AF360832

# Graphene for Electronics

# Graphene for Electronics

Editor

**Eugene Kogan**

MDPI • Basel • Beijing • Wuhan • Barcelona • Belgrade • Manchester • Tokyo • Cluj • Tianjin

*Editor*
Eugene Kogan
Physics
Bar-Ilan University
Ramat-Gan
Israel

*Editorial Office*
MDPI
St. Alban-Anlage 66
4052 Basel, Switzerland

This is a reprint of articles from the Special Issue published online in the open access journal *Nanomaterials* (ISSN 2079-4991) (available at: www.mdpi.com/journal/nanomaterials/special_issues/graphene_electronic).

For citation purposes, cite each article independently as indicated on the article page online and as indicated below:

LastName, A.A.; LastName, B.B.; LastName, C.C. Article Title. *Journal Name* **Year**, *Volume Number*, Page Range.

ISBN 978-3-0365-6168-4 (Hbk)
ISBN 978-3-0365-6167-7 (PDF)

# Contents

# About the Editor

**Eugene Kogan**

Eugene Kogan is affiliated with Bar-Ilan University. He has published more than 20 papers in the field of graphene. He has studied in particular RKKY interaction in graphene, the band structure of graphene and plasmons in graphene.

 *nanomaterials*



*Editorial*

# Graphene for Electronics

Eugene Kogan

Department of Physics, Bar-Ilan University, Ramat-Gan 52900, Israel; eugene.kogan@biu.ac.il

Graphene is an allotrope of carbon consisting of a single layer of atoms arranged in a two-dimensional (2D) honeycomb lattice. Graphene's unique properties of thinness and conductivity have led to global research into its applications as a semiconductor. With the ability to conduct electricity well at room temperature, graphene semiconductors could easily be implemented into the existing semiconductor technologies and, in some cases, successfully compete with the traditional ones, such as silicon. Research has already shown that graphene chips are much faster than existing ones made from silicon. The world's smallest transistor was manufactured using graphene. Flexible, wearable electronics may take advantage of graphene's mechanical properties, as well as its conductivity, to create bendable touch screens for phones and tablets, for example.

On the other hand, the physics of graphene and graphene-based systems has inspired the application (and development) of many advanced theoretical methods, including those outside the scope of traditional condensed matter physics. Graphene thus turned into the favorite benchmark of theorists. Fundamental studies go hand in hand with the applied ones and, in some cases, the former even opened doors to possible applications.

Graphene has already led to substantial progress in the development of the current electronic systems due to its unique electronic and thermal properties, including its high conductivity, quantum Hall effect, Dirac fermions, high Seebeck coefficient and thermoelectric effects. It paves the way for advanced biomedical engineering, reliable human therapy, and environmental protection. This suggests substantial improvements in current electronic technologies and applications in healthcare systems.

This Special Issue of Nanomaterials covers recent studies, both theoretical and experimental, that advance our understanding of graphene and may be relevant to graphene electronics. With the growing number of flexible electronics applications, environmentally friendly ways of mass-producing graphene electronics are required. Kralj and coworkers [1] present a scalable mechanochemical route for the exfoliation of graphite in a planetary ball mill with melamine to form melamine-intercalated graphene nanosheets.

Field-effect transistors have attracted significant attention in chemical sensing and clinical diagnosis, due to their high sensitivity and label-free operation. Huang and coworkers [2] present the study of a scalable photolithographic process of fabrication of the graphene-based ion-sensitive field-effect transistor (ISFET) arrays.

The study of electronic transport in the lowest Landau level of disordered graphene sheets placed in a homogeneous perpendicular magnetic field, a long-standing and cumbersome problem which defied a conclusive solution for several years, is presented in the paper by Sinner and Tkachov [3].

The paper by Berman et al. [4] contains theoretical analysis of Bose-Einstein condensation and superfluidity of dipolar excitons, formed by electron-hole pairs in spatially separated gapped hexagonal layers.

The electrical properties of polycrystalline graphene grown by chemical vapor deposition are determined by grain-related parameters, such as average grain size, single-crystalline grain sheet resistance, and grain boundary resistivity. Park et al. [5] have observed that the material property, graphene sheet resistance, could depend on the device

**Citation:** Kogan, E. Graphene for Electronics. *Nanomaterials* **2022**, *12*, 4359. https://doi.org/10.3390/nano12244359

Received: 18 November 2022
Accepted: 6 December 2022
Published: 7 December 2022

**Publisher's Note:** MDPI stays neutral with regard to jurisdictional claims in published maps and institutional affiliations.

dimension and developed an analytical resistance model based on the cumulative distribution function of the gamma distribution, explaining the effect of the grain boundary density and distribution in the graphene channel.

Electric devices have evolved to become smaller, more multifunctional, and increasingly integrated. When the total volume of a device is reduced, insufficient heat dissipation may result in device failure. A microfluidic channel with a graphene solution may replace solid conductors for simultaneously supplying energy and dissipating heat in a light emitting diode (LED). Chung et al. [6] designed, using a graphene solution, an automated recycling system that reduces the necessity of the manual operation of the device.

Silkin and coworkers [7] present a detailed first-principles investigation of the response of a free-standing graphene sheet to an external perpendicular static electric field.

The tunneling of electrons and holes in quantum structures plays a crucial role in studying the transport properties of materials and the related devices. A new two-dimensional Dirac material, 8-Pmmn borophene, hosts tilted Dirac cone and chiral, and anisotropic massless Dirac fermions. Kong et al. [8] adopted the transfer matrix method to investigate the Klein tunneling of massless fermions across the smooth NP junctions and NPN junctions of 8-Pmmn borophene.

The process of formation of carbon nanoscrolls with non-uniform curvatures is worthy of a detailed investigation. Lin et al. [9] present the first-principles method suitable for studying the combined effects due to the finite-size confinement, the edge-dependent interactions, the interlayer atomic interactions, the mechanical strains, and the magnetic configurations.

In the paper by Krasovskii [10], angle-resolved photoemission from monolayer and bilayer graphene is studied based on an ab initio one-step theory.

In the paper by Do [11] and coworkers, by introducing a generalized quantum-kinetic model which is coupled self-consistently with Maxwell and Boltzmann transport equations, the authors elucidate the significance of using input from first-principles band-structure computations for an accurate description of ultra-fast dephasing and the scattering dynamics of electrons in graphene.

In summary, this Special Issue presents several examples of the latest advancements on graphene science. We hope the readers will enjoy these articles and find them useful for their research.

**Funding:** This research received no external funding.

**Conflicts of Interest:** The authors declare no conflict of interest.

# References

1. Kralj, M.; Krivacic, S.; Ivanisevi, I.; Zubak, M.; Supina, A.; Marcius, M.; Halasz, I.; Kassal, P. Conductive Inks Based on Melamine Intercalated Graphene Nanosheets for Inkjet Printed Flexible Electronics. *Nanomaterials* **2022**, *12*, 2936. [CrossRef] [PubMed]
2. Huang, T.; Yeung, K.K.; Li, J.; Sun, H.; Alam, M.M.; Gao, Z. Graphene-Based Ion-Selective Field-Effect Transistor for Sodium Sensing. *Nanomaterials* **2022**, *12*, 2620. [CrossRef] [PubMed]
3. Sinner, A.; Tkachov, G. Quantum Diffusion in the Lowest Landau Level of Disordered Graphene. *Nanomaterials* **2022**, *12*, 1675. [CrossRef] [PubMed]
4. Berman, O.L.; Gumbs, G.; Martins, G.P.; Fekete, P. Superfluidity of Dipolar Excitons in a Double Layer of $\alpha$—T3 with a Mass Term. *Nanomaterials* **2022**, *12*, 1437. [CrossRef] [PubMed]
5. Park, H.; Lee, J.; Lee, C.-J.; Kang, J.; Yun, J.; Noh, H.; Park, M.; Lee, J.; Park, Y.; Park, J.; et al. Simultaneous Extraction of the Grain Size, Single-Crystalline Grain Sheet Resistance, and Grain Boundary Resistivity of Polycrystalline Monolayer Graphene. *Nanomaterials* **2022**, *12*, 206. [CrossRef] [PubMed]
6. Chung, Y.-C.; Chung, H.-H.; Lin, S.-H. Improvement of Temperature and Optical Power of an LED by Using Microfluidic Circulating System of Graphene Solution. *Nanomaterials* **2021**, *11*, 1719. [CrossRef] [PubMed]
7. Silkin, V.M.; Kogan, E.; Gumbs, G. Screening in Graphene: Response to External Static Electric Field and an Image-Potential Problem. *Nanomaterials* **2021**, *11*, 1561. [CrossRef] [PubMed]
8. Kong, Z.; Li, J.; Zhang, Y.; Zhang, S.-H.; Zhu, J.-J. Oblique and Asymmetric Klein Tunneling across Smooth NP Junctions or NPN Junctions in 8-Pmmn Borophene. *Nanomaterials* **2021**, *11*, 1462. [CrossRef] [PubMed]
9. Lin, S.-Y.; Chang, S.-L.; Chiang, C.-R.; Li, W.-B.; Liu, H.-Y.; Lin, M.-F. Feature-Rich Geometric and Electronic Properties of Carbon Nanoscrolls. *Nanomaterials* **2021**, *11*, 1372. [CrossRef] [PubMed]

10. Krasovskii, E. Ab Initio Theory of Photoemission from Graphene. *Nanomaterials* **2021**, *11*, 1212. [CrossRef] [PubMed]
11. Do, T.-N.; Huang, D.; Shih, P.-H.; Lin, H.; Gumbs, G. Atomistic Band-Structure Computation for Investigating Coulomb Dephasing and Impurity Scattering Rates of Electrons in Graphene. *Nanomaterials* **2021**, *11*, 1194. [CrossRef] [PubMed]

 *nanomaterials*

*Article*

# Conductive Inks Based on Melamine Intercalated Graphene Nanosheets for Inkjet Printed Flexible Electronics

Magdalena Kralj [1,†], Sara Krivačić [2,†], Irena Ivanišević [2], Marko Zubak [2], Antonio Supina [3], Marijan Marciuš [4], Ivan Halasz [1] and Petar Kassal [2,*]

1 Division of Physical Chemistry, Ruđer Bošković Institute, Bijenička cesta 54, 10000 Zagreb, Croatia
2 Faculty of Chemical Engineering and Technology, University of Zagreb, Marulićev trg 19, 10000 Zagreb, Croatia
3 Institute of Physics, Bijenička cesta 46, 10000 Zagreb, Croatia
4 Division of Materials Chemistry, Ruđer Bošković Institute, Bijenička cesta 54, 10000 Zagreb, Croatia
* Correspondence: pkassal@fkit.hr
† These authors contributed equally to this work.

**Abstract:** With the growing number of flexible electronics applications, environmentally benign ways of mass-producing graphene electronics are sought. In this study, we present a scalable mechanochemical route for the exfoliation of graphite in a planetary ball mill with melamine to form melamine-intercalated graphene nanosheets (M-GNS). M-GNS morphology was evaluated, revealing small particles, down to 14 nm in diameter and 0.4 nm thick. The M-GNS were used as a functional material in the formulation of an inkjet-printable conductive ink, based on green solvents: water, ethanol, and ethylene glycol. The ink satisfied restrictions regarding stability and nanoparticle size; in addition, it was successfully inkjet printed on plastic sheets. Thermal and photonic post-print processing were evaluated as a means of reducing the electrical resistance of the printed features. Minimal sheet resistance values (5 kΩ/sq for 10 printed layers and 626 Ω/sq for 20 printed layers) were obtained on polyimide sheets, after thermal annealing for 1 h at 400 °C and a subsequent single intense pulsed light flash. Lastly, a proof-of-concept simple flexible printed circuit consisting of a battery-powered LED was realized. The demonstrated approach presents an environmentally friendly alternative to mass-producing graphene-based printed flexible electronics.

**Keywords:** mechanochemistry; graphene nanosheets; conductive ink; inkjet printing; printed electronics

**Citation:** Kralj, M.; Krivačić, S.; Ivanišević, I.; Zubak, M.; Supina, A.; Marciuš, M.; Halasz, I.; Kassal, P. Conductive Inks Based on Melamine Intercalated Graphene Nanosheets for Inkjet Printed Flexible Electronics. *Nanomaterials* **2022**, *12*, 2936. https://doi.org/10.3390/nano12172936

Academic Editor: Eugene Kogan

Received: 27 July 2022
Accepted: 20 August 2022
Published: 25 August 2022

**Publisher's Note:** MDPI stays neutral with regard to jurisdictional claims in published maps and institutional affiliations.

## 1. Introduction

Flexible electronic devices manufactured by printing techniques on various substrates, such as paper, polymers, and textiles, have recently gained tremendous attention [1,2]. Unlike traditional silicon-based production techniques—often described as costly and complicated—printing offers faster, simpler, as well as environmentally and economically beneficial production possibilities [3,4]. Numerous examples include printed electronic circuits [5], displays [6], radio frequency identification tags (RFIDs) [7], thin-film transistors (TFTs) [8], and sensors [9]. Among different printing techniques, inkjet printing offers several advantages in the publishing and graphics industries [10]. This non-contact additive manufacturing technique is based on the selective ejection of individual drops of a liquid material (ink) from the nozzle upon thermal or pressure pulse [11]; this makes it easily adaptable for mass production. The arrival of inkjet printing outside the scope of classical application came with the development of nanoparticle-based inks with functional properties, especially electrical conductivity [2,12].

Printable inks are based upon a careful selection of the ink components, including the functional material and (a combination of) solvents and stabilizers [13]. Metal nanoparticles [14–16], conductive polymers [17,18], and carbon nanomaterials [19–21] are the most used functional components in electrically conductive inks. When formulating inks, rheological properties must be carefully tailored to ensure proper jetting. In the case of nanoparticle-based inks, additional restrictions regarding nanoparticle size and ink stability are imposed; these all prevent nozzle clogging.

Even though metal nanoparticles—due to their excellent conductivity—are among the most commonly used materials, a growing interest is being devoted to carbon (nano)-materials [22]. Inkjet-printed carbon nanomaterials have therefore been used in the development of flexible and wearable electronics [23,24], sensors [25,26], and film heaters [27]. Considering its intriguing and unique physicochemical properties, such as large surface area, exceptional thermal stability, excellent electrical conductivity, high electron mobility, superior mechanical strength, flexibility, and undemanding chemical functionalization, graphene has attracted attention as a promising functional material in the production of flexible electronic components [28]. To achieve greater concentration in an ink, without the reaggregation of the (nano)particles of the functional material, a suitable solvent based on adequate solubility parameters should be selected; and/or a stabilizing agent should be employed [29,30]. Common solvents for carbon (nano)material-based inks, such as N-methyl-2-pyrrolidone (NMP), N-cyclo-2-pyrrolidone, dimethylformamide (DMF), and dimethylsulfoxide (DMSO), are either expensive, chemically harsh, toxic, and/or difficult to remove post-printing due to their high boiling points [30,31]; thus, their use is not recommended [31–33]. However, environmentally compatible solvents often require the additional use of stabilizers, such as polymers and surfactants. For example, graphene suspensions in cyclohexanone and terpineol have been stabilized with ethyl-cellulose [34,35]; in water [36] and ethylene glycol [37] with the stabilizing agent sodium dodecyl sulfate (SDS) [38]; and in ethanol, ethanediol, propanetriol, and deionized water along with sodium carboxymethyl cellulose (CMC) [39].

Different synthetic routes toward graphene have been thoroughly investigated since its discovery. These include the two main approaches: the top-down (TD) approach and the bottom-up (BU) approach [28,32,40,41]. BU approaches, based on the nucleation of a carbon precursor, are generally expensive and time-consuming. On the other hand, in the TD approach, carbonaceous materials (such as graphite) are cut into nano-sized particles by physicochemical processes, which pave the road to the mass production of graphene. TD approaches include the famous Hummers' method and liquid-phase exfoliation (LPE) of graphite [42–44]. Yet, these have major limitations, such as the need for harmful and complex pretreatments, high energy consumption, low yields, agglomeration tendency, low-stability in polar solvents, high precursor costs, or the need for special equipment [41,45].

Clearly, a facile, sustainable, reproducible, and low-cost route for the large-scale preparation of graphene nanosheets (GNS) with minimal surface defects is required to satisfy the growing industry requirements. For this reason, methods of mechanochemistry have become attractive as they often provide quick and quantitative reactions of solids, even on a large scale, while according with the principles of Green Chemistry [46]. There are numerous examples of mechanochemical synthesis and modification of monodisperse nanoparticle systems in a solvent-free environment [47–49]. As recently demonstrated, graphite can be exfoliated through non-covalent interactions with melamine (1,3,5-Triazine-2,4,6-triamine) in a ball milling process under solid, i.e., dry conditions [50]; this is in contrast to the exfoliation of graphite with melamine in aqueous media [51–54].

We present here a facile, scalable, and green method for the development of inkjet printable conductive graphene-based inks. We have adopted a mechanochemical route for the exfoliation of graphite with melamine to form melamine-intercalated graphene nanosheets (M-GNS). The M-GNS were used as the conductive material in the formulation of an inkjet printable ink, with the aid of polymeric dispersants in green solvents (water, ethanol, ethylene glycol). The electrical properties of the printed features were evaluated and post-print processing optimized, to yield flexible printed electronic circuits.

## 2. Materials and Methods

### 2.1. Materials

Graphite flakes (G) having particle sizes 200–300 μm were purchased from Graphenea, San Sebastian, Spain and melamine from Alfa Aesar, Kandel, Germany. Ethanol (absolute) and 2-propanol were obtained from Gram-Mol, Zagreb, Croatia, ethylene glycol from Sigma-Aldrich, St. Louis, MO, USA, and terpineol (mixture of isomers) from Alfa Aesar, Kandel, Germany. All the chemicals were of analytical grade and were used as received. Melamine intercalated graphene nanosheets (M-GNS, 1–2 sheets) were obtained by a mechanochemical route and can be used without additional purification. Aqueous solutions were prepared with deionized water (Millipore Milli-Q, specific conductivity 0.059 μS cm$^{-1}$). Commercial polymeric stabilizing agents Solsperse 12000S and Solsperse 20000 were supplied by Lubrizol, Wickliffe, OH, USA. Surface mount light-emitting diodes for the proof of concept experiment were from Kingbright Electronic Co, New Taipei City, Taiwan; they were glued to the printed conducting traces using a conductive glue, Wire Glue, Anders Products, Melrose, MA, USA.

### 2.2. Mechanochemical Synthesis of Melamine-Intercalated Graphene Nanosheets (M-GNS)

Single- and double-layer melamine-intercalated graphene nanosheets (M-GNS) were obtained by neat grinding in a ball to powder ratio $m_b$:$m_p$ = 1:12.7. The process was performed at room temperature using a planetary ball mill PULVERISETTE 6 operating at 500 rpm, in a 50 mL stainless steel jar equipped with 12 stainless steel balls ($m$ = 4 g; $d$ = 10 mm). Graphite flakes ($m$ = 0.5 g), melamine ($m$ = 2.5 g), and dry ice ($m$ = 0.8 g) were milled in a mass ratio $m$(G):$m$(I):$m$(M) = 1:1.6:5 for 48 h in periods of 1 h milling, followed by 10 min of resting. The product was a black free-flowing powder that could be easily collected from the jar using a spatula.

### 2.3. Preparation of M-GNS Inks

M-GNS inks were prepared by dispersing the powdered product in various solvents using a Sonopuls Serie 2000.2 tip-sonicator, with the addition of Solsperse stabilizers. The sonication was performed for 15 min at 25% amplitude of the initial power of 70 W. The physical properties of the as-prepared formulations, including viscosity and surface tension, were measured with a micro-Ostwald viscosimeter 516 13/Ic, SI Analytics GmbH, (Mainz, Germany) and KRÜSS K6 tensiometer (Hamburg, Germany), respectively. All the measurements were performed at room temperature (23 ± 2 °C).

### 2.4. Inkjet Printing of M-GNS Inks and Post-Printing Processing

Inkjet printing was performed using a Fujifilm Dimatix DMP-2850 (Tokyo, Japan) drop-on-demand printer, which utilizes 16 nozzles with a diameter of 21 μm and a nominal drop volume of 10 pL. The experimental printing parameters were optimized to achieve continuous conductive features of the deposited ink on the selected substrates: PI (Kapton, DuPont, Wilmington, NC, USA, $d$ = 25 μm); and clear PET (Melinex 505, DuPont, Wilmington, NC, USA, $d$ = 125 μm). The printed patterns for characterization were 8 mm × 8 mm squares designed in Dimatix Drop Manager Software 3.0, Fujifilm, Tokyo, Japan.

To improve electrical conductivity, the printed squares were processed both thermally and photothermally using intense pulsed light (IPL). For thermal processing, the specimens were placed in a furnace (Demiterm, Estherm d.o.o., Sveta Nedelja, Croatia) at different temperatures for 1 h. For IPL processing, the jetted patterns were set approximately 1 cm from the flash lamp (Xenon, Wilmington, NC, USA, LH-912) of a Xenon X-1100 IPL system. A series of experiments were performed to find the optimal energy at 2500 V.

### 2.5. Characterization

Powder X-ray diffraction data were collected on a Aeris bench-top diffractometer, Panalytical, Almelo, Netherlands, with Ni-filtered CuK$\alpha$ radiation obtained from an X-ray tube operating at 7.5 mA and 40 kV, in the 2$\theta$ range of 5–70° (step size of 0.027166°, 7.65 s per step). Thermogravimetric measurements were performed with a Shimadzu, Kyoto, Japan, DTG-60H analyzer at a heating rate of 10 °C min$^{-1}$ from room temperature to 1000 °C in a stream of nitrogen, for bulk M-GNS samples and the prepared ink; and from room temperature to 1000 °C in a stream of oxygen for the polymeric stabilizers. Scanning electron microscopy (SEM) imaging was performed on a Jeol, Tokyo, Japan, JSM-7000F field emission scanning electron microscope, operating at 10 kV; while energy-dispersive X-ray spectroscopy (EDX) analysis was performed on an Oxford Instruments, Abingdon, UK, INCA 350 spectrometer coupled with the FE-SEM. Atomic force microscopy (AFM) micrographs were obtained by NanoWizard 4 ULTRA AFM, Bruker, Billerica, MA, USA in AC mode. Samples for AFM were prepared by diluting the stock solutions to a given concentration of $10^{-3}$ mg/mL and spin coating on a freshly exfoliated mica substrate before drying at 70 or 150 °C in a Biobase Bov-30V Lab high-temperature vacuum oven for 2 h. Fourier-transform infrared attenuated total reflectance (FTIR-ATR) spectra in KBr tablets were recorded on a PerkinElmer, Waltham, MA, USA, SpectrumTwo L1600400 spectrometer equipped with a diamond cell in the range of 4000−450 cm$^{-1}$ with a resolution of 8 cm$^{-1}$. UV–Vis spectroscopy of the conductive ink was performed with a Shimadzu, Kyoto, Japan, UV-1280 UV–Vis spectrometer. The absorption spectra were recorded in the range 320–800 nm after diluting to 1:100 to assure a meaningful absorbance range. Particle size distribution (PSD) was determined using a Zetasizer Ultra (Malvern Panalytical, Malvern, UK) based on a He-Ne laser ($\lambda$ = 632.8 nm) and a thermostated sample cell. The sample dilution was $\varphi$ = 1:33, accounting for the graphene refractive index of 1.957. Before measurement, the sample was equilibrated for 120 s at 25 °C $\pm$ 0.1 °C. The intensity of the scattered light was converted into contribution per number of particles within the measured sample volume. Zeta-potential measurements of the M-GNS ink formulation were carried out using the aforementioned instrument and the same thermostated sample cell. The ZS Xplorer v1.00, Malvern Panalytical, Malvern, UK software was used for data analysis. The sheet resistance of the printed samples was measured before and after both thermal and IPL processing using a four-point probe (Ossila, UK).

### 3. Results and Discussion

*3.1. Synthesis and Characterization of Melamine-Intercalated Graphene Nanosheets*

For the synthesis of the conducting nanoparticles, we adopted melamine-induced exfoliation in a planetary ball-mill that produces melamine intercalated single and two-layered graphene nanosheets (M-GNS). The role of melamine is to aid the exfoliation of graphite by noncovalent interactions, and prevent re-aggregation of the graphene sheets into a graphitic structure. Melamine has an aromatic core that interacts with the π-system of graphene; however, multiple melamine molecules can form extended 2D networks via hydrogen bonding, and this improves the exfoliation and stabilization of GNS [50]. The synthesized M-GNS were thoroughly characterized to determine their composition, morphology, and thermal properties. The Fourier-transform infrared attenuated total reflectance (FTIR-ATR) spectrum of M-GNS exhibits bands characteristic of melamine, while the absence of any additional bands demonstrates that the sample was pure (Figure 1).

Scanning electron microscopy (SEM) was used to evaluate the formation, size distribution, and morphology of the M-GNS; while EDX analysis provided additional information about the elemental composition of the sample. Figure 2 shows the morphology and elemental structure of the raw M-GNS sample. It is noticeable that melamine, after undergoing a grinding process, is present in the sample (evidenced by the significant nitrogen amount). A wide particle size distribution is also observed. The morphology corresponds to previously examined mechanochemically treated carbon materials [55].

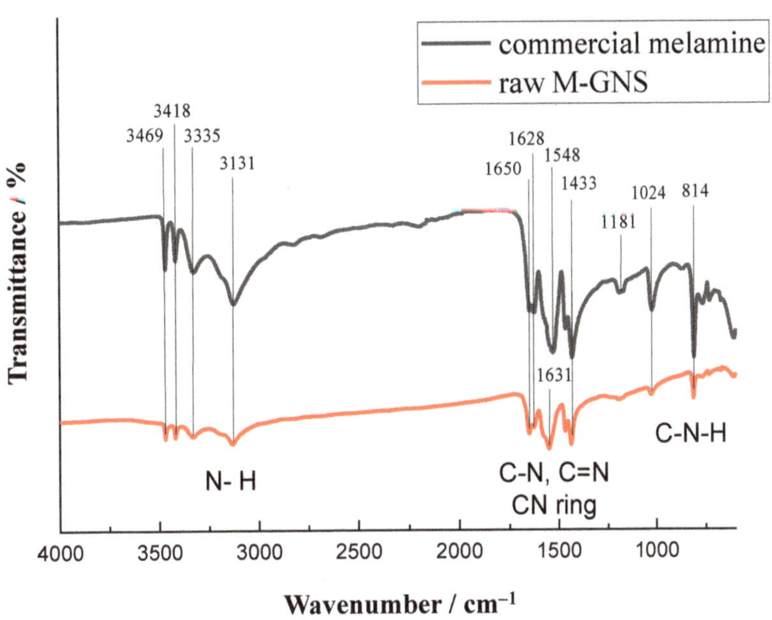



**Figure 1.** Fourier-transform infrared (FTIR) spectra for the raw sample of melamine-intercalated graphene nanosheets (M-GNS).

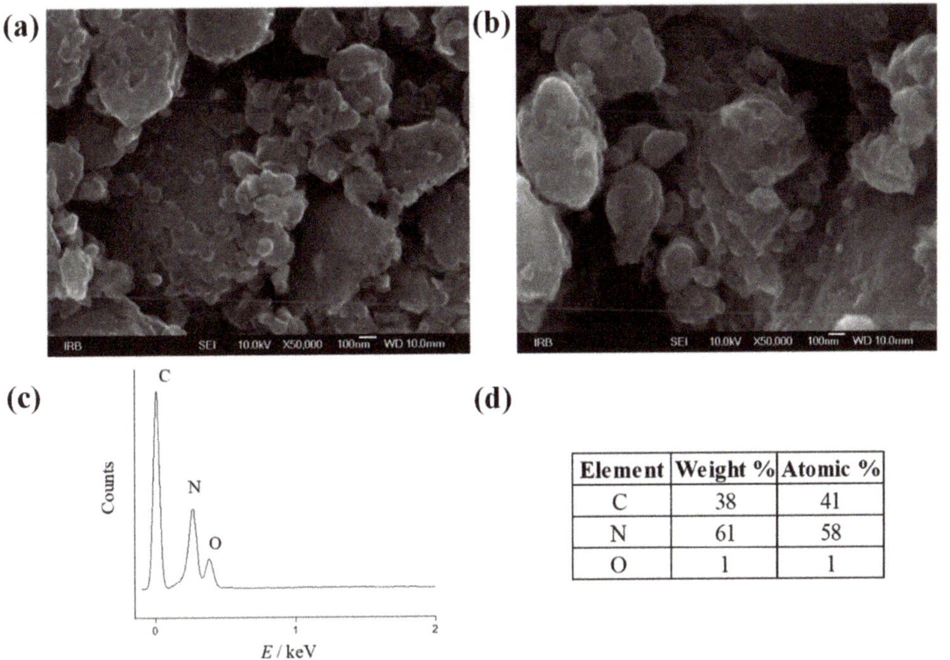



| Element | Weight % | Atomic % |
| --- | --- | --- |
| C | 38 | 41 |
| N | 61 | 58 |
| O | 1 | 1 |

**Figure 2.** (**a**,**b**) Scanning electron microscopy (SEM) images and (**c**,**d**) energy-dispersive X-ray (EDX) spectra of the raw M-GNS.

Topological and morphological studies via atomic force microscopy (AFM) were conducted primarily to visualize the surface structure of the M-GNS and height profiles with and without melamine present in the sample. The AFM image (Figure 3a) shows melamine layers with a lateral dimension of 200 nm. On the other hand, after thermal annealing at 130 °C in a vacuum oven, the larger melamine flakes were removed by sublimation and only smaller graphene nanosheets remained (Figure 3b,c). The average diameter of the GNS was determined to be around 14 nm. AFM height measurements revealed an average height of 0.3–0.65 nm corresponding to single- and double-layer GNS.

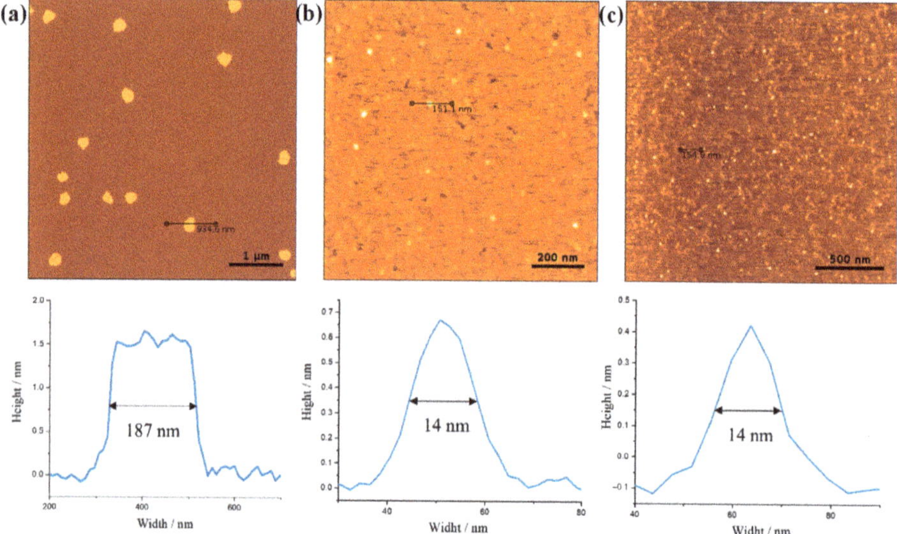


**Figure 3.** Atomic force microscopy (AFM) images and cross-section analysis of the M-GNS dispersed in a mixture of ethanol:water:EG = 0.50:0.45:0.05. The sample was spin-coated on freshly cleaved mica substrates and was dried in a vacuum oven for 2 h at (**a**) 70 °C and (**b**,**c**) 130 °C.

The prepared M-GNS were heated from room temperature to 1000 °C to determine their thermal stability. It has been reported that melamine decomposition takes place in three stages, undergoing progressive endothermic condensation during heating; with the release of ammonia and forms products, such as melam, melem, and melon [56]. Products of thermal decomposition of melamine are thermally more stable than melamine. Finally, graphitic carbon nitride, g-$C_3N_4$ is produced under further heating [57,58].

As expected, a typical differential weight loss in several regions was observed (Figure 4). The first stage covered the regions of maximum weight loss corresponding to the characteristic mass loss at the range of 300–400 °C. This is associated with melamine condensation to melam (a short-lived intermediate) and further condensation to melem [57]. At higher temperatures (around 400–600 °C), the condensation reaction slowly progresses; at first, it yields melon and then, graphitic carbon nitride [59,60]. Finally, thermal decomposition of graphitic carbon nitride takes place in the range of 600–750 °C [61]; whereas additional changes in weight loss were not observed, proving good thermal stability of GNS. The residual mass of GNS amounts to 18.6%, which is in good agreement with the initial mass ratio of graphite to melamine during the mechanochemical exfoliation.

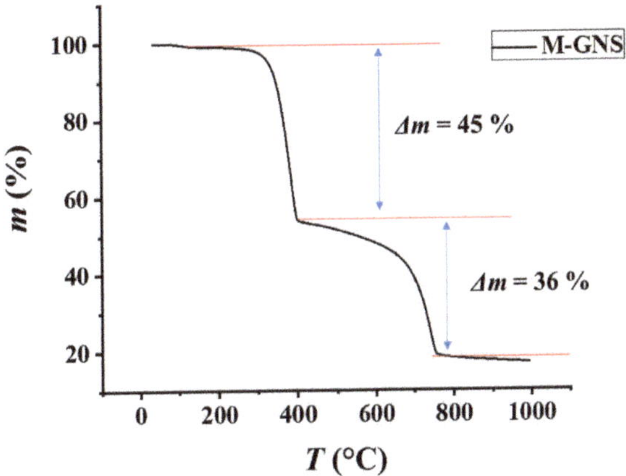


**Figure 4.** Thermogravimetric analysis (TGA) curve showing the mass loss profile of M-GNS.

### *3.2. Preparation and Characterization of M-GNS Inks*

The M-GNS were dispersed in green solvents using an ultrasonic probe to form conductive inks for inkjet printing. We focused on less harmful polar solvents (water and alcohols) and aimed to formulate an ink compatible with substrates commonly used in printed electronics—PET and PI. Graphene is known to form stable dispersions in solvents with a similar surface energy to itself [62]. When adopting solvents of incompatible surface energy, there is a requirement for stabilizers that adsorb to graphene nanosheets during the homogenization step [36]. In this way, the stability of the ink is improved and the shelf life is significantly extended. Commercially available Solsperse polymeric hyperdispersants—steric stabilizers with anchor groups optimized for strong adsorption to the particle surface—were used for this purpose. We evaluated the stability of the M-GNS in several solvents and their mixtures (see Supplementary Materials, Figure S1). Ultimately, the selected composition of the ink was 2 mg/mL of the M-GNS dispersed in a solvent mixture consisting of ethanol:water:ethylene glycol = 0.50:0.45:0.05 by volume; with the addition of 0.36 mg/mL of Solsperse 20000 and 0.04 mg/mL of Solsperse 12000S stabilizers. The prepared ink is shown in Figure 5a. This composition demonstrated good wetting of PET and PI substrates, with no observable coffee ring effect after drying.

The bottleneck of a piezoelectric drop-on-demand inkjet process is the development of stable, single droplets without the formation of satellite (secondary) droplets [63]. This can be achieved by tuning the inks composition and its physical properties, including viscosity, density, and surface tension. The droplet formation behavior is often characterized by $Z$, a dimensionless inverse Ohnesorge ($Oh$) number [11,64], $Z = \sqrt{\gamma \rho a}/\eta$; where $\rho$, $\eta$, $\gamma$, and $a$ are the density, dynamic viscosity, surface tension, and dimensional parameter of the printer, respectively. Low Z-values (<4) indicate possible difficulties in fluid ejection due to the high viscosity, whereas a higher Z-value (>13) suggests the formation of satellite droplets or, at least, uncontrollable ink leakage [2]. The requirements for inkjet printable fluids include low viscosity (4–30 mPa s) and relatively high surface tension (around ~35 mN m$^{-1}$) [11]. Our ink formulation had a measured Z-value of 7.7, indicating excellent suitability for jetting.

The second major requirement of nanoparticle-based inks is nanoparticle size and suspension stability. A maximum nanoparticle size of about 200–500 nm (1% of the nozzle diameter) is generally suggested, along with the necessary stability against aggregation and sedimentation [12]. Failure to meet either of these requirements can cause the clogging of printer nozzles. Dynamic light scattering (DLS) analysis was performed to determine the particle size distribution of the fabricated M-GNS ink based on the number of scattering particles, Figure 5b.

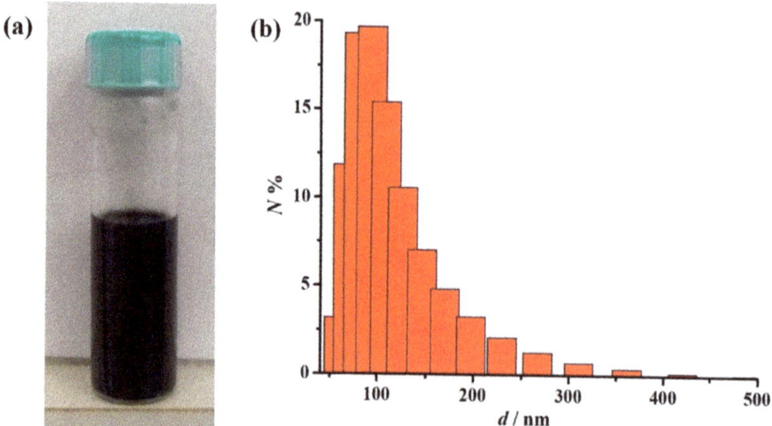



**Figure 5.** (**a**) The M-GNS ink after 24 h at rest after preparation; and (**b**) a histogram of the prepared M-GNS ink from Dynamic light scattering (DLS) measurements, recorded in an aqueous medium with a sample dilution of $\varphi$ = 1:33.

The histogram in Figure 5b shows that the sample contains particles of different sizes, resulting in an average hydrodynamic particle diameter of $d$ = 173.7 nm; this corresponds to the size of melamine sheets observed in the AFM measurements (Figure 3a) and indicates that the nanoparticles are small enough for printing without clogging. The stability of the conductive inks was evaluated using zeta-potential measurements. Particle dispersions with zeta-potential values of ±20–30 mV are generally assumed to be moderately stable [65]. The measured $\zeta$-potential of the M-GNS ink was –25.7 mV, indicating moderate stability of the prepared graphene particles in the solution phase [66]. The stability of real systems is determined by the relationship between attractive van der Waals forces (information that is not detectable with $\zeta$-potential measurements) and electrostatic repulsive forces between particles (provided by the zeta-potential). Accounting for that, dispersions with a lower absolute zeta-potential than that generally acknowledged should not be discarded in terms of colloidal stability [67].

The long-term stability of the ink was additionally evaluated by optical absorption spectroscopy. We collected UV–Vis spectra of freshly prepared ink and compared it to those taken up to 32 days post-formation. Graphene has an absorbance maximum at ~270 nm [68–70]; however, this part of the spectrum is affected by absorption of the Solsperse 20000 stabilizer. Solsperse 12000S, on the other hand, has very strong absorbance in the visible part of the spectrum (Figure S2); while its lowest absorbance is at $\lambda$ = 514 nm. Therefore, 514 nm was chosen as the wavelength for monitoring graphene absorbance reduction as a function of ink instability over time, Figure 6. As can be seen from Figure 6b, sedimentation is strongest within the first 6 h. However, the absorbance at 514 nm does not fall below 91% of the initial value; this indicates good short-term stability for single-day printing. In the following days, the absorbance decreases more slowly; it reaches a minimum at 66% of the initial value after 32 days. On the 34th day of the ink storage, we tip-sonicated the ink for 1 min (25% amplitude) and regained the initial absorbance value (100%). This indicates that although the ink shows moderate stability over prolonged periods, the maximum stability can be recovered after only one minute of ultrasonication.

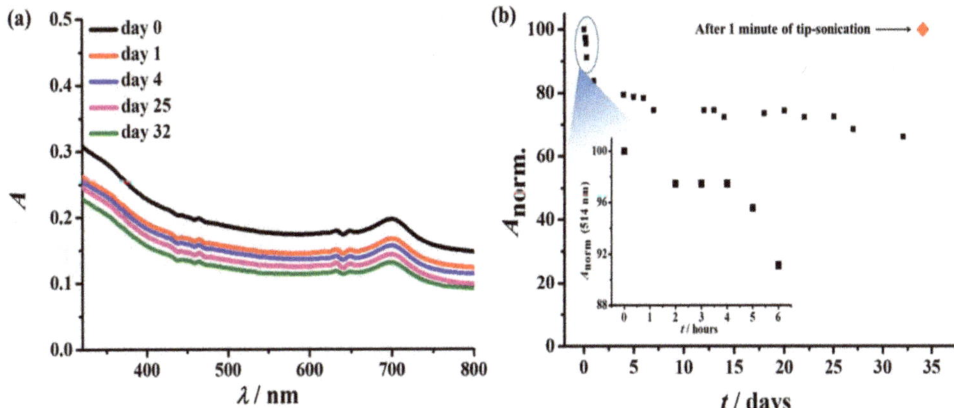

**Figure 6.** (**a**) A comparison of UV–Vis spectra recorded immediately after ink preparation (**day 0**), and **1, 4, 25** and **32 days** post-preparation (dilution: 100×); (**b**) the normalized view of the absorption at 514 nm (dilution: 100×); (**b**) **inset** sedimentation of the ink within the first six hours after ink preparation.

*3.3. Inkjet Printing and Post-Printing Processing*

The printing process starts with optimizing the printing parameters, which include the following: waveform, applied voltage, drop spacing, jetting frequency, cartridge height, number of printed layers, cartridge temperature, and platen temperature. The printing was performed at a moderate temperature of 55 °C and low jetting frequency. The other optimized printing parameters are shown in Table S1. For characterization, 8 × 8 mm squares were printed on polyethylene terephthalate (PET) and polyimide (PI) sheets; this is due to these substrates being commonly used in printed electronics [12,71]. Multiple layers of the conductive ink were printed (Figure 7), which is a practical way of increasing conductivity. The printed samples were characterized by sheet resistance measurements ($R_S$) with a four-point probe. The printed films were not electrically conductive up to three layers. At five layers, the measured sheet resistance was 4.27 ± 0.87 MΩ/sq (SD) and further decreased with additional layers. Nevertheless, such high sheet resistances are inadequate for most printed electronics applications; moreover, increasing the number of printed layers becomes pointless beyond a certain number of layers, since this greatly increases printing duration. Conductivity is instead commonly increased by removal of non-conducting ink components, usually by thermal post-print processing [72].

We exposed the printed squares to thermal annealing in order to improve the electrical conductivity via the removal of melamine and polymeric stabilizers. As presented in Figure 4, most of the melamine thermally decomposes at temperatures of up to 400 °C. The polymeric stabilizer Solsperse 20000 decomposes at somewhat lower temperatures (Figure S4); while Solsperse 12000S is more stable, but present in minuscule amounts. Therefore, the printed squares on PI were processed for 1 h at different temperatures, up to 400 °C (Figure 7b). The sheet resistance decreased gradually with temperature from the initial value of around 2.0 ± 0.9 MΩ/sq (SD), down to 44 ± 6 kΩ/sq (SD) at 400 °C. The thermal processing also benefited the homogeneity of the printed features, as evidenced by the decreasing standard deviations of measured sheet resistances.

To gain better insight into the morphology and topology of the printed features, SEM and AFM analysis were performed before and after thermal annealing at 400 °C (Figure 8). The surface morphology of the printed pattern before annealing is rough and inhomogeneous. We observed large melamine crystals (larger than 10 μm in diameter), which disrupt the electrical conductivity (Figure 8a). The SEM picture of the printed pattern after thermal annealing confirms a significant enhancement of the film quality and removal of melamine crystals due to thermal decomposition (Figure 8d,e). Accordingly, AFM measurements revealed a decrease in film thickness after annealing, along with a decrease in surface roughness (Figure 8c,f). The surface roughness parameter $R_a$ (average) decreased from 580.6 nm to 216.0 nm, while the $R_q$ (quadratic average) decreased from 789.1 nm to 289.1 nm.

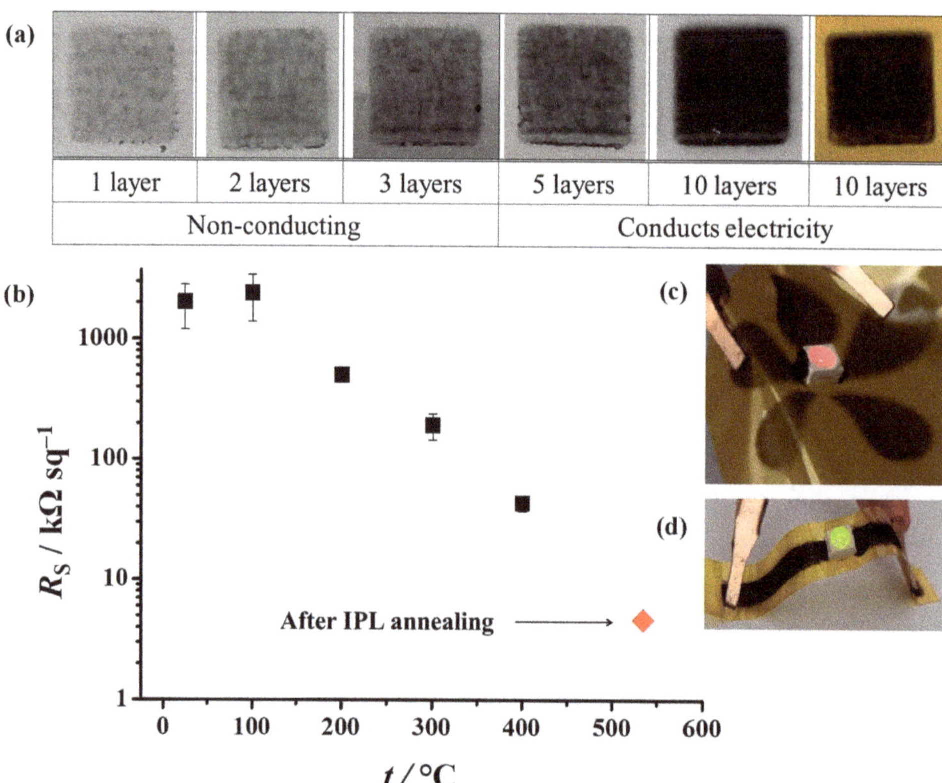


**Figure 7.** (**a**) The different number of printing passes of the conductive ink on a PET and PI substrate; (**b**) the sheet resistance of printed squares on PI, after thermal annealing at different temperatures. Error bars represent one standard deviation (*n* = 6); red diamond indicates the sheet resistance value after intense pulsed light (IPL) annealing at the energy of 700 J; examples of flexible printed electronics using 10 layers (**c**) and 20 layers (**d**) of the M-GNS ink.

In addition to thermal processing, we evaluated intense pulsed light (IPL) as a way of photothermal processing. IPL uses very short high-energy pulses of visible light; it thereby diminishes the thermal stress on the sensitive polymeric substrate, which makes it highly compatible with printed flexible electronics [73]. Graphene-based materials are great candidates for IPL annealing due to their high absorption coefficient in the visible part of the spectrum [74,75]. Nevertheless, the sheet resistance of printed squares was reduced only to around 43% of the initial value after exposure to 600 J (Figure S5). A further increase in IPL energy caused an increase in resistance, suggesting that the conductive film was damaged during the annealing process. This can be attributed to the formation of gaseous products (ammonia) of melamine decomposition [58] in very short time intervals, leading to the removal of graphene from the substrate. Therefore, IPL in itself is not an optimal processing technology for this kind of conducting ink containing melamine. However, we exposed the previously thermally annealed samples (at 400 °C) to IPL energies of

700 J. This combined processing procedure resulted in the lowest sheet resistances of only $5.0 \pm 0.3$ kΩ/sq (SD) for 10 printed layers (shown in Figure 7b) and $626 \pm 106$ Ω/sq (SD) for 20 printed layers. As can be seen from Table S2, the measured sheet resistances are comparable to, or better than, those obtained in similar studies and for a comparable number of printing passes. While printed metal nanoparticle inks can yield the lowest sheet resistances, in some cases less than 1 Ω/sq [12], in the case of graphene inks, sheet resistances are usually larger than 1 kΩ/sq for a single digit number of printed layers. Increasing the number of printing passes reduces the sheet resistance below the value of 1 kΩ/sq, which is usually observed at 20 passes or more (Table S2). Such resistivities are sufficient for different printed electronics applications [23]. Finally, as a proof-of-concept experiment, we constructed simple flexible LED circuits; we constructed the circuits by printing 10 and 20 layers of the conductive ink on PI, either as plain 2 mm wide conducting traces or in the shape of our institution logo (Figure 7c,d). The printed traces were annealed in the same way as previously optimized, by combining thermal and IPL processing. The attached surface mount LEDs were successfully powered from a single 9 V battery.

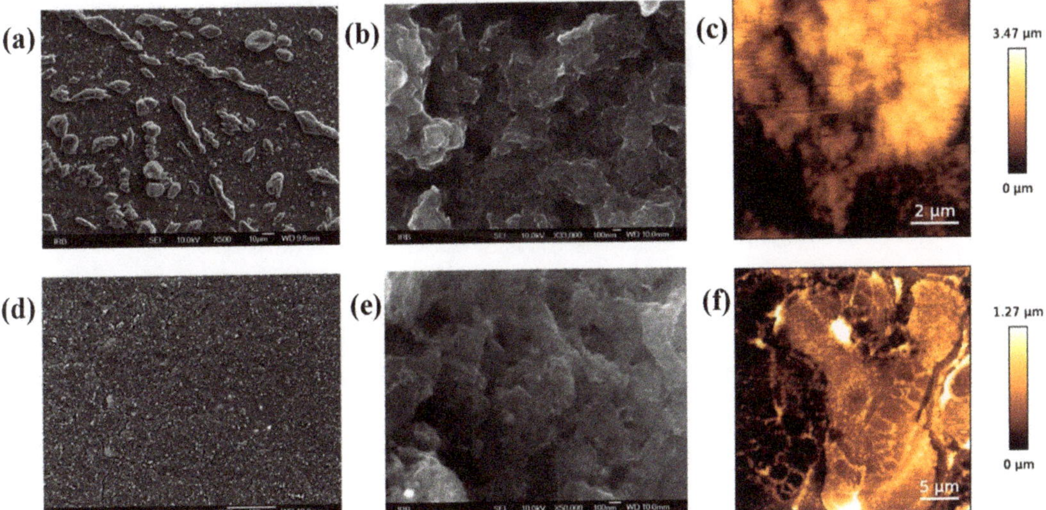

**Figure 8.** SEM (under different magnification) and AFM images of the M-GNS film on a PI substrate: (**a**–**c**) before annealing; and (**d**–**f**) after annealing at 400 °C.

## 4. Conclusions

We described here a novel mechanochemical synthesis of melamine-intercalated graphene nanosheets; we suggest it as a potential approach for large-scale preparation of graphene, which would comply with the basic principles of green chemistry. The prepared M-GNS were used as a functional material for the formulation of a stable graphene-based ink, suitable for inkjet printing. The printing process was optimized to generate electrically conductive patterns on flexible PET and PI substrates. A combination of thermal and photonic (IPL) annealing reduced the electrical resistance of the printed patterns by three orders of magnitude. The presented procedure is both scalable and environmentally friendly; in addition, it represents a starting point in the development of graphene-based printed flexible electronics.

**Supplementary Materials:** The following supporting information can be downloaded at: https://www.mdpi.com/article/10.3390/nano12172936/s1, Figure S1: Stability in different solvents; Figure S2: Absorption spectra of stabilizers and ink formulation; Figure S3: Unsuccessful printing example; Figure S4: Thermogravimetric analysis; Figure S5: Effect of intense pulsed light on sheet resistance; Table S1: Optimized printing parameters; Table S2: An overview of relevant literature. References [31,35–37,39,76–80] were cited in Supplementary Materials.

**Author Contributions:** Conceptualization, P.K.; formal analysis, M.K., S.K., I.I., M.Z., A.S. and M.M.; funding acquisition, I.H. and P.K.; investigation, M.K., S.K., I.I., M.Z., A.S. and M.M.; methodology, M.K., S.K. and P.K.; project administration, P.K.; resources, I.H. and P.K.; supervision, I.H. and P.K.; writing—original draft, M.K., S.K. and I.I.; writing—review and editing, I.H. and P.K. All authors have read and agreed to the published version of the manuscript.

**Funding:** This research was funded by the Croatian Science Foundation, grant numbers UIP-2020-02-9139 and DOK-2021-02-2362; the Ministry of Environment and Energy; the Ministry of Science and Education; the Environmental Protection and Energy Efficiency Fund; the Croatian Science Foundation under the project "New Materials for Energy Storage", in the total amount of 1,962,100 HRK (PKP-2016-06-4480); and by the Centre of Excellence for Advanced Materials and Sensing Devices, a project financed by the European Union through the European Regional Development Fund—the Competitiveness and Cohesion Operational Programme (KK.01.1.1.01.0001).

**Institutional Review Board Statement:** Not applicable.

**Informed Consent Statement:** Not applicable.

**Data Availability Statement:** The data presented in this study are available in the article and supplementary material.

**Acknowledgments:** We kindly thank Lubrizol for supplying the Solsperse stabilizers.

**Conflicts of Interest:** The authors declare no conflict of interest.

## References


1. Raut, N.C.; Al-Shamery, K. Inkjet printing metals on flexible materials for plastic and paper electronics. *J. Mater. Chem. C* **2018**, *6*, 1618–1641. [CrossRef]
2. Nayak, L.; Mohanty, S.; Nayak, S.K.; Ramadoss, A. A review on inkjet printing of nanoparticle inks for flexible electronics. *J. Mater. Chem. C* **2019**, *7*, 8771–8795. [CrossRef]
3. Wu, W. Inorganic nanomaterials for printed electronics: A review. *Nanoscale* **2017**, *9*, 7342–7372. [CrossRef] [PubMed]
4. Khan, Y.; Thielens, A.; Muin, S.; Ting, J.; Baumbauer, C.; Arias, A.C. A New Frontier of Printed Electronics: Flexible Hybrid Electronics. *Adv. Mater.* **2020**, *32*, 1905279. [CrossRef]
5. Huang, L.; Huang, Y.; Liang, J.; Wan, X.; Chen, Y. Graphene-based conducting inks for direct inkjet printing of flexible conductive patterns and their applications in electric circuits and chemical sensors. *Nano Res.* **2011**, *4*, 675–684. [CrossRef]
6. Pietsch, M.; Schlisske, S.; Held, M.; Strobel, N.; Wieczorek, A.; Hernandez-Sosa, G. Biodegradable inkjet-printed electrochromic display for sustainable short-lifecycle electronics. *J. Mater. Chem. C* **2020**, *8*, 16716–16724. [CrossRef]
7. Ali, Z.; Perret, E.; Barbot, N.; Siragusa, R.; Hely, D.; Bernier, M.; Garet, F. Authentication Using Metallic Inkjet-Printed Chipless RFID Tags. *IEEE Trans. Antennas Propag.* **2020**, *68*, 4137–4142. [CrossRef]
8. Baby, T.T.; Rommel, M.; von Seggern, F.; Friederich, P.; Reitz, C.; Dehm, S.; Kübel, C.; Wenzel, W.; Hahn, H.; Dasgupta, S. Sub-50 nm Channel Vertical Field-Effect Transistors using Conventional Ink-Jet Printing. *Adv. Mater.* **2017**, *29*, 1603858. [CrossRef]
9. Ivanisevic, I.; Milardović, S.; Ressler, A.; Kassal, P. Fabrication of an All-Solid-State Ammonium Paper Electrode Using a Graphite-Polyvinyl Butyral Transducer Layer. *Chemosensors* **2021**, *9*, 333. [CrossRef]
10. Cummins, G.; Desmulliez, M.P.Y. Inkjet printing of conductive materials: A review. *Circuit World* **2012**, *38*, 193–213. [CrossRef]
11. Derby, B. Inkjet Printing of Functional and Structural Materials: Fluid Property Requirements, Feature Stability, and Resolution. *Ann. Rev. Mater. Res.* **2010**, *40*, 395–414. [CrossRef]
12. Kamyshny, A.; Magdassi, S. Conductive nanomaterials for 2D and 3D printed flexible electronics. *Chem. Soc. Rev.* **2019**, *48*, 1712–1740. [CrossRef] [PubMed]
13. Kamyshny, A.; Magdassi, S. Conductive Nanomaterials for Printed Electronics. *Small* **2014**, *10*, 3515–3535. [CrossRef] [PubMed]
14. Zea, M.; Moya, A.; Fritsch, M.; Ramon, E.; Villa, R.; Gabriel, G. Enhanced Performance Stability of Iridium Oxide-Based pH Sensors Fabricated on Rough Inkjet-Printed Platinum. *Acs Appl. Mater. Interfaces* **2019**, *11*, 15160–15169. [CrossRef] [PubMed]
15. Sjoberg, P.; Määttänen, A.; Vanamo, U.; Novell, M.; Ihalainen, P.; Andrade, F.J.; Bobacka, J.; Peltonen, J. Paper-based potentiometric ion sensors constructed on ink-jet printed gold electrodes. *Sens. Actuators B-Chem.* **2016**, *224*, 325–332. [CrossRef]
16. Milardovic, S.; Ivanišević, I.; Rogina, A.; Kassal, P. Synthesis and Electrochemical Characterization of AgNP Ink Suitable for Inkjet Printing. *Int. J. Electrochem. Sci.* **2018**, *13*, 11136–11149. [CrossRef]

17. Park, J.; Yoon, H.; Kim, G.; Lee, B.; Lee, S.; Jeong, S.; Kim, T.; Seo, J.; Chung, S.; Hong, Y. Highly Customizable All Solution-Processed Polymer Light Emitting Diodes with Inkjet Printed Ag and Transfer Printed Conductive Polymer Electrodes. *Adv. Funct. Mater.* **2019**, *29*, 1902412. [CrossRef]

18. Kulkarni, M.V.; Apte, S.K.; Naik, S.D.; Ambekar, J.; Kale, B. Ink-jet printed conducting polyaniline based flexible humidity sensor. *Sens. Actuators B-Chem.* **2013**, *178*, 140–143. [CrossRef]

19. Gbaguidi, A.; Madiyar, F.; Kim, D.; Namilae, S. Multifunctional inkjet printed sensors for MMOD impact detection. *Smart Mater. Struct.* **2020**, *29*, 085052. [CrossRef]

20. Tortorich, R.P.; Song, E.; Choi, J.W. Inkjet-Printed Carbon Nanotube Electrodes with Low Sheet Resistance for Electrochemical Sensor Applications. *J. Electrochem. Soc.* **2014**, *161*, B3044–B3048. [CrossRef]

21. Vasiljevic, D.Z.; Mansouri, A.; Anzi, L.; Sordan, R.; Stojanovic, G.M. Performance Analysis of Flexible Ink-Jet Printed Humidity Sensors Based on Graphene Oxide. *IEEE Sens. J.* **2018**, *18*, 4378–4383. [CrossRef]

22. Huang, Q.J.; Zhu, Y. Printing Conductive Nanomaterials for Flexible and Stretchable Electronics: A Review of Materials, Processes, and Applications. *Adv. Mater. Technol.* **2019**, *4*, 1800546. [CrossRef]

23. Htwe, Y.Z.N.; Mariatti, M. Printed graphene and hybrid conductive inks for flexible, stretchable, and wearable electronics: Progress, opportunities, and challenges. *J. Sci. Adv. Mater. Devices* **2022**, *7*, 100435. [CrossRef]

24. Htwe, Y.Z.N.; Mariatti, M. Surfactant-assisted water-based graphene conductive inks for flexible electronic applications. *J. Taiwan Inst. Chem. Eng.* **2021**, *125*, 402–412. [CrossRef]

25. Maribou, K.; Gil, W.; Al Ghaferi, A.; Saadat, I.; Alhammadi, K.; Khair, A.M.; Younes, H. Assessing the Stability of Inkjet-Printed Carbon Nanotube for Brine Sensing Applications. *J. Nanosci. Nanotechnol.* **2020**, *20*, 7644–7652. [CrossRef]

26. Jelbuldina, M.; Younes, H.; Saadat, I.; Tizani, L.; Sofela, S.; Al Ghaferi, A. Fabrication and design of CNTs inkjet-printed based micro FET sensor for sodium chloride scale detection in oil field. *Sens. Actuators A Phys.* **2017**, *263*, 349–356. [CrossRef]

27. Xu, L.; Wang, H.; Wu, Y.; Wang, Z.; Wu, L.; Zheng, L. A one-step approach to green and scalable production of graphene inks for printed flexible film heaters. *Mater. Chem. Front.* **2021**, *5*, 1895–1905. [CrossRef]

28. Agudosi, E.S.; Abdullah, E.C.; Numan, A.; Mubarak, N.M.; Khalid, M.; Omar, N. A Review of the Graphene Synthesis Routes and its Applications in Electrochemical Energy Storage. *Crit. Rev. Solid State Mater. Sci.* **2020**, *45*, 339–377. [CrossRef]

29. Wang, C.Y.; Xia, K.; Wang, H.; Liang, X.; Yin, Z.; Zhang, Y. Advanced Carbon for Flexible and Wearable Electronics. *Adv. Mater.* **2019**, *31*, 1801072. [CrossRef]

30. Backes, C.; Higgins, T.M.; Kelly, A.; Boland, C.; Harvey, A.; Hanlon, D.; Coleman, J.N. Guidelines for Exfoliation, Characterization and Processing of Layered Materials Produced by Liquid Exfoliation. *Chem. Mater.* **2017**, *29*, 243–255. [CrossRef]

31. Capasso, A.; Castillo, A.D.R.; Sun, H.; Ansaldo, A.; Pellegrini, V.; Bonaccorso, F. Ink-jet printing of graphene for flexible electronics: An environmentally-friendly approach. *Solid State Commun.* **2015**, *224*, 53–63. [CrossRef]

32. Backes, C.; Abdelkader, A.M.; Alonso, C.; Andrieux-Ledier, A.; Arenal, R.; Azpeitia, J.; Balakrishnan, N.; Banszerus, L.; Barjon, J.; Bartali, R.; et al. Production and processing of graphene and related materials. *2D Mater.* **2020**, *7*, 022001. [CrossRef]

33. Kamarudin, S.F.; Mustapha, M.; Kim, J.K. Green Strategies to Printed Sensors for Healthcare Applications. *Polym. Rev.* **2021**, *61*, 116–156. [CrossRef]

34. He, Q.; Das, S.R.; Garland, N.T.; Jing, D.; Hondred, J.A.; Cargill, A.A.; Ding, S.; Karunakaran, C.; Claussen, J.C. Enabling Inkjet Printed Graphene for Ion Selective Electrodes with Postprint Thermal Annealing. *Acs Appl. Mater. Interfaces* **2017**, *9*, 12719–12727. [CrossRef] [PubMed]

35. Pandhi, T.; Cornwell, C.; Fujimoto, K.; Barnes, P.; Cox, J.; Xiong, H.; Davis, P.H.; Subbaraman, H.; Koehne, J.E.; Estrada, D. Fully inkjet-printed multilayered graphene-based flexible electrodes for repeatable electrochemical response. *Rsc Adv.* **2020**, *10*, 38205–38219. [CrossRef]

36. Parvez, K.; Worsley, R.; Alieva, A.; Felten, A.; Casiraghi, C. Water-based and inkjet printable inks made by electrochemically exfoliated graphene. *Carbon* **2019**, *149*, 213–221. [CrossRef]

37. Majee, S.; Song, M.; Zhang, S.-L.; Zhang, Z.-B. Scalable inkjet printing of shear-exfoliated graphene transparent conductive films. *Carbon* **2016**, *102*, 51–57. [CrossRef]

38. Kudr, J.; Zhao, L.; Nguyen, E.P.; Arola, H.; Nevanen, T.K.; Adam, V.; Zitka, O.; Merkoçi, A. Inkjet-printed electrochemically reduced graphene oxide microelectrode as a platform for HT-2 mycotoxin immunoenzymatic biosensing. *Biosens. Bioelectron.* **2020**, *156*, 112109. [CrossRef]

39. Ji, A.; Chen, Y.; Wang, X.; Xu, C. Inkjet printed flexible electronics on paper substrate with reduced graphene oxide/carbon black ink. *J. Mater. Sci.-Mater. Electron.* **2018**, *29*, 13032–13042. [CrossRef]

40. Kumar, N.; Salehiyan, R.; Chauke, V.; Botlhoko, O.J.; Setshedi, K.; Scriba, M.; Masukume, M.; Ray, S.S. Top-down synthesis of graphene: A comprehensive review. *Flatchem* **2021**, *27*, 100224. [CrossRef]

41. Kang, S.; Jeong, Y.K.; Jung, K.H.; Son, Y.; Choi, S.-C.; An, G.S.; Han, H.; Kim, K.M. Simple preparation of graphene quantum dots with controllable surface states from graphite. *RSC Adv.* **2019**, *9*, 38447–38453. [CrossRef]

42. Hummers, W.S.; Offeman, R.E. Preparation of Graphitic Oxide. *J. Am. Chem. Soc.* **1958**, *80*, 1339. [CrossRef]

43. Munuera, J.; Britnell, L.; Santoro, C.; Cuéllar-Franca, R.; Casiraghi, C. A review on sustainable production of graphene and related life cycle assessment. *2D Mater.* **2022**, *9*, 012002. [CrossRef]

44. Bonaccorso, F.; Lombardo, A.; Hasan, T.; Sun, Z.; Colombo, L.; Ferrari, A.C. Production and processing of graphene and 2d crystals. *Mater. Today* **2012**, *15*, 564–589. [CrossRef]

45. Yan, Y.X.; Nashath, F.Z.; Chen, S.; Manickam, S.; Lim, S.S.; Zhao, H.; Lester, E.; Wu, T.; Pang, C.H. Synthesis of graphene: Potential carbon precursors and approaches. *Nanotechnol. Rev.* **2020**, *9*, 1284–1314. [CrossRef]
46. Do, J.L.; Friscic, T. Mechanochemistry: A Force of Synthesis. *Acs Cent. Sci.* **2017**, *3*, 13–19. [CrossRef]
47. Chang, D.W.; Choi, H.-J.; Jeon, I.-Y.; Seo, J.-M.; Dai, L.; Baek, J.-B. Solvent-free mechanochemical reduction of graphene oxide. *Carbon* **2014**, *77*, 501–507. [CrossRef]
48. Jeon, I.Y.; Shin, Y.-R.; Sohn, G.-J.; Choi, H.-J.; Bae, S.-Y.; Mahmood, J.; Jung, S.-M.; Seo, J.-M.; Kim, M.-J.; Chang, D.W.; et al. Edge-carboxylated graphene nanosheets via ball milling. *Proc. Natl. Acad. Sci. USA* **2012**, *109*, 5588–5593. [CrossRef]
49. Yan, L.; Lin, M.; Zeng, C.; Chen, Z.; Zhang, S.; Zhao, X.; Wu, A.; Wang, Y.; Dai, L.; Qu, J.; et al. Electroactive and biocompatible hydroxyl-functionalized graphene by ball milling. *J. Mater. Chem.* **2012**, *22*, 8367–8371. [CrossRef]
50. Leon, V.; Rodriguez, A.M.; Prieto, P.; Prato, M.; Vázquez, E. Exfoliation of Graphite with Triazine Derivatives under Ball-Milling Conditions: Preparation of Few-Layer Graphene via Selective Noncovalent Interactions. *Acs Nano* **2014**, *8*, 563–571. [CrossRef]
51. Chen, C.H.; Yang, S.-W.; Chuang, M.-C.; Woon, W.-Y.; Su, C.-Y. Towards the continuous production of high crystallinity graphene via electrochemical exfoliation with molecular in situ encapsulation. *Nanoscale* **2015**, *7*, 15362–15373. [CrossRef] [PubMed]
52. Ma, H.; Shen, Z. Exfoliation of graphene nanosheets in aqueous media. *Ceram. Int.* **2020**, *5*, 1895–1905. [CrossRef]
53. Shi, G.; Araby, S.; Gibson, C.T.; Meng, Q.; Zhu, S.; Ma, J. Graphene Platelets and Their Polymer Composites: Fabrication, Structure, Properties, and Applications. *Adv. Funct. Mater.* **2018**, *28*, 1706705. [CrossRef]
54. Samoechip, W.; Pattananuwat, P.; Potiyaraj, P. Synthesis of Graphene Functionalized Melamine and its Application for Supercapacitor Electrode. *Key Eng. Mater.* **2018**, *773*, 128–132. [CrossRef]
55. Deng, M.J.; Cao, X.; Guo, L.; Cao, H.; Wen, Z.; Mao, C.; Zuo, K.; Chen, X.; Yu, X.; Yuan, W. Graphene quantum dots: Efficient mechanosynthesis, white-light and broad linear excitation-dependent photoluminescence and growth inhibition of bladder cancer cells. *Dalton Trans.* **2020**, *49*, 2308–2316. [CrossRef]
56. Wirnhier, E.; Mesch, M.B.; Senker, J.; Schnick, W. Formation and Characterization of Melam, Melam Hydrate, and a Melam-Melem Adduct. *Chem.-A Eur. J.* **2013**, *19*, 2041–2049. [CrossRef]
57. Lotsch, B.V.; Schnick, W. New Light on an Old Story: Formation of Melam during Thermal Condensation of Melamine. *Chem.-A Eur. J.* **2007**, *13*, 4956–4968. [CrossRef]
58. Liu, X.; Hao, J.W.; Gaan, S. Recent studies on the decomposition and strategies of smoke and toxicity suppression for polyurethane based materials. *Rsc Adv.* **2016**, *6*, 74742–74756. [CrossRef]
59. Costa, L.; Camino, G. Thermal behaviour of melamine. *J. Therm. Anal.* **1988**, *34*, 423–429. [CrossRef]
60. May, H. Pyrolysis of melamine. *J. Appl. Chem.* **1959**, *9*, 340–344. [CrossRef]
61. Fang, L.; Ohfuji, H.; Shinmei, T.; Irifune, T. Experimental study on the stability of graphitic $C_3N_4$ under high pressure and high temperature. *Diam. Relat. Mater.* **2011**, *20*, 819–825. [CrossRef]
62. Hernandez, Y.; Nicolosi, V.; Lotya, M.; Blighe, F.M.; Sun, Z.; De, S.; McGovern, I.T.; Holland, B.; Byrne, M.; Gun'Ko, Y.K.; et al. High-yield production of graphene by liquid-phase exfoliation of graphite. *Nat. Nanotechnol.* **2008**, *3*, 563–568. [CrossRef] [PubMed]
63. Hu, G.H.; Kang, J.; Ng, L.W.T.; Zhu, X.; Howe, R.C.T.; Jones, C.G.; Hersam, M.C.; Hasan, T. Functional inks and printing of two-dimensional materials. *Chem. Soc. Rev.* **2018**, *47*, 3265–3300. [CrossRef]
64. Li, D.D.; Lai, W.-Y.; Zhang, Y.-Z.; Huang, W. Printable Transparent Conductive Films for Flexible Electronics. *Adv. Mater.* **2018**, *30*, 1704738. [CrossRef] [PubMed]
65. Bhattacharjee, S. DLS and zeta potential—What they are and what they are not? *J. Control. Release* **2016**, *235*, 337–351. [CrossRef]
66. Krishnamoorthy, K.; Veerapandian, M.; Yun, K.; Kim, S.-J. The Chemical and structural analysis of graphene oxide with different degrees of oxidation. *Carbon* **2013**, *53*, 38–49. [CrossRef]
67. Lyklema, J.; van Leeuwen, H.P.; Minor, M. DLVO-theory, a dynamic re-interpretation. *Adv. Colloid Interface Sci.* **1999**, *83*, 33–69. [CrossRef]
68. Backes, C.; Paton, K.R.; Hanlon, D.; Yuan, S.; Katsnelson, M.I.; Houston, J.; Smith, R.J.; McCloskey, D.; Donegan, J.F.; Coleman, J.N. Spectroscopic metrics allow in situ measurement of mean size and thickness of liquid-exfoliated few-layer graphene nanosheets. *Nanoscale* **2016**, *8*, 4311–4323. [CrossRef]
69. Lotya, M.; Hernandez, Y.; King, P.J.; Smith, R.J.; Nicolosi, V.; Karlsson, L.S.; Blighe, F.M.; De, S.; Wang, Z.; McGovern, I.T.; et al. Liquid Phase Production of Graphene by Exfoliation of Graphite in Surfactant/Water Solutions. *J. Am. Chem. Soc.* **2009**, *131*, 3611–3620. [CrossRef]
70. Tiwari, S.K.; Huczko, A.; Oraon, R.; De Adhikari, A.; Nayak, G.C. Facile electrochemical synthesis of few layered graphene from discharged battery electrode and its application for energy storage. *Arab. J. Chem.* **2017**, *10*, 556–565. [CrossRef]
71. Zang, D.Y.; Tarafdar, S.; Tarasevich, Y.Y.; Choudhury, M.D.; Dutta, T. Evaporation of a Droplet: From physics to applications. *Phys. Rep.-Rev. Sect. Phys. Lett.* **2019**, *804*, 1–56. [CrossRef]
72. Ivanisevic, I.; Kassal, P.; Milinković, A.; Rogina, A.; Milardović, S. Combined Chemical and Thermal Sintering for High Conductivity Inkjet-printed Silver Nanoink on Flexible Substrates. *Chem. Biochem. Eng. Q.* **2019**, *33*, 377–384. [CrossRef]
73. Wunscher, S.; Abbel, R.; Perelaer, J.; Schubert, U.S. Progress of alternative sintering approaches of inkjet-printed metal inks and their application for manufacturing of flexible electronic devices. *J. Mater. Chem. C* **2014**, *2*, 10232–10261. [CrossRef]

74. Arapov, K.; Jaakkola, K.; Ermolov, V.; Bex, G.; Rubingh, E.; Haque, S.; Sandberg, H.; Abbel, R.; De With, G.; Friedrich, H.H. Graphene screen-printed radio-frequency identification devices on flexible substrates. *Phys. Status Solidi-Rapid Res. Lett.* **2016**, *10*, 812–818. [CrossRef]
75. Secor, E.B.; Gao, T.Z.; Dos Santos, M.H.; Wallace, S.G.; Putz, K.W.; Hersam, M.C. Combustion-Assisted Photonic Annealing of Printable Graphene Inks via Exothermic Binders. *Acs Appl. Mater. Interfaces* **2017**, *9*, 29418–29423. [CrossRef]
76. Secor, E.B.; Ahn, B.Y.; Gao, T.Z.; Lewis, J.A.; Hersam, M.C. Rapid and Versatile Photonic annealing of Graphene Inks for Flexible Printed Electronics. *Adv. Mat.* **2015**, *27*, 6683. [CrossRef]
77. Sui, Y.K.; Hess-Dunning, A.; Wei, P.; Pentzer, E.; Sankaran, R.M.; Zorman, C.A. Electrically Conductive, Reduced Graphene Oxide Structures Fabricated by Inkjet Printing and Low Temperature Plasma Reduction. *Adv. Mat. Tech.* **2019**, *4*, 1900834. [CrossRef]
78. Pei, L.M.; Li, Y.F. Rapid and efficient intense pulsed light reduction of graphene oxide inks for flexible printed electronics. *RSC Adv.* **2017**, *7*, 51711–51720. [CrossRef]
79. McManus, D.; Vranic, S.; Withers, F.; Sanchez-Romaguera, V.; Macucci, M.; Yang, H.; Sorrentino, R.; Parvez, K.; Son, S.-K.; Iannaccone, G.; et al. Water-based and biocompatible 2D crystal inks for all-inkjet-printed heterostructures. *Nat. Nanotechnol.* **2017**, *12*, 343–350. [CrossRef]
80. Romagnoli, M.; Lassinantti Gualtieri, M.; Cannio, M.; Barbieri, F.; Giovanardi, R. Preparation of an aqueous graphitic ink for thermal drop-on-demand inkjet printing. *Mat. Chem. Phys.* **2016**, *182*, 263–271. [CrossRef]

 *nanomaterials*

*Article*

# Graphene-Based Ion-Selective Field-Effect Transistor for Sodium Sensing

Ting Huang [1], Kan Kan Yeung [1,2], Jingwei Li [1,3], Honglin Sun [1], Md Masruck Alam [1] and Zhaoli Gao [1,2,*]

1   Biomedical Engineering Department, The Chinese University of Hong Kong, Shatin, New Territories, Hong Kong, China; irishhh@link.cuhk.edu.hk (T.H.); kyeung@cuhk.edu.hk (K.K.Y.); jlidt@connect.ust.hk (J.L.); sunhl20@link.cuhk.edu.hk (H.S.); masruckalam@link.cuhk.edu.hk (M.M.A.)
2   CUHK Shenzhen Research Institute, Nanshan, Shenzhen 518057, China
3   Department of Chemical and Biological Engineering, The Hong Kong University of Science and Technology, Clear Water Bay, Kowloon, Hong Kong, China
*   Correspondence: zlgao@cuhk.edu.hk

**Abstract:** Field-effect transistors have attracted significant attention in chemical sensing and clinical diagnosis, due to their high sensitivity and label-free operation. Through a scalable photolithographic process in this study, we fabricated graphene-based ion-sensitive field-effect transistor (ISFET) arrays that can continuously monitor sodium ions in real-time. As the sodium ion concentration increased, the current–gate voltage characteristic curves shifted towards the negative direction, showing that sodium ions were captured and could be detected over a wide concentration range, from $10^{-8}$ to $10^{-1}$ M, with a sensitivity of 152.4 mV/dec. Time-dependent measurements and interfering experiments were conducted to validate the real-time measurements and the highly specific detection capability of our sensor. Our graphene ISFETs (G-ISFET) not only showed a fast response, but also exhibited remarkable selectivity against interference ions, including $Ca^{2+}$, $K^+$, $Mg^{2+}$ and $NH_4^+$. The scalability, high sensitivity and selectivity synergistically make our G-ISFET a promising platform for sodium sensing in health monitoring.

**Keywords:** ion-selective field-effect transistor; graphene; sodium ions; real-time monitoring

**Citation:** Huang, T.; Yeung, K.K.; Li, J.; Sun, H.; Alam, M.M.; Gao, Z. Graphene-Based Ion-Selective Field-Effect Transistor for Sodium Sensing. *Nanomaterials* **2022**, *12*, 2620. https://doi.org/10.3390/nano12152620

Academic Editor: Eugene Kogan

Received: 30 June 2022
Accepted: 26 July 2022
Published: 29 July 2022

**Publisher's Note:** MDPI stays neutral with regard to jurisdictional claims in published maps and institutional affiliations.

## 1. Introduction

Sodium ions are important indicators for monitoring and evaluating health status owing to their important role in homeostasis and maintaining the proper functions of the nervous system [1–3]. For instance, the total sodium level in cognitively normal brain tissues is around 35–45 mM, and 12–21 mM in healthy muscle tissue [4–6]. Deviation of sodium concentrations in the human body is related to its hydration status, which can be used as an indicator for health monitoring [7,8]. Thus, rapid, reliable and real-time monitoring of sodium ions has been an increasing interest in the fields of precision medicine and personalized healthcare [1,9,10]. To date, solid-contact ion-selective electrodes (ISE) are the most commonly used platforms for ion sensing, due to their low cost, accuracy, and simple operation [11–15]. However, ISEs have drawbacks, including the relatively high detection limit and narrow detection range, e.g., $10^{-4}$ or $10^{-5}$ M for specific ions [16–19].

Recently, field-effect transistors (FET) have gained increasing attention in ion sensing, offering the prospect of simple, rapid, cost-effective, and label-free detection [20,21]. The FET biosensors hold tremendous promise for label-free detection of target molecules with high accuracy and selectivity, without the usage of fluorescent, isotopic, or electrochemical labeling [22,23]. In combination with an ion-selective membranes (ISM), ion-sensitive field-effect transistors (ISFETs) are promising for ion sensing with enhanced sensitivity, and reduced sensor sizes and response times, providing the possibility to integrate them with flexible electronics [24–28].

Graphene is a 2D material with unique material properties, such as high carrier mobility (up to $10^6$ cm$^2$/V·s) [29], high conductivity [30], excellent mechanical strength, etc. [30,31]. Taking advantage of all these features combined, we have fabricated G-ISFETs that offer high sensitivity, selectivity and real-time monitoring of sodium ions. The graphene channel was grown by atmospheric pressure chemical vapor deposition (CVD), and transferred to pre-patterned electrodes, followed by a scalable photolithographic process. The graphene FETs (GFETs) were then functionalized with a sodium ionophore to specifically capture the target sodium ions. A broad range of sodium concentrations, from $10^{-8}$ to $10^{-1}$ M, which covers the sodium concentration in tissues, was detected, with a sensitivity of 152.4 mV/dec. We further conducted time-dependent measurements and control experiments to demonstrate the capability of real-time monitoring with high selectivity. The high performance of our G-ISFET makes it a promising platform for the real-time monitoring of sodium ions for health monitoring through physiological liquids.

## 2. Materials and Methods

### 2.1. Graphene Synthesis

The monolayer graphene film was synthesized using a chemical vapor deposition system (Lindberg/Blue M™ Mini-Mite™ Thermo Scientific Co., Waltham, MA, USA). The copper foil (Alfa Aesar, #13382, Haverhill, MA, USA) was cleaned by sonication in 5.4% $HNO_3$ for 1 min and then rinsed in DI water twice, followed by drying with high-pressure nitrogen gas. The cleaned foil was then transferred into the quartz tube. The furnace was heated to 1050 °C with a constant flow of 500 sccm Ar and 30 sccm $H_2$ and then annealed for 5 min. The 5 sccm-diluted $CH_4$ (0.5% in Ar) was introduced as a carbon source, and the growth time was 1 h. Lastly, the furnace was rapidly cooled to room temperature under the $H_2$ and Ar atmosphere.

### 2.2. GFET Sensor Array Fabrication

The sensor fabrication process was summarized in Figure S1. First, the electrode pattern was defined on a 4-inch p-doped $SiO_2$ (285 nm)/Si wafer by standard photolithography. The contact metallization was 8 nm Cr/45 nm Au, deposited by e-beam evaporation. Monolayer graphene was then transferred onto the pre-patterned $SiO_2$/Si chip using a "bubbling" transfer method. Briefly, a layer of polymethylmethacrylate (PMMA) was spin-coated on the graphene-Cu foil, followed by baking at 105 °C for 2 min and then slowly immersed into a 50 mM NaOH aqueous solution [32]. By applying a 15 V voltage, the graphene/PMMA film was peeled off from Cu foil by the hydrogen bubbles formed on the copper surface. The film was washed with DI water thrice and transferred onto the electrode chip. The chip was air-dried and then baked at 150 °C for 2 min before removing the PMMA with acetone. The graphene/electrode chip was then spin-coated with PMGI (Micro Chem Corp., Newton, MA, USA) and a S1813 (Shipley) photoresist bilayer and exposed using an ABM aligner. Graphene outside the channels was removed by $O_2$ plasma etching. The remaining photoresist on graphene channels was stripped by Remover PG (Micro Chem Corp., Newton, MA, USA), acetone, and IPA. Finally, the GFET arrays were annealed in Ar/$H_2$ forming gas at 225 °C to remove photoresist residues.

### 2.3. Ionophore Membrane Preparation

Selectophore grade sodium ionophore X (4-tertbutylcalix [4]arene-tetraacetic acid tetraethyl ester), sodium tetrakis [3,5-bis(trifluoromethyl) phenyl] borate (Na-TFPB), 2-nitrophenyl octyl ether (2-NPOE), tetrahydrofuran (THF), and poly (vinyl chloride) (PVC) were purchased from Sigma-Aldrich. The ionophore membrane was prepared by mixing 1 mg sodium ionophore X, 47.2 mg PVC, 90.7 µL 2-NPOE, and 0.29 mg Na-TFPB [33]. The mixture was dissolved in 1 mL THF and sonicated for 1 h, then stored at 4 °C for further usage.

### 2.4. Material Characterization

Micro-Raman measurements were performed by using WiTec Alpha 300 system with a laser excitation wavelength of 532 nm. An atomic force microscope (AFM, Icon Bruker, Tucson, AZ, USA) was used to characterize the height increase during the fabrication process.

### 2.5. Solution Preparation

Sodium chloride (NaCl), potassium chloride (KCl), magnesium chloride ($MgCl_2$), calcium chloride ($CaCl_2$) and ammonium chloride ($NH_4Cl$) anhydrous salts with >99% purity were obtained from Sigma Aldrich. The desired concentrations were carefully prepared and diluted with de-ionized water (18.2 M$\Omega$ cm, Milli-Q$^®$ 3 UV Water Purification System). The sweat sample was collected from a cycling volunteer at different sporting times, and stored in $-20\,°C$ refrigerator before testing.

### 2.6. Electrical Measurement

The 285 nm-thick $SiO_2$ served as the gate dielectric, and the highly p-doped silicon substrate acted as the back-gate electrode. No liquid gate was applied in this study. The I-$V_g$ characteristic measurements were performed after each functionalization step. The probe station (FormFactor MPS 150, Livermore, CA, USA) was equipped with a customized probe card, allowing 100 devices to be measured simultaneously. The Keithley 2400 source meter was used to apply a bias voltage (V = 0.1 V), and the gate voltage was applied using the Keithley 6517 model. A Python program was developed to conduct the measurement and collect data.

### 3. Results and Discussion

Figure 1a shows an optical image of a GFET fabricated by the photolithographic process. The monolayer graphene film was synthesized on a copper foil using chemical vapor deposition, followed by a hydrolysis bubble transfer onto a $SiO_2$/Si chip with prefabricated Cr/Au electrodes to create an array of 100 GFETs. The graphene channel, as shown in Figure 1b, was defined by photolithography and oxygen plasma etching. The GFET chip was then annealed in an Ar/$H_2$ atmosphere to remove any photoresist residues on the graphene channels [34]. The high quality of the as-fabricated GFETs was verified by the negligible D peak (~1345 cm$^{-1}$) in the Raman spectrum (Figure 1c) [35]. The height of the GFET channel was ~0.5 nm, and there was a ~5 $\mu$m height increase after the immobilization of the sodium ionophore membrane. The Raman spectrum and AFM image together confirm the high quality of the as-grown CVD graphene, even after the photolithographic process.

As seen in Figure 2, the current-back gate voltage (I-$V_g$) measurements show good device-to-device uniformity across the 100 arrays. The Dirac voltage and carrier mobility were extracted by fitting the hole branch of the I-$V_g$ curve to the following equation [36,37]:

$$\sigma^{-1}(V_g) = \left[\mu c_g (V_D - V_g)\right]^{-1} + \sigma_s^{-1} \tag{1}$$

where $c_g$ is the gate capacitance per unit area (12.1 nF cm$^{-2}$ for the 285 nm thick $SiO_2$), $\mu$ is the hole carrier mobility, $\sigma_s$ is the saturation conductivity when $V_g$ approaches $-\infty$. The narrow distribution of the Dirac point voltage ($6.3 \pm 4$ V) and hole carrier mobility ($2400 \pm 600$ cm$^2$ V$^{-1}$ s$^{-1}$) indicates a low doping effect induced by the fabrication process.

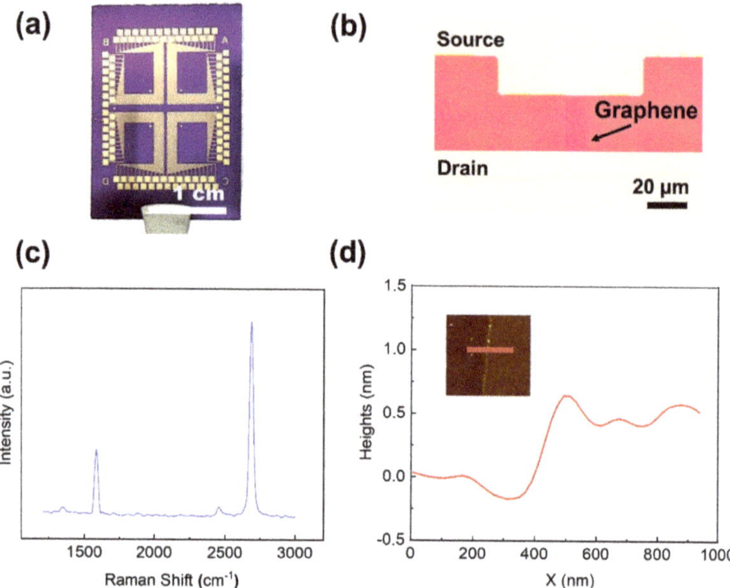



**Figure 1.** (**a**) Optical image of as-fabricated GFETs, (**b**) Optical image of the graphene channel and source/drain electrode. (**c**) Raman spectrum of a graphene channel after the fabrication process. Two characteristic peaks were found: G peak at ~1580 $cm^{-1}$ and 2D peak at ~2700 $cm^{-1}$ (**d**) The line scan profile of the as-annealed GFET, Inset: AFM image with scan line indicated. The thickness of the graphene channel is ~0.5 nm.

The as-fabricated GFETs were then functionalized with the prepared sodium selective membrane, as shown in Figure 3. Briefly, the sodium ionophore X was dissolved and mixed with ion-selective membrane (ISM) cocktails (see Materials and Methods). An amount of 25 µL of the solution was drop-cast on the GFET surface, followed by air-drying overnight, to obtain the G-ISFET. The I-$V_g$ characteristics was measured after the ionophore deposition, where the deposition of the ionophore leads to a negative Dirac point shift (Figure 2d). During sensing, the intrinsic structure of ionophore X, namely the calix [4] arenes, provides a scaffold with an optimum cavity for the complexation of sodium ions [38,39]. The captured ion in the sodium-selective membrane resulted in a surface potential change and the Dirac voltage shift in the characteristics curve.

A real-time measurement of the drain-source current through the ISM without the graphene channel against different sodium solutions ($10^{-5}$, $10^{-3}$, and $10^{-1}$ M) was conducted, as shown in Figure S2, and the leaking current between the source and drain electrodes was found at the sub-nA level, which did not affect our study. The G-ISFET was tested against a series of sodium concentrations, from $10^{-8}$ to $10^{-1}$ M, to confirm the sensor response. The ion sensitive membrane provided a cation exchange site and created a barrier that prevented nonspecific ions from reaching the sensing surface. As a result, only sodium ions were able to permeate and pass through the selective membrane to reach the ISM–graphene interface. Accordingly, the sodium ion accumulation on the graphene surface caused a doping effect. This G-ISFET response is shown in Figure 4. A fixed bias voltage of 100 mV was applied during the sensing measurements. As the sodium concentrations increased, there was a consistent trend of negative shifts in the transport curves. This Dirac point shift was attributed to the increase in the electron concentration on the graphene's surface, due to the accumulation of positively charged $Na^+$ ions, thereby driving the Fermi level closer to the charge neutrality point through chemical gating, and consequently decreasing the Dirac point. The dependence of $V_D$ on varying $Na^+$ values is plotted in Figure 4b, where the dotted line represents a linear fit. The slope of calibration fitting reflects the sensitivity of the G-ISFET, i.e., 152.4 mV/dec. The sensitivity is comparable to that of recent reports (see Table S1) [20,36,40], presumably attributed to the atomically thin nature of the graphene and the scalable fabrication of high-quality sensor arrays based on CVD graphene.

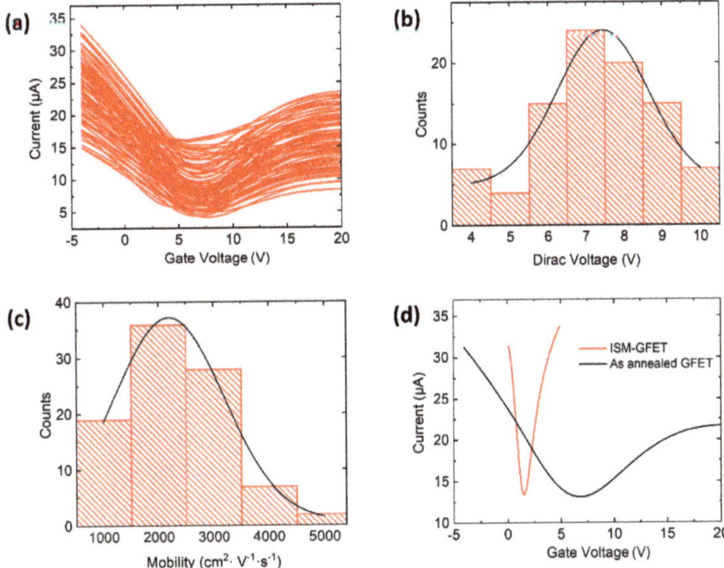

**Figure 2.** (a) Current-gate voltage characteristic of an array of 100 graphene field-effect transistors. Histograms and Gaussian fits (black lines) of (**b**) Dirac voltage and (**c**) hole mobility extracted from the curves in panel **a**. (**d**) Current-gate voltage characteristic curves before and after ionophore deposition.

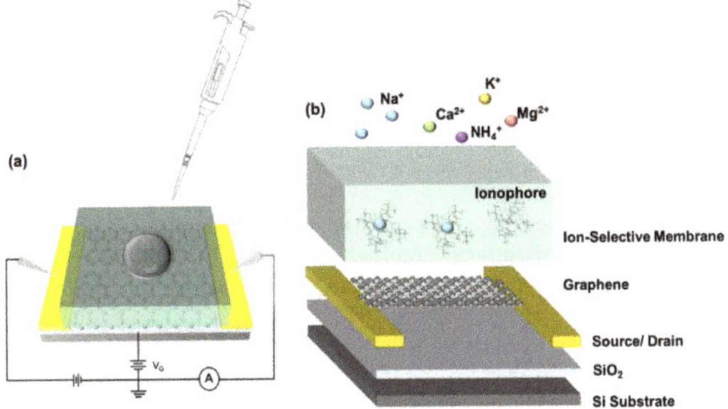


**Figure 3.** (a) Schematic of a back-gated G-ISFET. (**b**) Schematic of ionophore-functionalized GFET. The sodium ions captured in ionophores lead to a doping effect of the GFET.

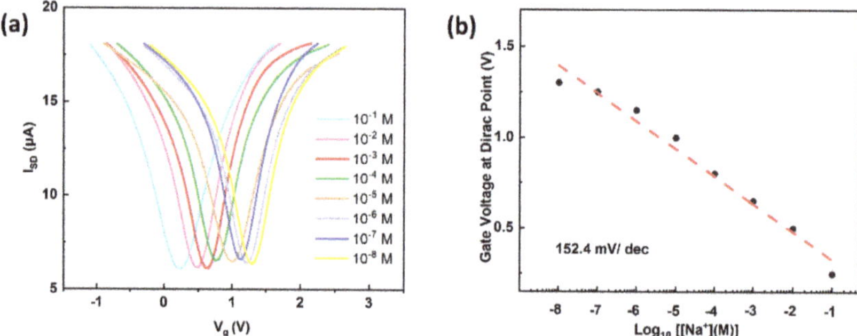

**Figure 4.** (**a**) Transport characteristic curves of G-ISFET against different $Na^+$ concentrations from $10^{-8}$ to $10^{-1}$ M with a bias voltage of 100 mV. (**b**) G-ISFET response as a function for different target sodium concentrations at the logarithmic scale. A response of 152.4 mV/dec was observed.

We next investigated the real-time response of G-ISFET against various sodium concentrations, by measuring $I_{DS}$ versus sensing time with a fixed gate voltage ($V_{ds}$ = 100 mV). As shown in Figure 5a, the source-drain current decreased with the increasing $Na^+$ concentration, in agreement with the n-doping effect by positively charged $Na^+$ ions. The linear response in $I_{DS}$ is plotted in Figure 5b, and the fitting indicates a response of $2.2 \pm 0.08$ µA/dec, consistent with previously reported ISFETs [41–43].

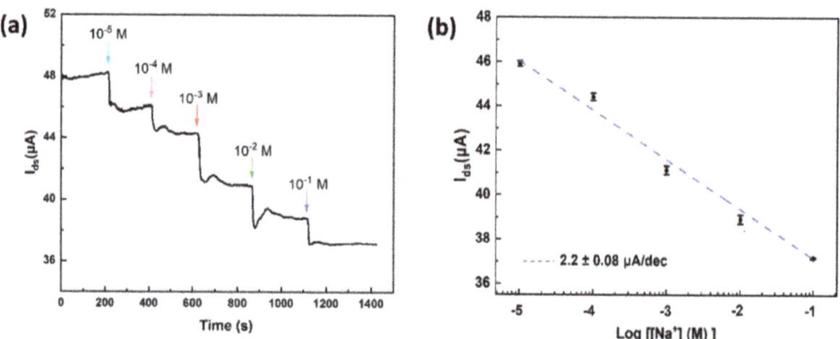

**Figure 5.** (**a**) Real-time response of the source–drain current against a series of $Na^+$ molar concentrations ($V_g$ = 0 V). (**b**) The linear response in $I_{DS}$ with different sodium concentrations from the real–time measurements in panel a.

Selectivity is a crucial factor in evaluating the performance of an ion sensor. We further carried out interference experiments to verify the effectiveness of our G-ISFET. As shown in Figure 6, several non-specific ions were tested, including $Ca^{2+}$, $K^+$, $Mg^{2+}$ and $NH_4^+$, and the relative Dirac point shift was plotted. In sharp contrast to the large Dirac voltage shift for sodium ions, the as-fabricated G-ISFET displays a negligible response to the interfering ions, indicating that the ion-selective membrane specifically captured the target ions, and possessed excellent selectivity against nonspecific ions. We also performed measurements with a real sample, i.e., human sweat. As shown in Figure 6b, the source-drain current decreased with the increasing concentration of sodium ions (from 47.91 mM to 49.62 mM). This result confirmed the high selectivity and rapid response of the G-ISFET, which offers a pathway toward health evaluation through sweat.

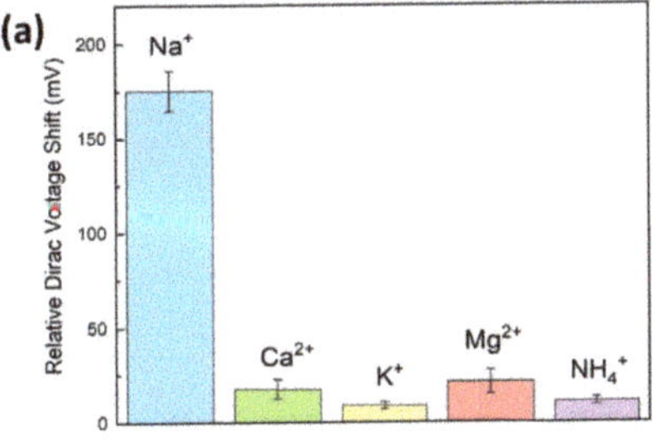
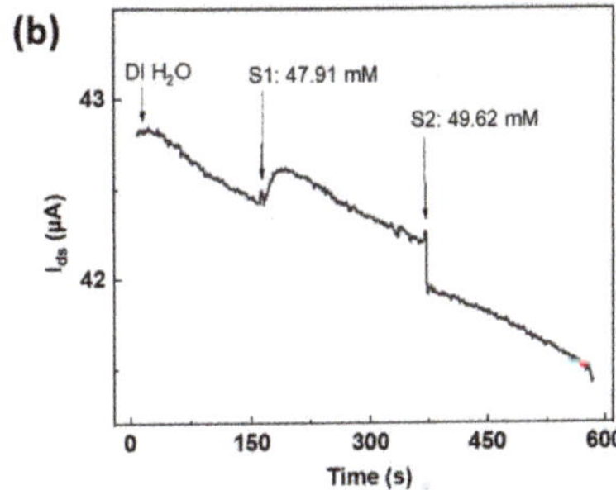



**Figure 6.** (**a**) The selectivity test against several interfering ions. Error bars are the standard deviation of the mean. (**b**) Real-time responses of G-ISFET against human sweat samples collected from a sporting volunteer.

## 4. Conclusions

We developed a graphene-based ISFET, incorporated with an ion-selective membrane, that can selectively detect sodium ions with high sensitivity. We grew large-area, high quality monolayer graphene by chemical vapor deposition, followed by a scalable photolithographic process, to fabricate the GFETs. The as-fabricated GFETs were then functionalized with sodium ionophore to sensitively capture sodium ions. We detected sodium ions with a wide range of concentrations, from $10^{-8}$ to $10^{-1}$ M, and achieved a sensitivity of 152.4 mV/dec, comparable to previously reported ISFET sensors. Nevertheless, the back-gate architecture of G-ISFET eliminates the usage of reference electrodes, offering a way to miniaturize the ISFET device. We further conducted time-dependent measurements and interfering experiments to demonstrate the real-time response and selectivity capabilities of our G-ISFETs, showing a fast response to changes in concentration, and exhibiting excellent selectivity against interference ions, including $Ca^{2+}$, $K^+$, $Mg^{2+}$ and $NH_4^+$. The scalability, sensitivity and selectivity synergistically make our G-ISFET a promising candidate for sodium sensing in health monitoring.

**Supplementary Materials:** The following supporting information can be downloaded at: https://www.mdpi.com/article/10.3390/nano12152620/s1, Figure S1: Illustration of the scalable fabrication process of G-ISFET; Figure S2: Real-time measurement of drain-source current with a bare ISM; Table S1: Comparison of ion sensitivities for sodium sensing.

**Author Contributions:** Supervision, Z.G.; Conceptualization, T.H.; Writing—original draft preparation, T.H. and K.K.Y.; writing—review and editing, J.L., H.S., M.M.A. and Z.G. All authors have read and agreed to the published version of the manuscript.

**Funding:** Z.G. thanks the support from the Key-Area Research and Development Program of Guangdong Province (Grant No. 2020B0101030002), the National Natural Science Foundation of China (Project No. 62101475), the Research Grant Council of Hong Kong (Project Nos. 24201020 and 14207421), and the Research Matching Grant Scheme of Hong Kong Government (Project No. 8601547).

**Institutional Review Board Statement:** Not applicable.

**Informed Consent Statement:** Not applicable.

**Data Availability Statement:** Not applicable.

**Conflicts of Interest:** The authors declare no conflict of interest.

# References

1. Lerch, M.M.; Hansen, M.J.; van Dam, G.M.; Szymanski, W.; Feringa, B.L. Emerging Targets in Photopharmacology. *Angew. Chem. Int. Edit.* **2016**, *55*, 10978–10999. [CrossRef] [PubMed]
2. Reynolds, R.M.; Padfield, P.L.; Seckl, J.R. Disorders of sodium balance. *BMJ* **2006**, *332*, 702–705. [CrossRef] [PubMed]
3. He, F.J.; MacGregor, G.A. Plasma sodium and hypertension. *Kidney Int.* **2004**, *66*, 2454–2466. [CrossRef] [PubMed]
4. Thulborn, K.; Lui, E.; Guntin, J.; Jamil, S.; Sun, Z.; Claiborne, T.C.; Atkinson, I.C. Quantitative sodium MRI of the human brain at 9.4 T provides assessment of tissue sodium concentration and cell volume fraction during normal aging. *NMR Biomed.* **2016**, *29*, 137–143. [CrossRef] [PubMed]
5. Inglese, M.; Madelin, G.; Oesingmann, N.; Babb, J.; Wu, W.; Stoeckel, B.; Herbert, J.; Johnson, G. Brain tissue sodium concentration in multiple sclerosis: A sodium imaging study at 3 tesla. *Brain* **2010**, *133*, 847–857. [CrossRef] [PubMed]
6. Gerhalter, T.; Gast, L.V.; Marty, B.; Uder, M.; Carlier, P.G.; Nagel, A.M. Assessing the variability of 23Na MRI in skeletal muscle tissue: Reproducibility and repeatability of tissue sodium concentration measurements in the lower leg at 3 T. *NMR Biomed.* **2020**, *33*, e4279. [CrossRef] [PubMed]
7. Titze, J.; Machnik, A. Sodium sensing in the interstitium and relationship to hypertension. *Curr. Opin. Nephrol. Hypertens.* **2010**, *19*, 385–392. [CrossRef] [PubMed]
8. Shirreffs, S. Markers of hydration status. *J. Sports Med. Phys. Fit.* **2000**, *40*, 80–84. [CrossRef]
9. Yang, Y.; Mis, M.A.; Estacion, M.; Dib-Hajj, S.D.; Waxman, S.G. NaV1. 7 as a pharmacogenomic target for pain: Moving toward precision medicine. *Trends Pharmacol. Sci.* **2018**, *39*, 258–275. [CrossRef]
10. Sun, W.; Lee, J.; Zhang, S.; Benyshek, C.; Dokmeci, M.R.; Khademhosseini, A. Engineering precision medicine. *Adv. Sci.* **2019**, *6*, 1801039. [CrossRef]
11. Freiser, H. *Ion-Selective Electrodes in Analytical Chemistry*; Springer Science & Business Media: Berlin/Heidelberg, Germany, 2012.
12. Mikhelson, K. Ion-selective electrodes with sensitivity in strongly diluted solutions. *J. Anal. Chem.* **2010**, *65*, 112–116. [CrossRef]
13. Arnold, M.A.; Meyerhoff, M.E. Ion-selective electrodes. *Anal. Chem.* **1984**, *56*, 20–48. [CrossRef]
14. De Marco, R.; Clarke, G.; Pejcic, B. Ion-selective electrode potentiometry in environmental analysis. *Electroanal. Int. J. Devoted Fundam. Pract. Asp. Electroanal.* **2007**, *19*, 1987–2001. [CrossRef]
15. Yeung, K.K.; Huang, T.; Hua, Y.; Zhang, K.; Yuen, M.M.; Gao, Z. Recent advances in electrochemical sensors for wearable sweat monitoring: A review. *IEEE Sens. J.* **2021**, *21*, 14522–14539. [CrossRef]
16. Bobacka, J.; Lindfors, T.; Mccarrick, M.; Ivaska, A.; Lewenstam, A. Single Piece All-Solid-State Ion-Selective Electrode. *Anal. Chem.* **1995**, *67*, 3819–3823. [CrossRef]
17. Cammann, K. Bio-sensors based on ion-selective electrodes. *Fresenius' Z. Für Anal. Chem.* **1977**, *287*, 1–9. [CrossRef]
18. Ezzat, S.; Ahmed, M.A.; Amr, A.G.E.; Al-Omar, M.A.; Kamel, A.H.; Khalifa, N.M. Single-Piece All-Solid-State Potential Ion-Selective Electrodes Integrated with Molecularly Imprinted Polymers (MIPs) for Neutral 2,4-Dichlorophenol Assessment. *Materials* **2019**, *12*, 2924. [CrossRef]
19. Cammann, K. *Working with Ion-Selective Electrodes: Chemical Laboratory Practice*; Springer Science & Business Media: Berlin/Heidelberg, Germany, 2012.
20. Fu, W.; Jiang, L.; van Geest, E.P.; Lima, L.M.; Schneider, G.F. Sensing at the surface of graphene field-effect transistors. *Adv. Mater.* **2017**, *29*, 1603610. [CrossRef]
21. Syu, Y.-C.; Hsu, W.-E.; Lin, C.-T. Field-effect transistor biosensing: Devices and clinical applications. *ECS J. Solid State Sci. Technol.* **2018**, *7*, Q3196. [CrossRef]
22. Xu, M.; Liang, T.; Shi, M.; Chen, H. Graphene-like two-dimensional materials. *Chem. Rev.* **2013**, *113*, 3766–3798. [CrossRef]
23. Akinwande, D.; Huyghebaert, C.; Wang, C.-H.; Serna, M.I.; Goossens, S.; Li, L.-J.; Wong, H.-S.P.; Koppens, F.H. Graphene and two-dimensional materials for silicon technology. *Nature* **2019**, *573*, 507–518. [CrossRef]
24. Cho, S.K.; Cho, W.J. Highly Sensitive and Selective Sodium Ion Sensor Based on Silicon Nanowire Dual Gate Field-Effect Transistor. *Sensors* **2021**, *21*, 4213. [CrossRef] [PubMed]
25. Lee, C.-S.; Kim, S.K.; Kim, M. Ion-sensitive field-effect transistor for biological sensing. *Sensors* **2009**, *9*, 7111–7131. [CrossRef]
26. Maehashi, K.; Sofue, Y.; Okamoto, S.; Ohno, Y.; Inoue, K.; Matsumoto, K. Selective ion sensors based on ionophore-modified graphene field-effect transistors. *Sens. Actuators B Chem.* **2013**, *187*, 45–49. [CrossRef]
27. Lim, H.R.; Lee, S.M.; Mahmood, M.; Kwon, S.; Kim, Y.S.; Lee, Y.; Yeo, W.H. Development of Flexible Ion-Selective Electrodes for Saliva Sodium Detection. *Sensors* **2021**, *21*, 1642. [CrossRef] [PubMed]
28. Spijkman, M.; Smits, E.C.P.; Cillessen, J.F.M.; Biscarini, F.; Blom, P.W.M.; de Leeuw, D.M. Beyond the Nernst-limit with dual-gate ZnO ion-sensitive field-effect transistors. *Appl. Phys. Lett.* **2011**, *98*, 6169. [CrossRef]
29. Neto, A.C.; Guinea, F.; Peres, N.M.; Novoselov, K.S.; Geim, A.K. The electronic properties of graphene. *Rev. Mod. Phys.* **2009**, *81*, 109. [CrossRef]
30. Novoselov, K.S.; Colombo, L.; Gellert, P.; Schwab, M.; Kim, K. A roadmap for graphene. *Nature* **2012**, *490*, 192–200. [CrossRef]
31. Ovid'Ko, I. Mechanical properties of graphene. *Rev. Adv. Mater. Sci.* **2013**, *34*, 1–11.
32. Gao, L.; Ren, W.; Xu, H.; Jin, L.; Wang, Z.; Ma, T.; Ma, L.-P.; Zhang, Z.; Fu, Q.; Peng, L.-M. Repeated growth and bubbling transfer of graphene with millimetre-size single-crystal grains using platinum. *Nat. Commun.* **2012**, *3*, 699. [CrossRef]

33. Bandodkar, A.J.; Molinnus, D.; Mirza, O.; Guinovart, T.; Windmiller, J.R.; Valdés-Ramírez, G.; Andrade, F.J.; Schöning, M.J.; Wang, J. Epidermal tattoo potentiometric sodium sensors with wireless signal transduction for continuous non-invasive sweat monitoring. *Biosens. Bioelectron.* **2014**, *54*, 603–609. [CrossRef] [PubMed]

34. Gao, Z.; Xia, H.; Zauberman, J.; Tomaiuolo, M.; Ping, J.; Zhang, Q.; Ducos, P.; Ye, H.; Wang, S.; Yang, X. Detection of sub-fM DNA with target recycling and self-assembly amplification on graphene field-effect biosensors. *Nano Lett.* **2018**, *18*, 3509–3515. [CrossRef] [PubMed]

35. Ferrari, A.C.; Meyer, J.C.; Scardaci, V.; Casiraghi, C.; Lazzeri, M.; Mauri, F.; Piscanec, S.; Jiang, D.; Novoselov, K.S.; Roth, S. Raman spectrum of graphene and graphene layers. *Phys. Rev. Lett.* **2006**, *97*, 187401. [CrossRef] [PubMed]

36. Morozov, S.; Novoselov, K.; Katsnelson, M.; Schedin, F.; Elias, D.C.; Jaszczak, J.A.; Geim, A. Giant intrinsic carrier mobilities in graphene and its bilayer. *Phys. Rev. Lett.* **2008**, *100*, 016602. [CrossRef] [PubMed]

37. Chen, F.; Xia, J.; Ferry, D.K.; Tao, N. Dielectric screening enhanced performance in graphene FET. *Nano Lett.* **2009**, *9*, 2571–2574. [CrossRef] [PubMed]

38. Bühlmann, P.; Pretsch, E.; Bakker, E. Carrier-based ion-selective electrodes and bulk optodes. 2. Ionophores for potentiometric and optical sensors. *Chem. Rev.* **1998**, *98*, 1593–1688. [CrossRef]

39. Luboch, E.; Jeszke, M.; Szarmach, M.; Łukasik, N. New bis (azobenzocrown) s with dodecylmethylmalonyl linkers as ionophores for sodium selective potentiometric sensors. *J. Incl. Phenom. Macrocycl. Chem.* **2016**, *86*, 323–335. [CrossRef]

40. Meric, I.; Han, M.Y.; Young, A.F.; Ozyilmaz, B.; Kim, P.; Shepard, K.L. Current saturation in zero-bandgap, top-gated graphene field-effect transistors. *Nat. Nanotechnol.* **2008**, *3*, 654–659. [CrossRef]

41. Garcia-Cordero, E.; Bellando, F.; Zhang, J.; Wildhaber, F.; Longo, J.; Guérin, H.; Ionescu, A.M. Three-dimensional integrated ultra-low-volume passive microfluidics with ion-sensitive field-effect transistors for multiparameter wearable sweat analyzers. *ACS Nano* **2018**, *12*, 12646–12656. [CrossRef] [PubMed]

42. Fakih, I.; Durnan, O.; Mahvash, F.; Napal, I.; Centeno, A.; Zurutuza, A.; Yargeau, V.; Szkopek, T. Selective ion sensing with high resolution large area graphene field effect transistor arrays. *Nat. Commun.* **2020**, *11*, 3226. [CrossRef] [PubMed]

43. Zhang, J.; Rupakula, M.; Bellando, F.; Garcia Cordero, E.; Longo, J.; Wildhaber, F.; Herment, G.; Guerin, H.; Ionescu, A.M. Sweat biomarker sensor incorporating picowatt, three-dimensional extended metal gate ion sensitive field effect transistors. *ACS Sens.* **2019**, *4*, 2039–2047. [CrossRef] [PubMed]

 *nanomaterials*

*Article*

# Quantum Diffusion in the Lowest Landau Level of Disordered Graphene

Andreas Sinner [1,2,*] and Gregor Tkachov [2,*]

[1]   Institute of Physics, University of Opole, 45-052 Opole, Poland
[2]   Institute of Physics, University of Augsburg, 86135 Augsburg, Germany
*   Correspondence: andreas.sinner@uni-a.de (A.S.); gregor1.tkachov@uni-a.de (G.T.)

**Abstract:** Electronic transport in the lowest Landau level of disordered graphene sheets placed in a homogeneous perpendicular magnetic field is a long-standing and cumbersome problem which defies a conclusive solution for several years. Because the modeled system lacks an intrinsic small parameter, the theoretical picture is infested with singularities and anomalies. We propose an analytical approach to the conductivity based on the analysis of the diffusive processes, and we calculate the density of states, the diffusion coefficient and the static conductivity. The obtained results are not only interesting from the purely theoretical point of view but have a practical significance as well, especially for the development of the novel high-precision calibration devices.

**Keywords:** low-dimensional semimetals; electronic transport in graphene; quantum hall effect

**Citation:** Sinner, A.; Tkachov, G. Quantum Diffusion in the Lowest Landau Level of Disordered Graphene. *Nanomaterials* **2022**, *12*, 1675. https://doi.org/10.3390/nano12101675

Academic Editors: Eugene Kogan and Vincenzo Carravetta

Received: 15 March 2022
Accepted: 11 May 2022
Published: 14 May 2022

**Publisher's Note:** MDPI stays neutral with regard to jurisdictional claims in published maps and institutional affiliations.

## 1. Introduction

Two-dimensional (2d) electronic systems in general and their transport properties in particular have been in the focus of intense research for several decades. In such systems, the effects due to quantum interference are strong and give rise to the interesting and rather unintuitive phenomena, as for instance various facets of the quantum Hall effect. Yet another effect on the transport that is supposed to be strong in 2d arises from the disorder which is always present in realistic materials. In conventional 2d electron systems, which are characterized by a parabolic and isomorphic spectrum, the presence of the disorder is widely believed to lead to the destructive interference of electronic quantum waves and consequently to the suppression of the electronic transport through the system on macroscopic scales. This phenomenon is usually called the Anderson localization of electronic wave functions, and it has received much of attention in the past [1–4]. This picture was challenged with the discovery of the unconventional behavior of electrons in the transition between Hall plateaux in quantum Hall systems. The experimental evidence from this observations points to the principal possibility for the existence of a metallic state in 2d systems under special conditions [5]. However, a real change of paradigm occured with the discovery of metallic states in graphene [6–8] and in a number of further low-dimensional systems, which is collectively known as the topological insulators [9–16]. A feature common to all these systems is the presence of the so-called nodes in the band structure and the linearity of the spectrum in the vicinity of these nodes. Despite being pristine 2d systems, they reveal a finite dc conductivity, which is very robust against disorder and thermal fluctuations.

The theoretical approach to the electronic transport of disordered electron gases is a rather formidable and cumbersome undertaking. Because the translational symmetry in the system is explicitly broken by the randomness, the usual methods of the theoretical analysis, which are mainly built around the duality between the position and momentum space representations and the special role of the Fourier transformation as the diagonalization tool for the quadratic Hamiltonians, no longer work. Therefore, the main idea behind every analytical approach to the macroscopic disordered systems is to reintroduce the translational invariance into the system by mapping the initial problem, which usually neglects the electron–electron and electron–phonon interactions from the outset, by a kind of an averaging operation on an effective interacting model in which the scattering of individual electrons on the randomly distributed potentials is approximated by the interaction operators expressed in terms of bilineals of second quantization operators. However, in practical terms, such an averaging procedure works well only under a weak disorder assumption, which guarantees a well-formed saddle-like shape of the free energy functional. In this case, the main effects caused by the disorder are taken into account by the summation of all contributions in partial diagrammatic channels [2–4,17–30].

In magnetic fields, the quantum mechanics of charge carriers with a linear spectrum specific for graphene is governed by an interplay of the intrinsic and magnetic-field-induced Berry curvatures. Several aspects of this physics remain widely untouched, though. For instance, relatively little is known about the role of disorder and its interplay with the magnetic field. The overall progress in this area has been slow because of the technical challenges, which are considerable even by the standards of the community [31–35]. A number of issues make the disordered electrons in the homogeneous perpendicular magnetic field look differently than the situation without a magnetic field. Due to the freedom of the gauge choice, the problem can be approached in a number of ways, which differ very much in details and in the outer appearance. The popular choice of the central gauge has the advantage that the solutions of the Schrödinger equation are states localized in the position space. Therefore, one can do computations in the position space in an exact manner.

The envisaged problem is notoriously difficult because the model lacks a small expansion parameter [36]. This inevitably leads to divergent expansion series. A powerful method developed to keep such divergences under control is the renormalization group. In the past, our understanding of the physics of disordered metals and semiconductors profited vastly from the various combinations of variational and perturbative techniques with the renormalization group, c.f. Refs. [1–3,25–27] and Refs. [37–39]. However, in the central gauge picture, there is no continuous variable to be sliced off by iterations in order to obtain the renormalization group equations. Of course, one can use a different gauge, which allows for a description in terms of states localized in one direction and propagating in the other. The price to pay is the loss of exactness, which is too costly to give up. In this paper, we develop a diagrammatic approach to the conductivity of the two-dimensional disordered electron gas in a strong magnetic field in a central gauge picture. While these series can still be wrapped up exactly for the single-particle propagators, as it was impressively demonstrated by Wegner in Ref. [32], additional technical issues make elusive every attempt of applying these techniques with the same success to the two-particles propagators. The available divergent series cannot be directly plugged into the Kubo formula without some not a priori obvious regularization or resummation. Hence, the usual way to approach the conductivity is via the Einstein relation and correspondingly via the notion of diffusion [40–42]. Because the corresponding statistical averages require normalization with respect to the vacuum fluctuations [43], this provides a tool of estimating the measurable quantities by means of some kind of analytical continuation [34,35,44,45].

To make our approach function, it relies on the information from the perturbative expansion. Therefore, we perform the exact computations of the perturbative series to the very high order. We identify the exact asymptotics of the two-particles propagator functions and approach the diffusion coefficient via the mean squared displacement using these asymptotics. It turns out that the behavior at longer time is dominated by the higher-order elements and tends toward a stationary state. On the sublaying time scales though, there is a large region with linear time dependence, which is characteristic of the diffusion. To approach this regime, we propose a self-consistent equation of motion for the mean squared displacement and extract the diffusion coefficient from there. With the obtained diffusion coefficient and density of states, we find via the Einstein relation a universal expression for the static conductivity in the lowest Landau level. All the system becomes metallic within a parameter window around the eigenvalues of the Hamiltonian of the clean system. With increasing disorder, this parameter window becomes broader. Numerically, the conductivity of disordered gapless and undoped graphene is very close to the experimentally established values.

The structure of this paper is as follows: In Section 2, we briefly discuss the main facts about the tight-binding Hamiltonian on the honeycomb lattice, its eigenvalues and eigenstates, and introduce the effective continuous model. In Section 3, we elaborate on the topological properties of the Hamiltonian and its eigenstates. In Section 4, we proceed with the consideration of the effective Hamiltonian, which describes the graphene in a strong external magnetic field and evaluates the single-particle propagator of the clean system in Section 5. In Section 6, we evaluate the Kubo–Greenwood formula for the dc conductivity of the clean gapless and chemically neutral graphene off and in an external magnetic field. In Section 7, we approach the single-particle propagator of the disordered system and discuss the Wegner's exact solution and the exact density of states. In Section 8, we give our result for the two-particles propagator and for the mean squared displacement. Finally, in Section 9, we extract the diffusion coefficient from the equation of motion of the mean squared displacement and with that the static conductivity.

## 2. Tight-Binding and Effective Hamiltonian of Graphene

First, we briefly review the main spectral and topological properties of the tight-binding Hamiltonian on a honeycomb lattice. In second quantization, it reads

$$H_{\text{TB}} = -t \sum_{\langle rr' \rangle} (c_r^\dagger d_{r'} + d_{r'}^\dagger c_r) , \tag{1}$$

where $c$ and $d$ ($c^\dagger$ and $d^\dagger$) denote the annihilation (creation) operators acting on the lattice sites of each sublattice of the honeycomb lattice, respectively. The nearest neighbor positions are $a_1 = a(0, -1)$, $a_{2,3} = \dfrac{a}{2}\left(\pm\sqrt{3}, 1\right)$, where $a$ denotes the lattice spacing. The tight-binding Hamiltonian Equation (1) is translationally invariant and is diagonalized by a Fourier transform, giving the eigenvalues $E_\pm = \pm E = \pm\sqrt{h_1^2 + h_2^2}$ and the respective eigenstates of the first-quantized Hamiltonian

$$|v_\pm\rangle = \pm\frac{1}{\sqrt{2E}}[(h_1 - ih_2) , \pm E]^{\text{T}}, \tag{2}$$

with $h_1 = -t\sum_{i=1}^{3}\cos(a_i \cdot k)$ and $h_2 = -t\sum_{i=1}^{3}\sin(a_i \cdot k)$. The eigenvalues of the tight-binding Hamiltonian vanish at nodal points at the corners of the hexagonal Brillouin zone. Each of the corners contributes with the fraction $1/3$ to the total number of cones, which therefore is 2. At chemical neutrality, i.e., for Fermi energy laying precisely at nodal points, there is no extended Fermi surface, and it became common to talk about Fermi points or semimetals. Close to the Fermi points, the fermion dispersion is linear and therefore describes massless Dirac particles, cf. Figure 1. The two Dirac cones are not exactly equivalent though, but they differ by a subtle notion of chirality. The states corresponding to each of two cones can be thought of as the chiral partners of each other. The total chirality of the tight-binding Hamiltonian is therefore zero. Being interested in the physics at low energies, it is usually sufficient to use the effective low-energy Hamiltonian

$$H = \Delta_0\Sigma_{03} + \epsilon_0\Sigma_{00} - iv(\mathcal{D}_+\nabla_- + \mathcal{D}_-\nabla_+), \tag{3}$$

where $\nabla_\pm = \partial_x \pm i\partial_y$. To describe the $4 \times 4$ matrix body of the Hamiltonian, it is useful to introduce the Dirac matrices $\Sigma_{ab} = \sigma_a \otimes \sigma_b$, $a, b = 0, 1, 2, 3$, with $\sigma_{a=1,2,3}$ denoting the Pauli matrices in their usual representation and $\sigma_{a=0}$ being the two-dimensional unity matrix. The first index refers to the valley and the second refers to the sublattice degree of freedom.

With that, $\mathcal{D}_\pm = 1/2[\Sigma_{01} \pm i\Sigma_{02}]$ follows. The band gap $\Delta_0$ in pristine graphene is usually attributed to the spin–orbit coupling and has the size of roughly $10^{-3}$ meV [11], but it can also be considered as a free parameter available for fine tuning. Finally, the chemical potential $\epsilon_0$ is an adjustable quantity.

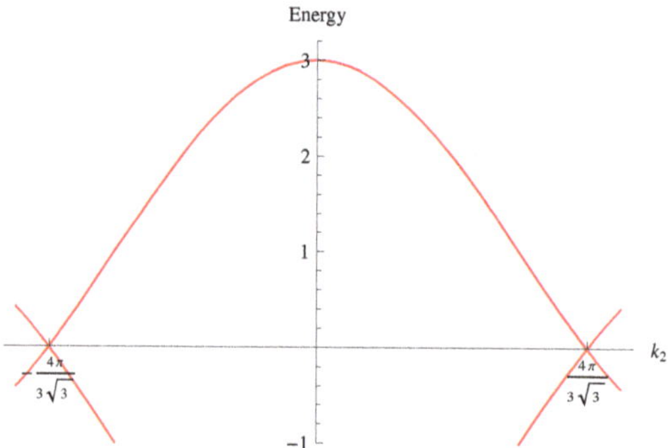


**Figure 1.** Spectrum of the tight-binding model along the line $k_1 = 0$ with two Dirac cones at the corners of the Brilloun zone. The energy axis is scaled in units of the hopping parameter between nearest-neighbors $t$.

### 3. Topological Chern Number

The phase of the wave function plays a crucial role for the properties of the related physical system. It is associated with a topological invariant called the Chern number and is ultimately responsible for the robust macroscopic properties, such as for instance the famous universal conductance. The Chern number is defined as a contour integral [46]

$$C = \frac{1}{2\pi} \oint_C d\vec{k} \cdot \vec{A}(k) \tag{4}$$

over the so-called Berry vector potential $\vec{A}(k) = -i\langle v_\pm | \vec{\nabla}_k | v_\pm \rangle$ along any closed path in the reciprocal space. Here, $|v_\pm\rangle$ denotes an eigenstate of the Hamiltonian defined in Equation (2). The Berry vector potential corresponding to the completely filled band of the full tight-binding Hamiltonian is shown in Figure 2. It appears to have the shape of a double vortex centered around the nodal points of the spectrum and demonstrates nicely the difference in chirality of the Dirac cones by whirling in opposite directions. The total Chern appears as the sum of Chern numbers from each eigenstate. Therefore, the total Chern number of the pristine graphene is zero, but this might change if a fundamental symmetry of the Hamiltonian is broken, e.g., by applying a magnetic field.

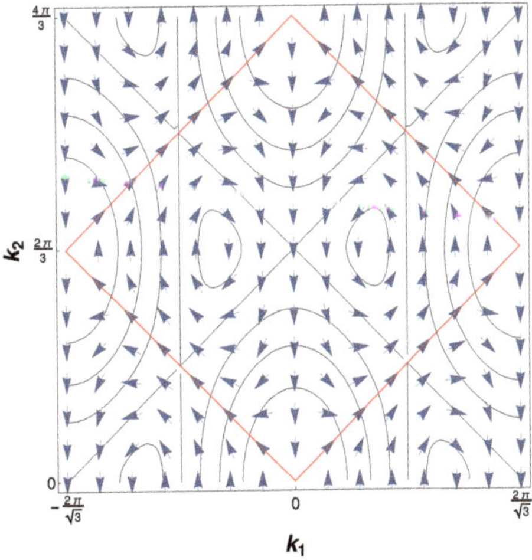



**Figure 2.** The circulation of the Berry vector potential corresponding to the occupied band of the full half filled tight-binding model in the reciprocal space with visible vortex-like structures around the position of the nodal points.

## 4. The Effective Hamiltonian in Strong Magnetic Field

In strong magnetic fields, we replace the usual derivatives by the covariant ones $\partial_\mu \to \partial_\mu + iA_\mu$, with the vector potential $A$ related to the magnetic field via $\nabla \times A = B$. Here, we use the central gauge $A = B/2(-y, x, 0)^{\mathrm{T}}$, the choice which makes analytical calculations particularly convenient. Introducing complex coordinates $z = x + iy$, $\bar{z} = x - iy$, and corresponding derivatives $\partial_z = (\partial_x - i\partial_y)/2$, $\partial_{\bar{z}} = (\partial_x + i\partial_y)/2$, with the properties $\partial_z z = \partial_{\bar{z}}\bar{z} = 1$, $\partial_z\bar{z} = \partial_{\bar{z}}z = 0$, we get $\nabla_- \to 2\partial_z + k^2\bar{z} = A$, $\nabla_+ \to 2\partial_{\bar{z}} - k^2 z = A^\dagger$, where

$$k^2 = \frac{eB}{2\hbar} = \frac{1}{\ell^2}. \tag{5}$$

where $\ell = 1/k$ is referred to as the magnetic length. The operator $A$ annihilates the functions

$$\varphi_n(r) = \frac{k}{\sqrt{\pi}}\frac{(k\bar{z})^n}{\sqrt{n!}}e^{-\frac{k^2}{2}z\bar{z}} \tag{6}$$

i.e., $A\varphi_n(r) = 0$, for every positive integer $n$. The Gaussian part of Equation (6) guarantees the localization in the position space and makes it possible to carry out an integration in the position space exactly. The holomorphic part of Equation (6) contains only powers of $\bar{z}$, and therefore, the wave function itself is manifestly chiral, which can be linked to the induced Berry curvature. The difference in the intrinsic Berry curvature discussed around Equation (4) is in the absence of the partner state with the opposite chirality, which reflects the explicit time-reversal symmetry breaking by an external magnetic field. The Hilbert space of the lowest Landau level is infinitely degenerate; i.e., $n$ can assume every positive integer value between zero and infinity. In this notation, the Hamiltonian becomes

$$H = \Delta_0\Sigma_{03} - \epsilon_0\Sigma_{00} - iv\left(\mathcal{D}_+ A + \mathcal{D}_- A^\dagger\right). \tag{7}$$

The ground state (i.e., the eigenstate in the lowest Landau level) suffices the condition

$$iv\left(\mathcal{D}_+ A + \mathcal{D}_- A^\dagger\right)\psi = 0, \tag{8}$$

which suggests two solutions:

$$\psi_{+,n}(r) = \varphi_n(r) \begin{pmatrix} 0 \\ 1 \\ 0 \\ 0 \end{pmatrix}, \quad \psi_{-,n}(r) = \varphi_n(r) \begin{pmatrix} 0 \\ 0 \\ 1 \\ 0 \end{pmatrix}, \tag{9}$$

which correspond to two valley polarization. The respective eigenvalues of the Hamiltonian for each spin projection in the lowest Landau level are found from the stationary Schrödinger equation

$$H\psi = E\psi, \tag{10}$$

which yields for both spectral branches (or Landau sublevels) [47]

$$E_\pm = -\epsilon_0 \pm \Delta_0, \tag{11}$$

i.e., the spectrum of both Dirac electron species consists of two flat bands irrespective of the strength of the magnetic field. Moreover, in chemically neutral and gapless graphene, the spectrum in the lowest Landau level is at zero [48].

## 5. Single-Particle Propagator in the Lowest Landau Level

The advanced $(+)$ or retarded $(-)$ Green's function in the lowest Landau level can be calculated using the spectral representation

$$G_{r,r'}^\pm \sim \sum_{n=0}^{\infty} \varphi_n(r)\bar{\varphi}_n(r') \sum_{s=\pm} \frac{\mathcal{P}_s}{E - E_s \pm 0^+}, \tag{12}$$

where $E_s$ are the eigenvalues of the Hamiltonian for each spin projection in the lowest Landau level, as shown in Equation (11), and the normalization will be fixed later. The projectors $\mathcal{P}_\pm$ on the spin space

$$\mathcal{P}_+ = \begin{pmatrix} 0 & 0 & 0 & 0 \\ 0 & 1 & 0 & 0 \\ 0 & 0 & 0 & 0 \\ 0 & 0 & 0 & 0 \end{pmatrix} \quad \text{and} \quad \mathcal{P}_- = \begin{pmatrix} 0 & 0 & 0 & 0 \\ 0 & 0 & 0 & 0 \\ 0 & 0 & 1 & 0 \\ 0 & 0 & 0 & 0 \end{pmatrix} \tag{13}$$

are idempotent and orthogonal matrices with properties $\mathcal{P}_+\mathcal{P}_- = 0$, $\mathcal{P}_+\mathcal{P}_+ = \mathcal{P}_+$, $\mathcal{P}_-\mathcal{P}_- = \mathcal{P}_-$. The summation over all $n$ yields

$$\sum_{n=0}^{\infty} \varphi_n(r)\bar{\varphi}_n(r') = \frac{k^2}{\pi} e^{-\frac{k^2}{2}(|z|^2+|z'|^2)} \sum_{n=0}^{\infty} \frac{(k^2\bar{z}z')^n}{n!} = \frac{k^2}{\pi} e^{-\frac{k^2}{2}(|z|^2+|z'|^2-2\bar{z}z')}, \tag{14}$$

which then gives for the Green's function [32,49]

$$G_{r,r'}^\pm(E) = \frac{k^2}{2\pi} e^{-\frac{k^2}{2}(|z|^2+|z'|^2-2\bar{z}z')} \sum_{s=\pm} \frac{\mathcal{P}_s}{E - E_s \pm 0^+}. \tag{15}$$

Notably, the local Green's function $(r = r')$ is a coordinate independent constant. The propagator is normalized this way in order to satisfy the usual sum rule

$$\mp \int_{-\infty}^{\infty} \frac{dE}{\pi} \, \text{Im tr} G_{r,r}^\pm(E) = \frac{k^2}{\pi} = \frac{eB}{h}, \tag{16}$$

where the trace operator acts only on the spin space. Equation (16) gives the number of the elementary flux quanta $\phi_0 = h/e$ per unit volume. In the real-time representation, the

Green's function represents a simple collection of undamped harmonic functions with the period determined by the eigenenergies of the lowest Landau level modes

$$G^{\pm}_{r,r'}(t) = \mp i \frac{k^2}{2\pi} e^{-\frac{k^2}{2}(|z|^2+|z'|^2-2\bar{z}z')} \sum_{s=\pm} \mathcal{P}_s e^{\pm i E_s t}, \tag{17}$$

and the initial time is assumed to be at zero. The Green's function is totally separable on the space-time.

For the case of chemically neutral gapless graphene, the Green's function becomes particularly simple [49]:

$$G^{\pm}_{rr'}(E) = \frac{k^2}{2\pi} \frac{1}{E \pm i0^+} e^{-\frac{k^2}{2}(|z|^2+|z'|^2-2\bar{z}z')} [\mathcal{P}_+ + \mathcal{P}_-], \tag{18}$$

i.e., in the real-time representation, it is just a step function $\theta(t)$.

## 6. Static Conductivity of the Pristine Graphene vs. the Lowest Landau Level

The static conductivity of the clean system can be evaluated from the Kubo–Greenwood formula [28,29,34,35,50]:

$$\sigma^{dc}_{\mu\mu} = \frac{e^2}{h} \lim_{E \to 0} E^2 \, \mathrm{tr} \int d^2r \, r^2_\mu \, G^+_{r,0}(E) G^-_{0,r}(E). \tag{19}$$

We first evaluate this expression for the pristine graphene without a magnetic field. The Green's function of such system reads

$$G^{\pm}_{r,r'}(E) = \int \frac{d^2q}{(2\pi)^2} e^{-iq\cdot(r-r')}[\pm iE\Sigma_{00} + q \cdot J]^{-1} = \int \frac{d^2q}{(2\pi)^2} e^{-iq\cdot(r-r')} G^{\pm}(q), \tag{20}$$

where

$$J_\mu = \frac{\partial H}{\partial q_\mu} \tag{21}$$

denotes the current operator, while the second power of the position operator can be written as

$$r^2_\mu = -\frac{\partial^2}{\partial q^2_\mu}\bigg|_{q=0} e^{-iq\cdot r}. \tag{22}$$

Therefore, the Kubo–Greenwood formula changes to

$$\sigma^{dc}_{\mu\mu} = \frac{e^2}{h} \lim_{E \to 0} E^2 \, \mathrm{tr} \int \frac{d^2q}{(2\pi)^2} J_\mu G^-_q(E) G^+_q(E) J_\mu G^+_q(E) G^-_q(E). \tag{23}$$

Taking into account

$$G^{\pm}_q(E) G^{\mp}_q(E) = \frac{1}{q^2 + E^2}, \tag{24}$$

we then get to

$$\sigma^{dc}_{\mu\mu} = \frac{e^2}{h} \lim_{E \to 0} \int \frac{d^2q}{(2\pi)^2} \frac{4E^2}{[q^2 + E^2]^2}, \tag{25}$$

with 4 being the trace of the unity matrix. Assuming an infinitely large upper cutoff, we finally get for the conductivity a universal number

$$\sigma^{dc}_{\mu\mu} = \frac{1}{\pi} \frac{e^2}{h}, \tag{26}$$

which is the famous universal conductivity of graphene [6]. Remarkably, this result is also valid for the case of weakly disordered Dirac electron gas [30,51–53].

For the calculation of the static conductivity in the lowest Landau level of clean gapless and undoped graphene, we employ the Green's function shown in Equation (18). Here, we can evaluate the Kubo–Greenwood formula directly in the position space

$$\sigma_{\mu\mu}^{dc} = \frac{e^2}{h}\left(\frac{k^2}{2\pi}\right)^2 \lim_{E\to 0} E^2 \int d^2r \, r_\mu^2 e^{-k^2 r^2} \frac{2}{E^2} = \frac{1}{4\pi}\frac{e^2}{h}, \tag{27}$$

where 2 is the trace of the matrix $\mathcal{P}_+ + \mathcal{P}_-$. In addition, here is the conductivity of a universal number, but its magnitude is only a quarter of the clean graphene. It is obvious that this result is solely due to the presence of the zero mode in the spectrum of the gapless and chemically neutral graphene. A slightest doping or a smallest spectral gap would destroy this dc conductivity. Because of this fragility, we can think of the resulting Equation (27) as an anomaly in the parametric space of infinitely small thickness. In analogy to the situation without a magnetic field, we expect the widening of this line by disorder [53].

## 7. Single-Particle Propagator Renormalization Due to the Disorder

The disorder is introduced in the form of the fluctuating chemical potential $v(r)$, which couples in the spin space to the unity matrix $\Sigma_{00}$, with the white noise correlator:

$$\langle v_r \rangle_g = 0, \quad \langle v_{r_1} v_{r_2} \rangle_g = g\delta_{r_1 r_2}. \tag{28}$$

The averaged propagator reads

$$\bar{G}_{r_1 r_2}^{\pm} = \langle [(G^{\pm})^{-1} + v\Sigma_{00}]_{r_1 r_2}^{-1} \rangle_g. \tag{29}$$

To perform the disorder average perturbative, we expand the propagator in powers of $v$. Because of the properties of the disorder correlator Equation (28), all terms with an odd number of potentials $v$ vanish. The series then becomes

$$\bar{G}_{r_1 r_2}^{\pm} = \langle G_{r_1 r_2}^{\pm} + G_{r_1 x_1}^{\pm} v_{x_1} G_{x_1 x_2}^{\pm} v_{x_2} G_{x_2 r_2}^{\pm} + G_{r_1 x_1}^{\pm} v_{x_1} G_{x_1 x_2}^{\pm} v_{x_2} G_{x_2 x_3}^{\pm} v_{x_3} G_{x_3 x_4}^{\pm} v_{x_4} G_{x_4 r_2}^{\pm}$$
$$+ G_{r_1 x_1}^{\pm} v_{x_1} G_{x_1 x_2}^{\pm} v_{x_2} G_{x_2 x_3}^{\pm} v_{x_3} G_{x_3 x_4}^{\pm} v_{x_4} G_{x_4 x_5}^{\pm} v_{x_5} G_{x_5 x_6}^{\pm} v_{x_6} G_{x_6 r_2}^{\pm} \cdots \rangle_g. \tag{30}$$

Here, the summation over repeating indices is understood.

The Green's function shown in Equation (15) is spanned by the spin projectors $\mathcal{P}_s$. Therefore, only the disorder diagonal in the spin space is of importance. In addition to the randomly fluctuating chemical potential considered here, these might include the randomly fluctuating gap, which couples to $\Sigma_{03}$, the random "chiral" chemical potential ($\Sigma_{30}$), or the random "chiral" mass ($\Sigma_{33}$). Each product of these matrices with $\mathcal{P}_s$ projects them bar the sign back onto $\mathcal{P}_s$ again. Therefore, the perturbative series shown in Equation (30) does not depend on a particular disorder type, and our analysis is generic and disorder type independent.

The exact Green's function of disordered electrons in the lowest Landau level was obtained by Wegner in Ref. [32] in the distinctly separable form

$$\bar{G}_{rr'}^{\pm}(E) = \frac{k^2}{\pi} e^{-\frac{k^2}{2}(|r|^2+|r'|^2-2\bar{r}r')} \sum_{s=\pm} \mathcal{F}_s^{\pm}(E)\mathcal{P}_s. \tag{31}$$

The frequency-dependent part of the Green's function evaluated to the order $g^3$ by evaluation of the diagrams shown in Figure 3 reads

$$\mathcal{F}_s^{\pm}(E) = \frac{1}{2}\frac{1}{E-E_s}\left[1 + \frac{E_g^2}{[E-E_s]^2} + \frac{5}{2}\frac{E_g^4}{[E-E_s]^4} + \frac{37}{4}\frac{E_g^6}{[E-E_s]^6} \cdots \right], \tag{32}$$

where $E_g^2 = \frac{gk^2}{4\pi}$. The expansion coefficients $1, 1, 5/2, 37/4...$ are precisely those of the Wegner's exact solution [32]. They are determined as expansion coefficients of the function

$$-\frac{\partial}{\partial a}\log\left[\frac{2\pi}{\sqrt{b}}e^{-\frac{a^2}{b}}\int_{\frac{a}{\sqrt{b}}}^{\infty}dt\, e^{-t^2}\right]. \tag{33}$$

in powers of $b/a^2$. Following [32], we find for the frequency-dependent part of the dressed single-particle propagator

$$\mathcal{F}_s^{\pm}(E) = \eta_s(E) \mp i\rho_s(E), \tag{34}$$

with the following explicit expressions for the real

$$\eta_s(E) = \frac{1}{E_g}\left(\frac{2}{\pi}\frac{e^{v_s^2}\int_0^{v_s}dt\, e^{t^2}}{1+\left(\frac{2}{\sqrt{\pi}}\int_0^{v_s}dt\, e^{t^2}\right)^2} - v_s\right), \tag{35}$$

and imaginary parts [33,34]

$$\rho_s(E) = \frac{1}{\sqrt{\pi}E_g}\frac{e^{v_s^2}}{1+\left(\frac{2}{\sqrt{\pi}}\int_0^{v_s}dt\, e^{t^2}\right)^2}. \tag{36}$$

They depend on the dimensionless energy

$$v_s = \frac{E - E_s}{E_g}, \quad \text{where}\quad E_g^2 = \frac{gk^2}{4\pi}. \tag{37}$$

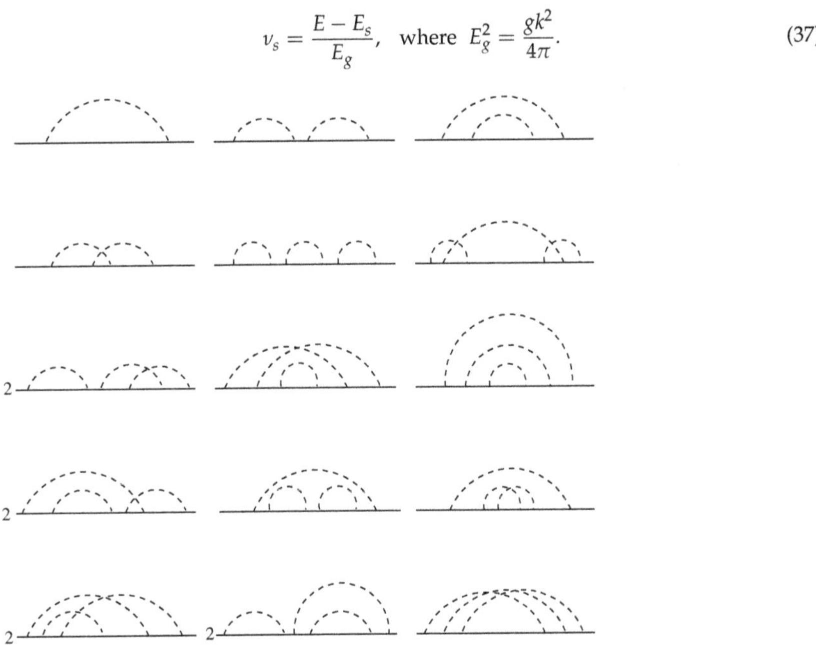

**Figure 3.** Perturbative processes contributing to the dressing of the single-particle propagator due to the disorder to order $g^1$ (one diagram), $g^2$ (three diagrams), and $g^3$ (fifteen diagrams). Some of the diagrams of order $g^3$ should be counted twice because of the degeneracy due to the mirror symmetry with respect to the imaginable vertical axis, which is accounted for by the factors 2 in front of them.

In the chosen units, the disorder-related energy $E_g$ is a dimensionless quantity, which is proportional to the ratio of two relevant lengths $E_g \sim l_\lambda / \ell$: the magnetic length $\ell \sim 1/k$ and the disorder related length $l_\lambda \sim \sqrt{g}$. The total density of states

$$\rho(E) = \mp \frac{1}{\pi} \mathrm{Im}\, \mathrm{tr} G_{rr}^{\pm}(E) = \frac{1}{\pi^{5/2}} \frac{k^2}{E_g} \sum_{s=\pm} \frac{e^{v_s^2}}{1 + \left( \frac{2}{\sqrt{\pi}} \int_0^{v_s} dt\, e^{t^2} \right)^2}. \tag{38}$$

is correctly normalized in accordance with Equation (16). Figure 4 shows the DOS from Equation (38). For weak disorder strength, the density of states that appears has the form of two sharp peaks placed symmetrically around the energy eigenvalues in the lowest Landau level. It is plotted in units of $\frac{1}{\pi^{5/2}} \frac{k^2}{E_g} \sim (\ell l_\lambda)^{-1}$, $\ell l_\lambda$ being the parametric volume constructed from the two specific lengths of the model. The peaks become broader with increasing disorder strength and overlap with each other until they merge to a single structure.

For the gapless and chemically neutral graphene, both peaks overlap and form a unique structure around the zero energy

$$\rho(E) = \frac{2}{\pi^{5/2}} \frac{k^2}{E_g} \frac{e^{v^2}}{1 + \left( \frac{2}{\sqrt{\pi}} \int_0^{v} dt\, e^{t^2} \right)^2}, \quad v = \frac{E}{E_g}. \tag{39}$$

Therefore, at the band center, we get

$$\rho(0) = \frac{2}{\pi^{5/2}} \frac{k^2}{E_g}. \tag{40}$$

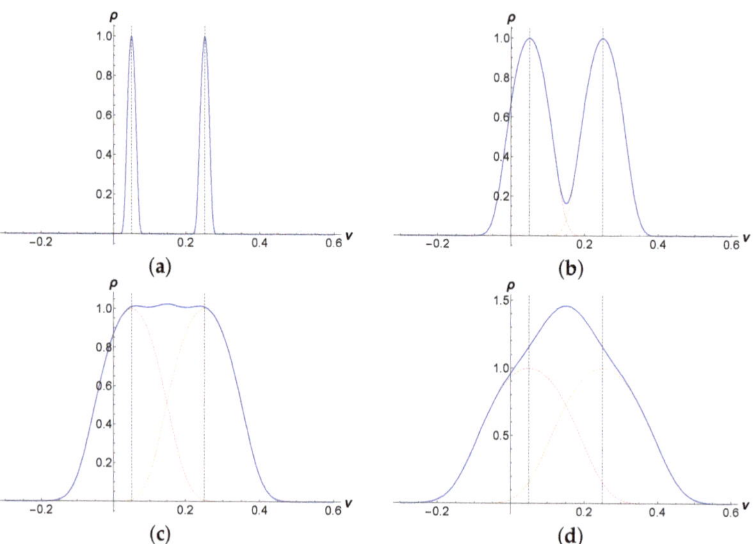

**Figure 4.** Evolution of the DOS of both Landau sublevels defined in Equation (11) (**a**–**d**) plotted in units of the DOS at each suband center $\frac{1}{\pi^{5/2}} \frac{k^2}{E_g}$ with increasing disorder strength as a function of the dimensionless energy $v$. The following quantities are used: $\epsilon_0/t = 0.15$, $\Delta_0/t = 0.1$ and $E_g/t = 0.01, 0.045, 0.073$, and $0.1$ in units of the hopping amplitude. Dashed lines emphasize the position of each eigenvalue.

### 8. Mean Squared Displacement of the Disordered System

The access to the diffusion goes via the mean squared displacement

$$\langle r_\mu^2(t)\rangle = \frac{\mathrm{tr}\sum_r r_\mu^2 P_{r0}(t)}{\mathrm{tr}\sum_r P_{r0}(t)},\tag{41}$$

where $r_\mu$ is the position operator and $P_{rr'}(t)$ is the return probability density defined as

$$P_{rr'}(t) = \int \frac{dE}{2\pi}\, e^{-iEt} P_{rr'}(E),\tag{42}$$

where

$$P_{rr'}(E) = \langle G_{rr'}^+(E)G_{r'r}^-(E)\rangle_g,\tag{43}$$

is the disorder averaged two-particles propagator. Notably, the numerator of Equation (41) appears to be essentially the Kubo–Greenwood formula shown in Equation (19). The relation between the mean squared displacement and diffusion is established via

$$\frac{d}{dt}\Big|_{t=0}\langle r_\mu^2(t)\rangle = 2D,\tag{44}$$

where $D$ is the diffusion coefficient. If we would be able to determine the diffusion coefficient of the disordered system through the direct evaluation of Equation (41), then it will be possible to compute the conductivity from the Einstein relation (in this particular form adopted from [29,50])

$$\sigma = \frac{e^2}{\hbar} D\rho(E),\tag{45}$$

where $\rho(E)$ is the density of states discussed in the previous paragraph.

A rigorous evaluation of the full perturbative series for the two-particles propagator $\langle G_{r,0}^+ G_{0,r}^-\rangle_g$ along the lines of Wegner's calculations for the single-particle propagator is principally impossible. Therefore, we need to consider the spatial averages. We evaluate both expressions from the numerator and denominator of Equation (41) perturbatively. Evaluation of all perturbative diagrams to order $g^3$, shown in Figure 5 yields for the expression in the numerator of Equation (41)

$$\mathrm{tr}\sum_r r_\mu^2 P_{r0}(E) = \frac{1}{4\pi}\frac{1}{E_g^2}\sum_{s=\pm}(2X_s)^2\Big[\tfrac{1}{2} + (2X_s)^2 + 2(2X_s)^4 + \tfrac{167}{36}(2X_s)^6 + \cdots$$
$$\left(\tfrac{3}{4}(2X_s)^4 + \tfrac{343}{72}(2X_s)^6 + \cdots\right)\cos 2\phi_s + \left(\tfrac{139}{72}(2X_s)^6 + \cdots\right)\cos 4\phi_s + \cdots\Big],\tag{46}$$

where

$$X_s^2(E) = E_g^2[\eta_s^2(E) + \rho_s^2(E)] \quad\text{and}\quad \phi_s(E) = \arctan\left[\frac{\rho_s(E)}{\eta_s(E)}\right].\tag{47}$$

According to Equations (35) and (36), $X_s^2(E)$ and $\phi_s(E)$ are dimensionless functions of the argument $v_s = (E - E_s)/E_g$. The analogous computation for the denominator of Equation (41) yields

$$\mathrm{tr}\sum_r P_{r0}(E) = \frac{k^2}{4\pi}\frac{1}{E_g^2}\sum_{s=\pm}(2X_s)^2\Big[1 + (2X_s)^2 + \tfrac{3}{2}(2X_s)^4 + \tfrac{13}{4}(2X_s)^6 + \cdots$$
$$\left((2X_s)^4 + \tfrac{9}{2}(2X_s)^6 + \cdots\right)\cos 2\phi_s + \left(\tfrac{5}{2}(2X_s)^6 + \cdots\right)\cos 4\phi_s + \cdots\Big].\tag{48}$$

A reasonable approximation for the two-particles propagator that leads beyond this partially rigorous result includes all diagrams of the so-called ladder channel. The four lowest order ladder diagrams are evaluated as [54]

$$\bigcirc = \frac{1}{4E_g^2}\left(\frac{k^2}{\pi}\right)^2 \sum_{s=\pm}(2X_s)^2 \exp\left[-k^2r^2\right],\tag{49}$$

$$\oplus = \frac{1}{4E_g^2}\left(\frac{k^2}{\pi}\right)^2 \sum_{s=\pm}\frac{(2X_s)^4}{2} \exp\left[-\frac{k^2r^2}{2}\right],\tag{50}$$

$$\oplus = \frac{1}{4E_g^2}\left(\frac{k^2}{\pi}\right)^2 \sum_{s=\pm}\frac{(2X_s)^6}{3} \exp\left[-\frac{k^2r^2}{3}\right],\tag{51}$$

$$\oplus = \frac{1}{4E_g^2}\left(\frac{k^2}{\pi}\right)^2 \sum_{s=\pm}\frac{(2X_s)^8}{4} \exp\left[-\frac{k^2r^2}{4}\right],\tag{52}$$

which suggests the following asymptotics of the two-particles propagator in the form of an infinite series:

$$P_{r0}^{\text{lad}}(E) \approx \frac{1}{4E_g^2}\left(\frac{k^2}{\pi}\right)^2 \sum_{s=\pm}\sum_{n=1}^{\infty}\frac{(2X_s)^{2n}}{n} \exp\left[-\frac{k^2r^2}{n}\right].\tag{53}$$

Using this expression, one can complement Equations (46) and (48) to any order.

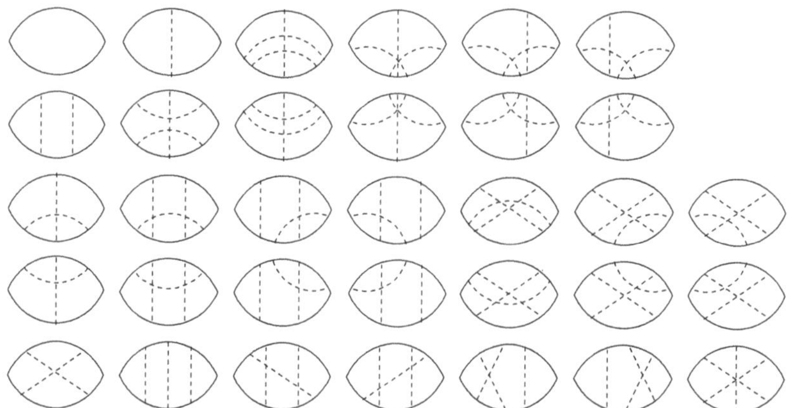

**Figure 5.** Perturbative processes contributing to the dressing of the two-particles propagator up to the third order in disorder strength. Solid lines denote the fully dressed Wegner's propagators and the dashed lines denote the disorder correlators.

## 9. Equation of Motion for the Mean Squared Displacement

The equation of motion for the mean squared displacement has the form of the second-order ordinary differential equation

$$k^2\frac{\partial^2}{\partial t^2}\langle r_\mu^2(t)\rangle = -E_g^2 I_N\left[\frac{N}{2}-k^2\langle r_\mu^2(t)\rangle\right],\tag{54}$$

where $N$ refers to the order of the perturbative expansion. The expression for the integral $I_N$ can be found in [54]. For our purposes, it is only important that it decreases with increasing $N$.

The large-time asymptotics ($t \gg t_c$, $t_c$ being some crossover time far in the past, which can be chosen zero) is given by the solution of the self-consistent equation

$$\langle r_\mu^2(t)\rangle^> \approx \frac{N}{2}\ell^2 + \ell^2 C \exp\left[-\frac{t}{\tau}\right],\tag{55}$$

where we introduced the scattering time as

$$\tau \approx \frac{1}{\sqrt{I_N} E_g}.$$ (56)

In the limit $t \to \infty$, the mean squared displacement approaches its upper bound

$$\lim_{t \to \infty} \langle r_\mu^2(t) \rangle^> \to \frac{N}{2} \ell^2,$$ (57)

which for $N \to \infty$ lies in the infinity and is therefore never reached. Hence, it can be only approached from below, which requires $\mathcal{C}$ to be negative. If $\sqrt{I_N}$ is small, then $\tau$ is large, and the regime with linear time dependence should be broad. The diffusion coefficient is then obtained from

$$\frac{\partial}{\partial t} \langle r_\mu^2(t) \rangle^> \Big|_{t=t_c} = |\mathcal{C}| \frac{\ell^2}{\tau}.$$ (58)

Formally, $|\mathcal{C}|$ should follow from the initial condition at $t = t_c$, but for this, we need to know $\langle r_\mu^2(t_c) \rangle^>$, which lies far in the past and is therefore forgotten. In order to be a physical quantity, we demand for $D$ an invariance with respect to $N$. This is similar to the version of the renormalization group typically used in the high-energy physics. This implies

$$\frac{\partial}{\partial N} (\sqrt{I_N} |\mathcal{C}|) = 0,$$ (59)

from where then follows

$$\sqrt{I_N} |\mathcal{C}| = \text{const.}$$ (60)

Even though this constrain might appear not entirely transparent, it has a natural analogy in the case of disordered electron gas without a magnetic field. Here, the diffusion coefficient is determined from the self-consistent Born approximation and appears unchanged in the partial series, e.g., cooperon or diffuson [22,23]. A comparison with Equation (44) suggests this constant to be 2. Then, the physical diffusion coefficient becomes

$$D \approx \frac{E_g}{k^2} \sim \ell l_\lambda,$$ (61)

i.e., it is proportional to the parametric volume of the model.

Inserting the density of states from Equation (38) and the diffusion coefficient Equation (61) into the Einstein relation Equation (45) yields the conductivity. The system is conducting within a parametric window located around each of the Landau sublevels. The width of the conducting window is determined by the parameters of the microscopic model and by the disorder. The transition $g \to 0$ is smooth, and the conductivity degenerates to two sharp peaks at the Landau sublevels. With increasing disorder, the peaks become broader and merge at some point to an amorphous structure. Simultaneously, the amplitude becomes smaller, signaling the suppression of the conductivity in the strong disorder limit.

We are now in the position to compute the conductivity at an arbitrary Landau sublevel defined in Equation (11). The corresponding density of the states is shown in Figure 4. For weak disorder, the contribution from the other sublevel is negligible, and we get for the density of states

$$\rho_{sl}(0) \approx \frac{1}{\pi^{5/2}} \frac{k^2}{E_g}.$$ (62)

Using the units of $e^2/h$ instead of $e^2/\hbar$ adds an extra factor of $2\pi$, i.e.,

$$\sigma_{sl} = 2\pi D \rho_{SL}(0) \frac{e^2}{h} \approx \frac{2\pi}{\pi^{5/2}} \frac{e^2}{h} \approx 0.36 \frac{e^2}{h}.$$ (63)

Numerically, it is close to the dc conductivity in clean graphene off the magnetic field $\frac{e^2}{\pi h} \approx 0.32\frac{e^2}{h}$ evaluated in Equation (26), which is valid even for the weakly disordered systems [30,51–53]. For us, the most interesting limit is the case of the neutral and gapless graphene, for which a rich empirical knowledge is available. Our estimation would give for this zero mode

$$\sigma_{\mathrm{zm}}(0) = 2\sigma_{\mathrm{sl}} \approx 0.72\frac{e^2}{h}. \tag{64}$$

Respected experimental studies of Ref. [55,56] determine the room-temperature longitudinal resistivity at the band center in the lowest Landau level as roughly 35 k$\Omega$ and 42 k$\Omega$, respectively, which corresponds to (with $h/e^2 \approx 25,812\ \Omega$) to

$$\sigma_{\mathrm{exp}}(0) \approx 0.614\frac{e^2}{h} \div 0.737\frac{e^2}{h}, \tag{65}$$

which is surprisingly close to our estimation. The comparison is justified, since the disorder can be considered as an effective temperature, cf. [57] and references therein.

## 10. Discussions

The diffusion of electrons in random environments confined to the lowest Landau level in two spatial dimensions is a long-standing and conceptually challenging problem of quantum statistical mechanics. Without disorder, the quantum mechanical description of the problem is simply that of the harmonic oscillator with a discrete, though highly degenerate spectrum, comprising of the so-called Landau levels. In the strong magnetic field, the gap between the lowest and the first Landau levels is very large and only the lowest Landau level is relevant. In this regime, the electrons are distributed between the stationary Landau orbits in the position space and should stay there forever, thus forbidding any transport across the sample. This is due to the disorder that the electrons can move from one Landau orbit to another, producing an observable current. While several analytical approaches have been developed in the past for systems of disorder electrons off a magnetic field, a meaningful formulation of the problem in the strong magnetic field is conceptually difficult, because the problem lacks a small parameter, and therefore, the perturbative expansions in powers of the disorder potential diverge. At the single-particle level, the problem has been solved by Wegner [32], who found an exact expression for the single-particle propagator. The Wegner propagator does not reveal any singularities and describes a state of the matter without pronounced resonances and consequently without clearly defined quasiparticles. However, the difficulties aggregate by far if one goes beyond the single-particle picture and considers processes involving two or more particles.

Our main intention is to calculate the static conductivity of disordered graphene in a strong magnetic field. In the lowest Landau level, the spectrum of the gapless and chemically neutral graphene Hamiltonian has zero energy, which are the consequence of its band topology and responsible for the exceptional transport properties under normal conditions. The spectral gap or fluctuations of the Fermi energy due to the hopping between second-nearest neighbors on the honeycomb lattice split this zero mode in two sublevels. Surprisingly, the static conductivity of the clean system evaluated from the Kubo–Greenwood formula gives a conductivity within infinitely thin parametric windows around the zero mode. One would expect that the disorder broadens this window to considerable sizes. However, because the perturbative series for the two-particles propagator diverges, a naive use of the Kubo formula fails. Therefore, we approach the conductivity via the Einstein relation, which requires the knowledge of the density of states and of the diffusion coefficient. While the former is known from the Wegner's solution, the latter is not. To deal with such divergences, we develop an analytical approach based on a self-consistent equation for the mean square displacement, which allows one to directly extract the diffusion coefficient and static conductivity.

Following the line of Wegner's exact considerations, we determine the general expression of the density of states of graphene. With a gap, the density of states of weakly disordered graphene represents two sharp peaks centered around each of the sublevels. For the case of gapless and chemically neutral graphene, both peaks coalesce to a single one with twice the height. The diffusion coefficient is extracted from the time evolution of the mean squared displacement. The latter tends toward a stationary state, which would reestablish the situation we observe in the clean system with all electrons distributed between stationary orbits. However, our findings suggest an infinitely large time needed for the system to arrive in this state. At the intermediary time scales, the mean squared displacement behaves lineary in time from which the diffusion coefficient is extracted. The combination of the density of states and diffusion coefficient, known as the Einstein relation, gives a universal, i.e., disorder independent value for the static conductivity. At the band center of the lowest Landau level, we find for the conductivity a universal value $\sim 0.72 \, e^2/h$, which is surprisingly close to the established results for the conductivity of the disordered Dirac electrons. In the subsequent work, we intend to extend our analysis to higher Landau levels and to address the Hall conductivity with the aim of arriving at an effective description of a kind of the Chern–Simons theories .

The quantum Hall effect has long become the standard tool for high-precision measurements and adjustments [58]. With our clearly laid out prediction for the dc conductivity in the lowest Landau level of graphene, we have provided a benchmark for prospective graphene-based metrological standardization devices.

**Author Contributions:** Conceptualization, A.S. and G.T.; methodology, A.S. and G.T.; supervision of the project, A.S. and G.T. All authors have read and agreed to the published version of the manuscript.

**Funding:** A.S. acknowledges the support by the grants of the Julian Schwinger Foundation for Physics Research and from the personal research grant PCI2021-122057-2B of the Agencia Estatal de Investigacion de Espana.

**Data Availability Statement:** The data presented in this study are available on request from the corresponding author.

**Acknowledgments:** The work was concluded at the IMDEA Nanotechnologia Madrid.

**Conflicts of Interest:** The authors declare no conflict of interest.

# References


1. Abrahams, E.; Anderson, P.W.; Licciardello, D.C.; Ramakrishnan, T.V. Scaling Theory of Localization: Absence of Quantum Diffusion in Two Dimensions. *Phys. Rev. Lett.* **1979**, *42*, 673–676. [CrossRef]
2. Gor'kov L.G.; Larkin A.I.; Khmel'nitskii, D.E. Particle conductivity in a two-dimensional random potential. *JETP Lett.* **1979**, *30*, 228–232.
3. Hikami, S.; Larkin, A.; Nagaoka, Y. Spin-Orbit Interaction and Magnetoresistance in the Two Dimensional Random System. *Prog. Theor. Phys.* **1980**, *63*, 707–710. [CrossRef]
4. Vollhardt, D.; Wölfle, P. Diagrammatic, self-consistent treatment of the Anderson localization problem in $d \leqslant 2$ dimensions. *Phys. Rev. B* **1980**, *22*, 4666–4679. [CrossRef]
5. Hanein, Y.; Meirav, U.; Shahar, D.; Li, C.C.; Tsui, D.C.; Shtrikman H. The metallic like conductivity of a two-dimensional hole system. *Phys. Rev. Lett.* **1998**, *80*, 1288–1291. [CrossRef]
6. Novoselov, K.S.; Geim, A.K.; Morozov, S.V.; Jiang, D.; Katsnelson, M.I.; Grigorieva, I.V.; Dubonos, S.V.; Firsov, A.A. Two-dimensional gas of massless Dirac fermions in graphene. *Nature* **2005**, *438*, 197–200. [CrossRef]
7. Tan, Y.-W.; Zhang, Y.; Bolotin, K.; Zhao, Y.; Adam, S.; Hwang, E.H.; Das Sarma, S.; Stormer, H.L.; Kim, P. Measurement of scattering rate and minimal conductivity in graphene. *Phys. Rev. Lett.* **2007**, *99*, 246803. [CrossRef]
8. Elias, D.C.; Nair R.R.; Mohiuddin, T.M.G.; Morozov, S.V.; Blake, P.; Halsall, M.P.; Ferrari, A.C.; Boukhvalov, D.W.; Katsnelson, M.I.; Geim, A.K.; et al. Control of graphene's properties by reversible hydrogenation: Evidence for graphane. *Science* **2009**, *323*, 610–613. [CrossRef]
9. Allen, M.J.; Tung, V.C.; Kaner, R.B. Honeycomb carbon: A review of graphene. *Chem. Rev.* **2010**, *110*, 132–145. [CrossRef]
10. Chen, L.; Liu, C.-C.; Feng, B.; He, X.; Cheng, P.; Ding, Z.; Meng, S.; Yao, Y.; Wu, K. Evidence for Dirac fermions in a honeycomb lattice based on silicon. *Phys. Rev. Lett.* **2012**, *109*, 056804. [CrossRef]
11. Castro Neto, A.H.; Guinea, F.; Peres, N.M.R.; Novoselov, K.S.; Geim, A.K. The electronic properties of graphene. *Rev. Mod. Phys.* **2009**, *81*, 109–162. [CrossRef]
12. Kotov, V.N.; Uchoa, B.; Pereira, V.M.; Guinea, F.; Castro Neto, A.H. Electron-Electron Interactions in Graphene: Current Status and Perspectives. *Rev. Mod. Phys.* **2012**, *84*, 1067–1125. [CrossRef]

13. Hasan, M.Z.; Kane, C.L. Colloquium: Topological insulators. *Rev. Mod. Phys.* **2010**, *82*, 3045–3067. [CrossRef]
14. Qi, X.-L.; Zhang, S.-C. Topological insulators and superconductors. *Rev. Mod. Phys.* **2011**, *83*, 1057–1110. [CrossRef]
15. Avsar, A.; Ochoa, H.; Guinea, F.; Özyilmaz, B.; Van Wees, B.J.; Vera-Marun, I.J. Colloquium: Spintronics in graphene and other two-dimensional materials. *Rev. Mod. Phys.* **2020**, *92*, 021003. [CrossRef]
16. Bernevig, B.A.; Hughes, T.L.; Zhang, S.-C. Quantum spin Hall effect and topological phase transition in HgTe quantum wells. *Science* **2006**, *314*, 1757–1761. [CrossRef] [PubMed]
17. Shon, N.H.; Ando, T. Quantum transport in two-dimensional graphite system. *J. Phys. Soc. Jpn.* **1998**, *67*, 2421–2429. [CrossRef]
18. Ando, T.; Zheng, Y.; Suzuura, H. Dynamical conductivity and zero-mode anomaly in honeycomb lattices. *J. Phys. Soc. Jpn.* **2002**, *71*, 1318–1324. [CrossRef]
19. Suzuura, H.; Ando, T. Crossover from symplectic to orthogonal class in a two-dimensional honeycomb lattice. *Phys. Rev. Lett.* **2002**, *89*, 266603. [CrossRef]
20. McCann, E.; Kechedzhi, K.; Fal'ko, V.I.; Suzuura, H.; Ando, T.; Altshuler, B.L. Weak-localization magnetoresistance and valley symmetry in graphene. *Phys. Rev. Lett.* **2006**, *97*, 146805. [CrossRef]
21. Altshuler, B.L.; Aronov, A.G.; Larkin, A.I.; Khmel'nitskii, D.E. Anomalous magnetoresistance in semiconductors. *Sov. Phys. JETP* **1981**, *54*, 411–419.
22. Altshuler, B.L.; Simons, B.D. Universalities: From Anderson localization to quantum chaos. In *Mesoscopic Quantum Physics, Les Houches 1994*; Akkermans, E., Montambaux, G., Pichard, J.-L., Zinn-Justin, J., Eds.; North Holland: Amsterdam, The Netherlands, 1995; pp. 1–98.
23. Efetov, K. *Supersymmetry in Disorder and Chaos*; Cambridge University Press: Cambridge, UK, 1997.
24. Lee, P.A. Localized states in a d-wave superconductor. *Phys. Rev. Lett.* **1993**, *71*, 1887–1890. [CrossRef] [PubMed]
25. Wegner, F.J. The mobility edge problem: Continuous symmetry and a conjecture. *Z. Physik B* **1979**, *35*, 207–210. [CrossRef]
26. Schäfer; L.; Wegner, F.J. Disordered system withn orbitals per site: Lagrange formulation, hyperbolic symmetry, and Goldstone modes. *Z. Physik B* **1980**, *38*, 113–126. [CrossRef]
27. Hikami, S. Anderson localization in a nonlinear-$\sigma$-model representation. *Phys. Rev. B* **1981**, *24*, 2671–2679. [CrossRef]
28. Wegner, F.J. Disordered system with n orbitals per site: $n = \infty$ limit. *Phys. Rev. B* **1979**, *19*, 783–792. [CrossRef]
29. McKane, A.J.; Stone, M. Localization as an alternative to Goldstone's theorem. *Ann. Phys.* **1981**, *131*, 36–55. [CrossRef]
30. Fradkin, E. Critical behavior of disordered degenerate semiconductors. II. Spectrum and transport properties in mean-field theory. *Phys. Rev. B* **1986**, *33*, 3263–3268. [CrossRef]
31. Ando, T. Theory of quantum transport in a two-dimensional electron system under magnetic field. III. Many-site approximation *J. Phys. Soc. Jpn.* **1974**, *37*, 622–630. [CrossRef]
32. Wegner, F.J. Exact density of states for lowest Landau level in white noise potential. Superfield representation for interacting systems. *Z. Phys. B Condens. Matter* **1983**, *51*, 279–285. [CrossRef]
33. Brézin, E.; Gross, D.J.; Itzykson, C. Density of states in the presence of a strong magnetic field and random impurities. *Nucl. Phys. B* **1984**, *235*, 24–44. [CrossRef]
34. Hikami, S. Borel-Padé analysis for the two-dimensional electron in a random potential under a strong magnetic field. *Phys. Rev. B* **1984**, *29*, 3726–3729. [CrossRef]
35. Hikami, S. Anderson Localization of the two-dimensional electron in a random potential under a strong magnetic field. *Prog. Theor. Phys.* **1984**, *72*, 722–735. [CrossRef]
36. Aoki, H. Quantised Hall effect. *Rep. Prog. Phys.* **1987**, *50*, 655–730. [CrossRef]
37. Tkachov, G. *Topological Insulators: The Physics of Spin Helicity in Quantum Transport*; Pan Stanford: Boca Raton, FL, USA, 2015.
38. Sinner, A.; Ziegler, K. Two-parameter scaling theory of transport near a spectral node. *Phys. Rev. B* **2014**, *90*, 174207. [CrossRef]
39. Sinner, A.; Ziegler, K. Finite-size scaling in a 2D disordered electron gas with spectral nodes. *J. Phys. Condens. Matter* **2016**, *28*, 305701. [CrossRef] [PubMed]
40. Goldenfeld, N. *Lectures on Phase Transitions and the Renormalization Group*; Perseus Books: Reading, MA, USA, 1992.
41. Huang, K. *Statistical Mechanics*, 2nd ed.; John Wiley: New York, NY, USA, 1987.
42. Chaikin, P.M.; Lubenski, T.C. *Principles of Condensed Matter Physics*; Cambridge University Press: Cambridge, UK, 1995.
43. Ziegler, K. Quantum diffusion in two-dimensional random systems with particle–hole symmetry. *J. Phys. A Math. Theor.* **2012**, *45*, 335001. [CrossRef]
44. Singh, R.R.P.; Chakravarty, S. A disordered two-dimensional system in a magnetic field: Borel-Padé analysis. *Nucl. Phys. B* **1986**, *265*, 265–292. [CrossRef]
45. Hikami, S.; Shirai, M.; Wegner, F.J. Anderson localization in the lowest Landau level for a two-subband model. *Nucl. Phys. B* **1993**, *408*, 415–426. [CrossRef]
46. Culcer, D.; Keser, A.C.; Li, Y.; Tkachov, G. Transport in two-dimensional topological materials: Recent developments in experiment and theory. *2D Mater.* **2020**, *7*, 022007. [CrossRef]
47. König, M.; Buhmann, H.; Molenkamp, L.W.; Hughes, T.; Liu, C.-X.; Qi, X.-L.; Zhang, S.-C. The quantum spin Hall effect: Theory and experiment. *J. Phys. Soc. Jpn.* **2008**, *77*, 031007. [CrossRef]
48. Li, G.; Andrei, E.Y. Observation of Landau levels of Dirac fermions in graphite. *Nat. Phys.* **2007**, *3*, 623–627. [CrossRef]
49. Goswami, P.; Jia, X.; Chakravarty, S. Quantum Hall plateau transition in the lowest Landau level of disordered graphene. *Phys. Rev. B* **2007**, *76*, 205408. [CrossRef]

50. Ludwig, A.W.W.; Fisher, M.P.A.; Shankar, R.; Grinstein, G. Integer quantum Hall transition: An alternative approach and exact results. *Phys. Rev. B* **1994**, *50*, 7526–7552. [CrossRef]
51. Ziegler, K. Robust transport properties in graphene. *Phys. Rev. Lett.* **2006**, *97*, 266802. [CrossRef]
52. Ziegler, K. Minimal conductivity of graphene: Nonuniversal values from the Kubo formula. *Phys. Rev. B* **2007**, *75*, 233407. [CrossRef]
53. Sinner, A.; Ziegler, K. Conductivity of disordered 2d binodal Dirac electron gas: Effect of internode scattering. *Philos. Mag.* **2018**, *98*, 1799. [CrossRef]
54. Sinner, A.; Tkachov, G. Diffusive transport in the lowest Landau level of disordered 2d semimetals: The mean-square-displacement approach. *Eur. Phys. J. B* **2022**, *submitted*.
55. Novoselov, K.S.; Jiang, Z.; Zhang, Y.; Morozov, S.V.; Stormer, H.L.; Zeitler, U.; Maan, J.C.; Boebinger, G.S.; Kim, P.; Geim, A.K. Room-temperature quantum Hall effect in graphene. *Science* **2007**, *315*, 1379. [CrossRef]
56. Jiang, Z.; Zhang, Y.; Tan, Y.-W.; Stormer, H.L.; Kim, P. Quantum Hall effect in graphene. *Solid State Comm.* **2007**, *143*, 14–19. [CrossRef]
57. Shemer, Z.; Barkai, E. Einstein relation and effective temperature for systems with quenched disorder. *Phys. Rev. E* **2009**, *80*, 031108. [CrossRef] [PubMed]
58. Jeckelmann, B.; Jeanneret, B. The quantum Hall effect as an electrical resistance standard. *Rep. Prog. Phys.* **2001**, *64*, 1603–1655. [CrossRef]

*nanomaterials*



*Article*

# Superfluidity of Dipolar Excitons in a Double Layer of $\alpha - T_3$ with a Mass Term

Oleg L. Berman [1,2,*], Godfrey Gumbs [2,3,4], Gabriel P. Martins [1,2,3] and Paula Fekete [5]

1   Physics Department, New York City College of Technology, City University of New York, New York, NY 11201, USA; gpimentamartins@gradcenter.cuny.edu
2   The Graduate School and University Center, City University of New York, New York, NY 10016, USA; ggumbs@hunter.cuny.edu
3   Department of Physics and Astronomy, Hunter College, City University of New York, New York, NY 10065, USA
4   Donastia International Physics Center (DIPC), P de Manuel Lardizabal, 4, 20018 San Sebastian, Spain
5   US Military Academy at West Point, 606 Thayer Road, West Point, NY 10996, USA; paula.fekete@westpoint.edu
*   Correspondence: oberman@citytech.cuny.edu

**Abstract:** We predict Bose-Einstein condensation and superfluidity of dipolar excitons, formed by electron-hole pairs in spatially separated gapped hexagonal $\alpha - T_3$ (GHAT3) layers. In the $\alpha - T_3$ model, the AB-honeycomb lattice structure is supplemented with C atoms located at the centers of the hexagons in the lattice. We considered the $\alpha - T_3$ model in the presence of a mass term which opens a gap in the energy-dispersive spectrum. The gap opening mass term, caused by a weak magnetic field, plays the role of Zeeman splitting at low magnetic fields for this pseudospin-1 system. The band structure of GHAT3 monolayers leads to the formation of two distinct types of excitons in the GHAT3 double layer. We consider two types of dipolar excitons in double-layer GHAT3: (a) "A excitons", which are bound states of electrons in the conduction band (CB) and holes in the intermediate band (IB), and (b) "B excitons", which are bound states of electrons in the CB and holes in the valence band (VB). The binding energy of A and B dipolar excitons is calculated. For a two-component weakly interacting Bose gas of dipolar excitons in a GHAT3 double layer, we obtain the energy dispersion of collective excitations, the sound velocity, the superfluid density, and the mean-field critical temperature $T_c$ for superfluidity.

**Keywords:** Bose-Einstein condensation; superfluidity; dipolar exitons

**Citation:** Berman, O.L.; Gumbs, G.; Martins, G.P.; Fekete, P. Superfluidity of Dipolar Excitons in a Double Layer of $\alpha - T_3$ with a Mass Term. *Nanomaterials* **2022**, *12*, 1437. https://doi.org/10.3390/nano12091437

Academic Editors: Yia-Chung Chang and Daniele Fazzi

Received: 13 January 2022
Accepted: 12 April 2022
Published: 22 April 2022

**Publisher's Note:** MDPI stays neutral with regard to jurisdictional claims in published maps and institutional affiliations.

## 1. Introduction

The many-particle systems of dipolar (indirect) excitons, formed by spatially separated electrons and holes, in semiconductor coupled quantum wells (CQWs) and novel two-dimensional (2D) materials have been the subject of numerous experimental and theoretical studies. These systems are attractive in large part due to the possibility of Bose-Einstein condensation (BEC) and superfluidity of dipolar excitons, which can be observed as persistent electrical currents in each quantum well, and also through coherent optical properties [1–5]. Recent progress in theoretical and experimental studies of BEC and superfluidity of dipolar excitons in CQWs have been reviewed in [6]. Electron-hole superfluidity in double layers can occur not only in the BEC regime, but also in the Bardeen-Cooper-Schrieffer (BCS)-BEC crossover regime [7].

A number of experimental and theoretical investigations have been devoted to the BEC of electron-hole pairs, formed by spatially separated electrons and holes in a double layer formed by parallel graphene layers. These investigations were reported in [8–13]. Both BEC and superfluidity of dipolar excitons in double layers of transition-metal dichalcogenides (TMDCs) [14–18] and phosphorene [19,20] have been discussed, because the exciton binding energies in novel 2D semiconductors are quite large. Possible BEC in a long-lived dark spin state of 2D dipolar excitons has been experimentally observed for GaAs/AlGaAs semiconductor CQWs [21].

Recently, the electronic properties of the $\alpha - T_3$ lattice have been the subject of the intensive theoretical and experimental investigations due to its surprising fundamental physical properties as well as its promising applications in solid-state devices [22–35]. For a review of artificial flat band systems, see [36]. Raoux, et al. [22] proposed that an $\alpha - T_3$ lattice could be assembled from cold fermionic atoms confined to an optical lattice by means of three pairs of laser beams for the optical dice lattice ($\alpha = 1$) [37]. This structure consists of an AB-honeycomb lattice (the rim) like that in graphene which is combined with C atoms at the center/hub of each hexagon. A parameter $\alpha$ represents the ratio of the hopping integral between the rim and the hub to that around the rim of the hexagonal lattice. By dephasing one of the three pairs of laser beams, one could vary the parameter $0 \leq \alpha = \tan \varphi \leq 1$. Optically induced dressed states [38], and their tunneling, transport [33,39], and collective properties [40], as well as $\alpha - T_3$ based nanoribbons [41] have been analyzed. The BEC and superfluidity of dipolar magnetoexcitons in $\alpha - T_3$ double layers in a strong uniform perpendicular magnetic field were proposed in [42].

We present the conditions for BEC and superfluidity of a two-component weakly interacting Bose gas of dipolar excitons, formed by electron-hole pairs in spatially separated GHAT3 layers. An applied weak magnetic field to this pseudospin-1 monolayer system results in a Zeeman-type splitting of the energy subbands [43]. This dispersion relation consists of three bands: CB, IB, and VB. We consider two types of dipolar excitons in a double-layer of GHAT3: (a) "A excitons", formed as bound states of electrons in CB and holes in IB, and (b) "B excitons", formed as bound states of electrons in CB and holes in VB. The binding energy of A and B dipolar excitons is calculated. For a two-component weakly interacting Bose gas of dipolar excitons in a GHAT3 double layer, we obtain the energy dispersion of collective excitations, the sound velocity, the superfluid density, and the mean-field critical temperature $T_c$ for superfluidity.

Our paper is organized in the following way. In Section 2, the two-body problem for an electron and a hole, spatially separated in two parallel GHAT3 monolayers, is formulated, and the effective masses and binding energies are obtained for two types of dipolar excitons. The spectrum of collective excitations and the sound velocity for the two-component weakly interacting Bose gas of dipolar excitons in the double layer of GHAT3 are derived in Section 3. In Section 4 the superfluidity of the weakly interacting Bose gas of dipolar excitons in the double layer of GHAT3 is predicted, and the mean-field critical temperature of the phase transition is obtained. The results of our calculations are discussed in Section 5. In Section 6 our conclusions are reported.

## 2. Dipolar Excitons in a Double Layer of $\alpha - T_3$ with a Mass Term

We will consider charge carriers in the conduction band, valence band, and the intermediate band, which corresponds to the flat band in an $\alpha - T_3$ layer without a mass term. In the presence of a weak magnetic field, the low-energy Hamiltonian of the charge carriers in a GHAT3 monolayer at the K and K′ points are given by [43]

$$\hat{\mathcal{H}}_\lambda = \begin{pmatrix} \Delta & f(\mathbf{k})\cos\phi & 0 \\ f^*(\mathbf{k})\cos\phi & 0 & f(\mathbf{k})\sin\phi \\ 0 & f^*(\mathbf{k})\sin\phi & -\Delta \end{pmatrix}, \tag{1}$$

where the origin in **k**-space is defined to be around the K point, $\mathbf{k} = (k_x, k_y)$ and $\tan\theta_\mathbf{k} = k_y/k_x$, $\phi = \tan^{-1}\alpha$, $f(\mathbf{k}) = \hbar v_F(\lambda k_x - i k_y) = \lambda\hbar v_F k e^{-i\lambda\theta_\mathbf{k}}$, with $\lambda = \pm 1$ being the valley index at the K and K′ points, $2\Delta$ is the gap in the energy spectrum of a GHAT3 layer due to the mass term in the Hamiltonian. In an $\alpha - T_3$ layer honeycomb lattice, there is an added fermionic hub atom C at the center of each hexagon. Let the hopping integral be $t_1$ between the hub atom and either an A or B atom on the rim and $t_2$ between nearest neighbors on the rim of the hexagon. The ratio of these two nearest neighbor hopping terms is denoted as $t_2/t_1 = \alpha$, where the parameter $\alpha$ satisfies $0 \leq \alpha \leq 1$. The largest value when $\alpha$ is 1 is for the dice lattice, whereas its value of 0 corresponds to graphene for decoupled hub from rim atoms [43].

At small momenta near K and K' points, the dispersion for the charge carriers in the conduction band $\epsilon_{CB}(k)$ is given by the relation [43]

$$\epsilon_{CB}(k) \approx \Delta + \frac{\hbar^2 k^2}{2m_{CB}},$$ 

(2)

where $\mathbf{k} = \mathbf{p}/\hbar$ and $\mathbf{p}$ are the wave vector and momentum of a quasiparticle, $m_{CB}$ is the effective mass of the charge carriers in the conduction band, given by

$$m_{CB} = \frac{(1+\alpha^2)\Delta}{2v_F^2},$$

(3)

where $v_F$ is the Fermi velocity in a GHAT3 layer, and $\varphi = \tan^{-1}\alpha$ [43]. At small momenta near K and K' points, the dispersion for the charge carriers in the valence band $\epsilon_{VB}(k)$ is given by the relation [43]

$$\epsilon_{VB}(k) \approx -\Delta - \frac{\hbar^2 k^2}{2m_{VB}},$$

(4)

with $m_{VB}$ the effective mass of the charge carriers in the valence band, given by

$$m_{VB} = \frac{(1+\alpha^2)\Delta}{2v_F^2\alpha^2}.$$

(5)

At small momenta near K and K' points, the dispersion for the charge carriers in the intermediate band, corresponding to the flat band in an $\alpha - T_3$ layer without a mass term, $\epsilon_{IB}(k)$ is given by the relation [43]

$$\epsilon_{IB}(k) \approx -\frac{\hbar^2 k^2}{2m_{IB}},$$

(6)

where $m_{IB}$ is the effective mass of the charge carriers in the intermediate band, given by

$$m_{IB} = \frac{(1+\alpha^2)\Delta}{2v_F^2(1-\alpha^2)}.$$

(7)

It is worth noting that there are spin degeneracy and valley degeneracy for the energy of the charge carriers in a GHAT3 layer.

In the system under consideration in this paper, electrons are confined in a 2D GHAT3 monolayer, while an equal number of positive holes are located in a parallel GHAT3 monolayer at a distance $D$ away as demonstrated in Figure 1. This electron-hole system in two parallel GHAT3 layers is treated as a 2D system without interlayer hopping. Due to the absence of tunneling of electrons and holes between different GHAT3 monolayers, electron-hole recombination is suppressed by a dielectric barrier with dielectric constant $\epsilon_d$ that separates the GHAT3 monolayers. Therefore, the dipolar excitons, formed by electrons and holes, located in two different GHAT3 monolayers, have a longer lifetime than direct excitons. The electron and hole are attracted via electromagnetic interaction $V(r_{eh})$, where $r_{eh}$ is the distance between the electron and hole, and they could form a bound state, i.e., an exciton, in three-dimensional (3D) space. Therefore, to determine the binding energy of the exciton a two-body problem in restricted 3D space has to be solved. However, if one projects the electron position vector onto the GHAT3 plane with holes and replaces the relative coordinate vector $\mathbf{r}_{eh}$ by its projection $\mathbf{r}$ on this plane, the potential $V(r_{eh})$ may be expressed as $V(r_{eh}) = V(\sqrt{r^2 + D^2})$, where $r$ is the relative distance between the hole and the projection of the electron position vector onto the GHAT3 plane with holes. A schematic illustration of the dipolar exciton in a GHAT3 double layer is presented in Figure 1. By introducing in-plane coordinates $\mathbf{r}_1 = (x_1, y_1)$ and $\mathbf{r}_2 = (x_2, y_2)$ for the electron and the projection vector of the hole, respectively (where $\mathbf{r} = \mathbf{r}_1 - \mathbf{r}_2$), the dipolar exciton can be described by employing a two-body 2D Schrödinger equation with potential $V(\sqrt{r^2 + D^2})$. So that the restricted 3D two-body problem can be reduced to a 2D two-body problem on a GHAT3 layer with the holes.

The dipolar excitons with spatially separated electrons and holes in two parallel GHAT3 monolayers can be created by laser pumping with an applied external voltage. While an electron in the conduction band and a hole in the valence or intermediate band are excited due to absorption of a photon, voltages are applied with opposite signs to confine electrons on one layer and holes on another so that dipoles point in one direction only.

In our case, "both" the energy bands and the exciton modes referred to the K-point, not one to the $\Gamma$ point and the other to the K point. We note that in the dispersion equations appearing in [43–46] the origin of the k-space was specified to be around the K point, (and not the $\Gamma$ point) as did several authors investigating $\alpha - T_3$. So, our choice of origin not being the center of the Brillouin zone has precedence. For graphene, the plasmon dispersion relation and low-energy bands, presented by [47] were both consistently measured from the K point taken as the origin and not the center of the Brillouin zone.

We consider excitons, formed by an electron and a hole from the same valley, because an electron and a hole from different valleys cannot be excited by absorption of photon due to conservation of momentum. The reason is that photons carry momenta much smaller than the difference between K and K' in reciprocal space.

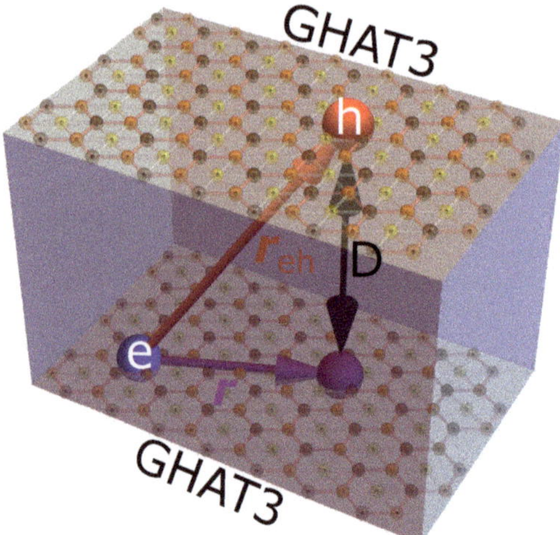



**Figure 1.** Schematic illustration of a dipolar excitonin a pair of GHAT3 double layers embedded in an insulating material.

The effective Hamiltonian of an electron and a hole, spatially separated in two parallel GHAT3 monolayers with the interlayer distance $D$ has the following form

$$\hat{H}_{ex} = -\frac{\hbar^2}{2m_e}\Delta_{\mathbf{r}_1} - \frac{\hbar^2}{2m_h}\Delta_{\mathbf{r}_2} + V(r), \tag{8}$$

where $\Delta_{\mathbf{r}_1}$ and $\Delta_{\mathbf{r}_2}$ are the Laplacian operators with respect to the components of the vectors $\mathbf{r}_1$ and $\mathbf{r}_2$, respectively, and $m_e$ and $m_h$ are the effective masses of the electron and hole, respectively. For CV excitons $m_e = m_{CB}$ and $m_h = m_{VB}$; and for CI excitons $m_e = m_{CB}$ and $m_h = m_{IB}$, where $m_{CB}$, $m_{VB}$, and $m_{IB}$ are given by Equations (3), (5) and (7), correspondingly. The problem of the in-plane motion of an interacting electron and hole forming the exciton in a GHAT3 double layer can be reduced to that of one particle with the reduced mass $\mu = m_e m_h / (m_e + m_h)$ in a $V(r)$ potential and motion of the center-of-mass of the exciton with the mass $M = m_e + m_h$. We introduce the coordinates of the center-of-mass $\mathbf{R}$ of an exciton and the coordinate of the relative motion $\mathbf{r}$ of an electron and hole as $\mathbf{R} = (m_e \mathbf{r}_1 + m_h \mathbf{r}_2)/(m_e + m_h)$ and $\mathbf{r} = \mathbf{r}_1 - \mathbf{r}_2$, correspondingly. The Hamiltonian $\hat{H}_{ex}$ can be represented in the form: $\hat{H}_{ex} = \hat{H}_R + \hat{H}_r$, where the Hamiltonian of the motion of the center-of-mass is $\hat{H}_R$ and that of the relative motion of electron and a hole is $\hat{H}_r$. The solution of the Schrödinger equation for the center-of-mass of an exciton $\hat{H}_R \psi(\mathbf{R}) = \mathcal{E}\psi(\mathbf{R})$ is the plane wave $\psi(\mathbf{R}) = e^{i\mathbf{P}\cdot\mathbf{R}/\hbar}$ with the quadratic energy spectrum $\mathcal{E} = P^2/(2M)$, where $\mathbf{P}$ is the momentum of the center-of-mass of an exciton.

We consider electrons and holes to be located in GHAT3 parallel layers, embedded in a dielectric with the dielectric constant $\epsilon_d$. The potential energy of electron-hole Coulomb attraction is

$$V(r) = -\frac{\kappa e^2}{\epsilon_d \sqrt{r^2 + D^2}}, \tag{9}$$

where $\kappa = 9 \times 10^9 \ N \times m^2/C^2$, $\epsilon_d$ is the dielectric constant of the insulator (SiO$_2$ or $h$-BN), surrounding the electron and hole GHAT3 monolayers, forming the double layer. For the $h$-BN barrier we substitute the dielectric constant $\epsilon_d = 4.89$, while for the SiO$_2$ barrier we substitute the dielectric constant $\epsilon_d = 4.50$. For $h$-BN insulating layers, $\epsilon_d = 4.89$ is the effective dielectric constant, defined as $\epsilon_d = \sqrt{\varepsilon^\perp}\sqrt{\varepsilon^\parallel}$ [14], where $\varepsilon^\perp = 6.71$ and $\varepsilon^\parallel = 3.56$ are the components of the dielectric tensor for $h$-BN [48]. Assuming $r \ll D$, we approximate $V(r)$ by the first two terms of the Taylor series and obtain

$$V(r) = -V_0 + \gamma r^2, \quad \text{where } V_0 = \frac{\kappa e^2}{\epsilon_d D}, \quad \gamma = \frac{\kappa e^2}{2\epsilon_d D^3}. \tag{10}$$

The similar approach has been applied for excitons in TMDC double layers [16,17]. The solution of the Schrödinger equation for the relative motion of an electron and a hole $\hat{H}_r \Psi(\mathbf{r}) = E\Psi(\mathbf{r})$ with the potential (10) is reduced to the problem of a 2D harmonic oscillator with the exciton reduced mass $\mu$. Following [49,50] one obtains the radial Schrödinger equation and the solution for the eigenfunctions for the relative motion of an electron and a hole in a GHAT3 double layer in terms of associated Laguerre polynomials, which can be written as

$$\Psi_{NL}(\mathbf{r}) = \frac{N!}{a^{|L|+1}\sqrt{\tilde{n}!\tilde{n}'!}} 2^{-|L|/2}\mathrm{sgn}(L)^L r^{|L|} e^{-r^2/(4a^2)} \times L_N^{|L|}(r^2/(2a^2)) \frac{e^{-iL\varphi}}{(2\pi)^{1/2}}, \tag{11}$$

where $N = \min(\tilde{n}, \tilde{n}')$, $L = \tilde{n} - \tilde{n}'$, $\tilde{n}, \tilde{n}' = 0,1,2,3,\ldots$ are the quantum numbers, $\varphi$ is the polar angle, and $a = \left[\hbar/(2\sqrt{2\mu\gamma})\right]^{1/2}$ is a Bohr radius of a dipolar exciton. The corresponding energy spectrum is given by

$$E_{NL} \equiv E_{e(h)} = -V_0 + (2N + 1 + |L|)\hbar\left(\frac{2\gamma}{\mu}\right)^{1/2}. \tag{12}$$

At the lowest quantum state $N = L = 0$ as it follows from Equation (12) the ground state energy for the exciton is given by

$$E_{00} = -V_0 + \hbar\left(\frac{2\gamma}{\mu}\right)^{1/2}. \tag{13}$$

The important characteristic of the exciton is the square of the in-plane gyration radius $r_X^2$. It allows one to estimate the condition when the excitonic gas is dilute enough. One can obtain the square of the in-plane gyration radius $r_X$ of a dipolar exciton [14], which is expressed as the average squared projection of an electron-hole separation onto the plane of a GHAT3 monolayer

$$r_X^2 \equiv \left\langle r^2 \right\rangle = \int \Psi_{00}^*(\mathbf{r}) r^2 \Psi_{00}(\mathbf{r}) d^2 r = \frac{2\pi}{2\pi a^2} \int_0^{+\infty} r^2 e^{-\frac{r^2}{2a^2}} r\, dr = 2a^2. \tag{14}$$

We consider dipolar excitons, formed by an electron in the conduction band and a hole in the valence band (CV excitons) and formed by and electron in the conduction band and a hole in the intermediate valence band (CI excitons). For CV excitons one has

$$\mu_{CV} = \frac{m_{CB} m_{VB}}{m_{CB} + m_{VB}} = \frac{\Delta}{2v_F^2}; \qquad M_{CV} = m_{CB} + m_{VB} = \frac{(1+\alpha^2)^2 \Delta}{2v_F^2 \alpha^2}. \tag{15}$$

For CI excitons one has

$$\mu_{CI} = \frac{m_{CB} m_{IB}}{m_{CB} + m_{IB}} = \frac{(1+\alpha^2)\Delta}{2v_F^2(2-\alpha^2)}; \qquad M_{CI} = m_{CB} + m_{IB} = \frac{(1+\alpha^2)(2-\alpha^2)\Delta}{2v_F^2(1-\alpha^2)}. \tag{16}$$

### 3. The Collective Excitations Spectrum and Superfluidity for the Two-Component System of Dipolar Excitons

We consider the dilute limit for dipolar exciton gas in a GHAT3 double layer, when $n_A a_{B\ A}^2 \ll 1$ and $n_B a_{B\ B}^2 \ll 1$, where $n_{A(B)}$ and $a_{B\ A(B)}$ are the concentration and effective exciton Bohr radius for A(B) dipolar excitons, correspondingly. In the dilute limit, dipolar A and B excitons are formed by electron-hole pairs with the electrons and holes spatially separated in two different GHAT3 layers. We will treat the two-component weakly interacting Bose gas of dipolar excitons in a GHAT3 double layer by applying the approach analogous to the one used for dipolar excitons in a transition metal dichalcogenide (TMDC) double layer [16,17].

Since the dipolar excitons, formed by the charge carriers in different valleys, are characterized by the same energy, the exciton states are degenerate with respect to the valley degree of freedom. Therefore, we consider the Hamiltonian of the weakly interacting Bose gas of dipolar excitons, formed in a single valley. We will take into account the degeneracy of the exciton states with respect to spin and valley degrees of freedom by the introducing the spin and valley degeneracy factor $s = 16$ below. The Hamiltonian $\hat{H}$ of the 2D A and B weakly interacting dipolar excitons can be written as

$$\hat{H} = \hat{H}_A + \hat{H}_B + \hat{H}_I, \tag{17}$$

where $\hat{H}_{A(B)}$ are the Hamiltonians of A(B) excitons defined as

$$\hat{H}_{A(B)} = \sum_{\mathbf{k}} E_{A(B)}(k) a_{\mathbf{k}A(B)}^\dagger a_{\mathbf{k}A(B)} + \frac{g_{AA(BB)}}{2S} \sum_{\mathbf{klm}} a_{\mathbf{k}A(B)}^\dagger a_{\mathbf{l}A(B)}^\dagger a_{A(B)\mathbf{m}} a_{A(B)\mathbf{k+l-m}}, \tag{18}$$

and $\hat{H}_I$ is the Hamiltonian of the interaction between A and B excitons presented as

$$\hat{H}_I = \frac{g_{AB}}{S} \sum_{\mathbf{klm}} a_{\mathbf{k}A}^\dagger a_{\mathbf{l}B}^\dagger a_{B\mathbf{m}} a_{A\mathbf{k+l-m}}, \tag{19}$$

where $a_{\mathbf{k}A(B)}^\dagger$ and $a_{\mathbf{k}A(B)}$ are Bose creation and annihilation operators for A(B) dipolar excitons with the wave vector $\mathbf{k}$, correspondingly, $S$ is the area of the system, $E_{A(B)}(k) \equiv \epsilon_{A(B)} = \varepsilon_{(0)A(B)}(k) + \mathcal{A}_{A(B)}$ is the energy spectrum of non-interacting A(B) dipolar excitons, respectively, $\varepsilon_{(0)A(B)}(k) = \hbar^2 k^2 / (2M_{A(B)})$, $M_{A(B)}$ is an effective mass of non-interacting dipolar excitons, $\mathcal{A}_{A(B)}$ is the constant, which depends on A(B) dipolar exciton binding energy and the corresponding gap, $g_{AA(BB)}$ and $g_{AB}$ are the interaction constants for the repulsion between two A dipolar excitons, two B dipolar excitons and for the interaction between A and B dipolar excitons, respectively.

In dilute system with large interlayer separation $D$, two dipolar excitons, located at distance $R$, repel each other via the dipole-dipole interaction potential $U(R) = \kappa e^2 D^2/(\epsilon_d R^3)$. Following the procedure described in [51], the interaction parameters for the exciton-exciton repulsion in very dilute systems can be obtained implying the exciton-exciton dipole-dipole repulsion exists only at the distances between excitons greater than the distance from the exciton to the classical turning point.

The many-particle Hamiltonian for a weakly interacting Bose gas can be diagonalized within the Bogoliubov approximation [52], replacing the product of four operators in the interaction term with the product of two operators. The Bogoliubov approximation is valid if one assumes that most of the particles belong to BEC. In this case, in the Hamiltonian one can keep only the terms responsible for the interactions between the condensate and non-condensate particles, while the terms describing the interactions between non-condensate particles are neglected.

Following the procedure, described in [16,17], applying the Bogoliubov approximation [52], generalized for a two-component weakly interacting Bose gas [53,54] and introducing the following notation,

$$
\begin{aligned}
G_{AA} &= g_{AA}n_A = g n_A,\ G_{BB} = g_{BB}n_B = g n_B,\ G_{AB} = g_{AB}\sqrt{n_A n_B} = g\sqrt{n_A n_B}, \\
\omega_A(k) &= \sqrt{\epsilon_{(0)A}^2(k) + 2G_{AA}\epsilon_{(0)A}(k)}, \\
\omega_B(k) &= \sqrt{\epsilon_{(0)B}^2(k) + 2G_{BB}\epsilon_{(0)B}(k)},
\end{aligned}
\tag{20}
$$

one obtains two modes of the spectrum of Bose collective excitations $\epsilon_j(k)$

$$
\epsilon_j(k) = \sqrt{\frac{\omega_A^2(k) + \omega_B^2(k) + (-1)^{j-1}\sqrt{\left(\omega_A^2(k) - \omega_B^2(k)\right)^2 + (4G_{AB})^2 \epsilon_{(0)A}(k)\epsilon_{(0)B}(k)}}{2}},
\tag{21}
$$

where $j = 1, 2$. In our approach, the condition $G_{AB}^2 = G_{AA}G_{BB}$ holds.

At small momenta $p = \hbar k$, when $\epsilon_{(0)A}(k) \ll G_{AA}$ and $\epsilon_{(0)B}(k) \ll G_{BB}$, expanding the spectrum of collective excitations $\epsilon_j(k)$ up to the first order with respect to the momentum $p$, one obtains two sound modes in the spectrum of the collective excitations $\epsilon_j(p) = c_j p$, where $c_j$ is the sound velocity written as

$$
c_j = \sqrt{\frac{G_{AA}}{2M_A} + \frac{G_{BB}}{2M_B} + (-1)^{j-1}\sqrt{\left(\frac{G_{AA}}{2M_A} - \frac{G_{BB}}{2M_B}\right)^2 + \frac{G_{AB}^2}{M_A M_B}}},
\tag{22}
$$

At $j = 1$, the spectrum of collective excitations is determined by the non-zero sound velocity $c_1$, while at $j = 2$ the sound velocity vanishes with $c_2 = 0$. At large momenta, for the conditions when $\epsilon_{(0)A}(k) \gg G_{AA}$ and $\epsilon_{(0)B}(k) \gg G_{BB}$, one obtains two parabolic modes of collective excitations with the spectra $\epsilon_1(k) = \epsilon_{(0)A}(k)$ and $\epsilon_2(k) = \epsilon_{(0)B}(k)$, if $M_A < M_B$ and if $M_A > M_B$ with the spectra $\epsilon_1(k) = \epsilon_{(0)B}(k)$ and $\epsilon_2(k) = \epsilon_{(0)A}(k)$.

## 4. Superfluidity of the Weakly-Interacting Bose Gas of Dipolar Excitons

Since when $j = 2$ the sound velocity vanishes, below we take into account only the branch of the spectrum of collective excitations at $j = 1$, neglecting the branch at $j = 2$. According to [52,55], it is clear that we need a finite sound velocity for superfluidity. Since the branch of the collective excitations at zero sound velocity for the collective excitations corresponds to the zero energy of the quasiparticles (which means that no quasiparticles are created with zero sound velocity), this branch does not lead to the dissipation of energy resulting in finite viscosity and, therefore, does not influence the Landau critical velocity.

This is the reason for eliminating the zero sound velocity case in our considerations here. The weakly-interacting gas of dipolar excitons in the double layer of GHAT3 satisfies the Landau criterion for superfluidity [52,55], because at small momenta, the energy spectrum of the quasiparticles in the weakly-interacting gas of dipolar excitons at $j = 1$ is sound-like with the finite sound velocity, $c_1$. In the moving weakly-interacting gas of dipolar excitons the quasiparticles are created at velocities above the velocity of sound, and the critical velocity for superfluidity reads as $v_c = c_1$. The difference between the ideal Bose gas and two-component weakly interacting Bose gas of dipolar excitons is that while the spectrum of ideal Bose gas has no branch with finite sound velocity, the dipolar exciton system under consideration has one branch in the spectrum of collective excitations with finite sound velocity at $j = 1$ due to exciton-exciton interaction. Therefore, at low temperatures, the two-component system of dipolar excitons exhibits superfluidity due to exciton-exciton interactions, while the ideal Bose gas does not demonstrate superfluidity.

We defined the density of the superfluid component $\rho_s(T)$ as $\rho_s(T) = \rho - \rho_n(T)$, where $\rho = M_A n_A + M_B n_B$ is the total 2D density of the dipolar excitons and $\rho_n(T)$ denotes the density of the normal component. The density $\rho_n(T)$ of the normal component can be defined using standard procedure [56]. The assumption that the dipolar exciton system moves with a velocity $\mathbf{u}$ implies that the superfluid component moves with the velocity $\mathbf{u}$. The energy dissipation at nonzero temperatures $T$ is characterized by the occupancy of quasiparticles in this system. Since the density of quasiparticles is small at low temperatures, the gas of quasiparticles can be treated as an ideal Bose gas. In order to obtain the density of the superfluid component, one can define the total mass flow for a Bose gas of quasiparticles in the frame, in which the superfluid component is assumed to be at rest, as

$$\mathbf{J} = s \int \frac{d^2p}{(2\pi\hbar)^2} \mathbf{p} f[\varepsilon_1(p) - \mathbf{p} \cdot \mathbf{u}], \tag{23}$$

where $s = 16$ is the spin and valley degeneracy factor, $f[\varepsilon_1(p))] = (\exp[\varepsilon_1(p)/(k_B T)] - 1)^{-1}$ is the Bose-Einstein distribution function for the quasiparticles with the dispersion $\varepsilon_1(p)$, and $k_B$ is the Boltzmann constant. Expanding the expression under the integral in Equation (23) up to the first order with respect to $\mathbf{p} \cdot \mathbf{u}/(k_B T)$, one has:

$$\mathbf{J} = -s\frac{\mathbf{u}}{2} \int \frac{d^2p}{(2\pi\hbar)^2} p^2 \frac{\partial f[\varepsilon_1(p)]}{\partial \varepsilon_1(p)}. \tag{24}$$

The density $\rho_n$ of the normal component in the moving weakly-interacting Bose gas of dipolar excitons is defined as [56]

$$\mathbf{J} = \rho_n \mathbf{u}. \tag{25}$$

Employing Equations (24) and (25), one derives the normal component density as

$$\rho_n(T) = -\frac{s}{2} \int \frac{d^2p}{(2\pi\hbar)^2} p^2 \frac{\partial f[\varepsilon_1(p)]}{\partial \varepsilon_1(p)}. \tag{26}$$

At low temperatures $k_B T \ll M_{A(B)} c_j^2$, the small momenta ($\varepsilon_{(0)A}(k) \ll G_{AA}$ and $\varepsilon_{(0)B}(k) \ll G_{BB}$) make the dominant contribution to the integral on the right-hand side of Equation (26). The quasiparticles with such small momenta are characterized by the sound spectrum $\varepsilon_1(k) = c_1 k$ with the sound velocity defined by Equation (22). By substituting $\varepsilon_1(k) = c_1 k$ into Equation (26), we obtain

$$\rho_n(T) = \frac{3s\zeta(3)}{2\pi\hbar^2 c_1^4} k_B^3 T^3, \tag{27}$$

where $\zeta(z)$ is the Riemann zeta function ($\zeta(3) \simeq 1.202$).

The mean field critical temperature $T_c$ of the phase transition at which the superfluidity occurs, implying neglecting the interaction between the quasiparticles, is obtained from the condition $\rho_s(T_c) = 0$ [56]:

$$\rho_n(T_c) = \rho = M_A n_A + M_B n_B .\tag{28}$$

At low temperatures $k_B T \ll M_{A(B)} c_1^2$ by substituting Equation (27) into Equation (28), one derives

$$T_c = \left[\frac{2\pi\hbar^2 \rho c_1^4}{3\zeta(3)sk_B^3}\right]^{1/3} .\tag{29}$$

While Bose-Einstein condensation occurs at absolute zero even in a two-dimensional (2D) system, it is well known that in a 2D bosonic system, Bose-Einstein condensation does not occur at finite temperature, and only the quasi-long-range order appears. In this paper, we have obtained the mean-field critical temperature $T_c$ of the phase transition at which superfluidity appears without claiming BEC in a 2D system at finite temperature. In this work, we have considered BEC only at absolute zero temperature. The similar approach has been applied for excitons in TMDC double layers [16,17].

## 5. Discussion

In this section we now discuss the results of our calculations. In Figure 2, we present the results for the exciton binding energy $\mathcal{E}_b(\alpha, \Delta, D)$ for CV and CI excitons as functions of the gap $\Delta$ for chosen parameter $\alpha = 0.6$ and interlayer separations $D = 25$ nm. According to Figure 2, $\mathcal{E}_b(\alpha, \Delta, D)$ is an increasing function of $\Delta$, whereas for CV excitons the exciton binding energy is slightly larger than that for CI excitons.

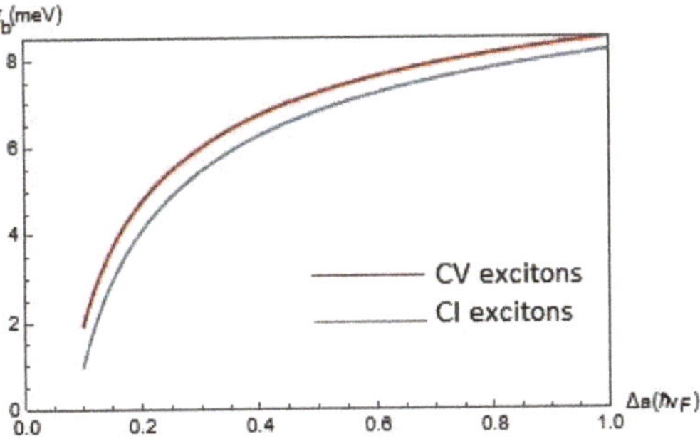


**Figure 2.** The exciton binding energy $\mathcal{E}_b(\alpha, \Delta, D)$ for CV and CI excitons as functions of the gap $\Delta$ for chosen parameter $\alpha = 0.6$ and interlayer separations $D = 25$ nm. The lattice constant of $\alpha - T_3$ is $a = 2.46$ .

In Figure 3, we present our results for the exciton binding energy $\mathcal{E}_b(\alpha, \Delta, D)$ for CV and CI excitons as functions of the parameter $\alpha$ for chosen gap $\Delta = 0.5\,\hbar v_F/a$ and interlayer separations $D = 25$ nm. According to Figure 3, $\mathcal{E}_b(\alpha, \Delta, D)$ does not depend on $\alpha$ for CV excitons, whereas it is an increasing function of $\alpha$ for CI excitons. At $\alpha \lesssim 0.7\,\mathcal{E}_b(\alpha, \Delta, D)$ for CV excitons is larger than for CI excitons, while at $\alpha \gtrsim 0.7\,\mathcal{E}_b(\alpha, \Delta, D)$ for CI excitons is larger than for CV excitons.

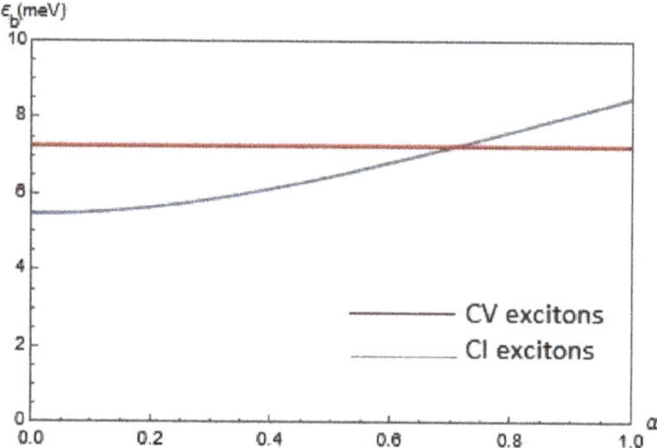

**Figure 3.** The exciton binding energy $\mathcal{E}_b(\alpha, \Delta, D)$ for CV and CI excitons as functions of the parameter $\alpha$ for chosen gap $\Delta = 0.5\ \hbar v_F/a$ and interlayer separations $D = 25$ nm.

In Figure 4, we present the results of our calculations for the exciton binding energy $\mathcal{E}_b(\alpha, \Delta, D)$ for CV and CI excitons as functions of the interlayer separation $D$ for chosen parameter $\alpha = 0.6$ and gap $\Delta = 0.5\ \hbar v_F/a$. According to Figure 4, $\mathcal{E}_b(\alpha, \Delta, D)$ is a decreasing function of $D$, whereas for CV excitons the exciton binding energy is slightly larger than for CI excitons.

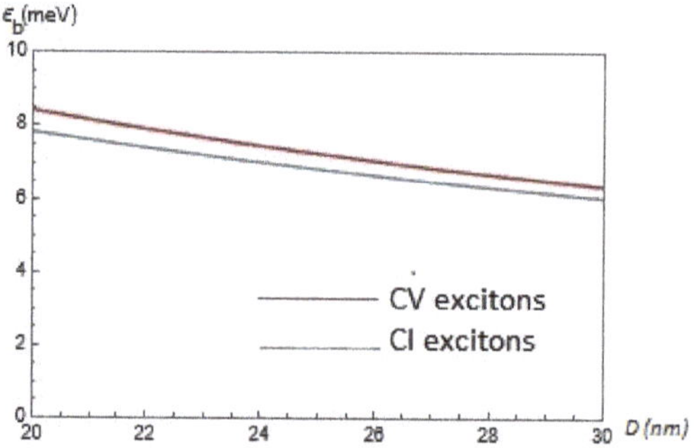

**Figure 4.** The exciton binding energy $\mathcal{E}_b(\alpha, \Delta, D)$ for CV and CI excitons as functions of the interlayer separation $D$ for chosen parameter $\alpha = 0.6$ and gap $\Delta = 0.5\ \hbar v_F/a$.

In Figure 5, we present plots of the effective masses for CV and CI dipolar excitons as functions of the gap $\Delta$ for chosen $\alpha = 0.6$ for (a) center-of-mass exciton mass $M$ on the left-hand side and (b) reduced exciton mass $\mu$, on the right. According to Figure 5, both $M$ and $\mu$ for the CV and CI excitons are increasing functions of $\Delta$, while for CV excitons both $M$ and $\mu$ are slightly larger than for CI excitons.

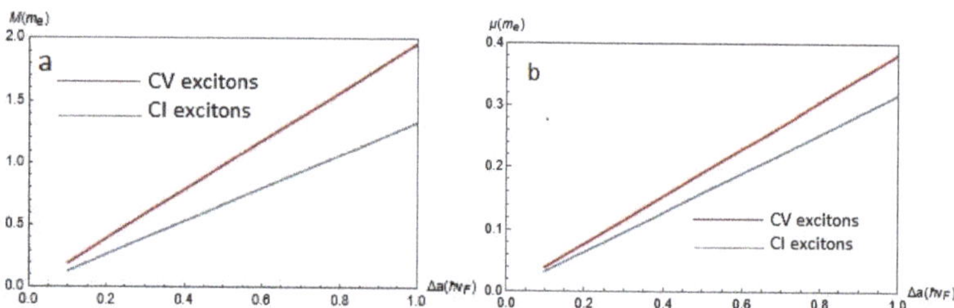

**Figure 5.** The effective masses of a dipolar exciton for CV and CI excitons as functions of the gap $\Delta$ for chosen $\alpha = 0.6$ for (a) center-of-mass exciton mass $M$ on the left panel and (b) reduced exciton mass $\mu$, on the right.

Figure 6 shows the effective masses of a dipolar exciton for CV and CI excitons as functions of $\alpha$ for chosen $\Delta = 0.5\,\hbar v_F/a$ for (a) center-of-mass exciton mass $M$ in the left panel and (b) reduced exciton mass $\mu$, on the right. According to Figure 6, for CV excitons $M$ is a decreasing function of $\alpha$, whereas $\mu$ does not depend on $\alpha$. For CI excitons, both $M$ and $\mu$ increase as $\alpha$ is increased. For $\alpha \lesssim 0.7$, both $M$ and $\mu$ for CV excitons are larger than for CI excitons, but when $\alpha \gtrsim 0.7$ $\mathcal{E}_b(\alpha, \Delta, D)$ both $M$ and $\mu$ for CV excitons are smaller than for CI excitons.

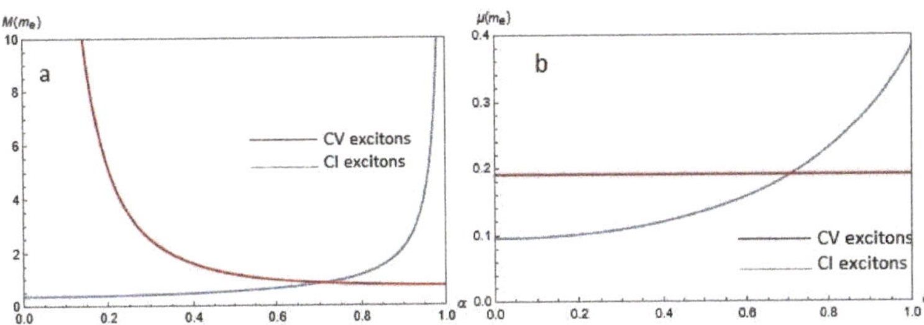


**Figure 6.** The effective masses of dipolar excitons for CV and CI excitons as functions of the hopping parameter $\alpha$ for chosen gap $\Delta = 0.5\,\hbar v_F/a$ for (a) center-of-mass exciton mass $M$ in the left panel and (b) reduced exciton mass $\mu$, on the right.

Figure 7 demonstrates the dependence of the sound velocity $c \equiv c_1$ on the hopping parameter $\alpha$ for chosen $\Delta = \hbar v_F/a$, interlayer separations $D = 25$ nm at fixed concentrations $n_A = 50 \times 10^{11}$ cm$^{-2}$ and $n_B = 50 \times 10^{11}$ cm$^{-2}$ of A and B excitons, respectively. According to Figure 7, $c$ does not depend much on $\alpha$ when $\alpha \lesssim 0.5$, while for $\alpha \gtrsim 0.5$, the sound velocity $c$ is a decreasing function of $\alpha$.

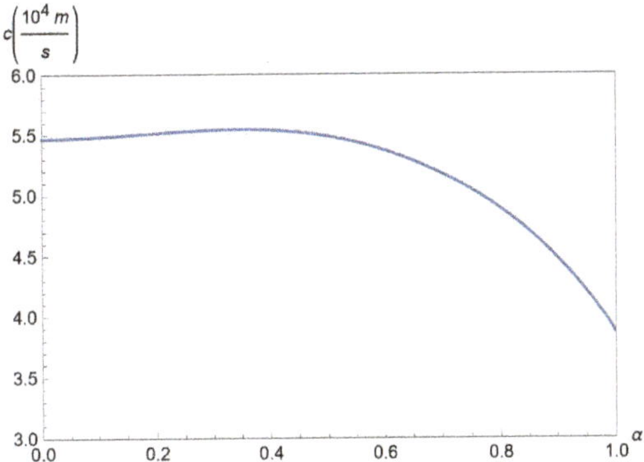


**Figure 7.** Plot of the sound velocity $c \equiv c_1$ versus $\alpha$ for chosen gap $\Delta = \hbar v_F/a$, interlayer separations $D = 25$ nm at fixed concentrations $n_A = 50 \times 10^{11}$ cm$^{-2}$ and $n_B = 50 \times 10^{11}$ cm$^{-2}$ of A and B excitons, respectively.

In Figure 8, we plot the sound velocity $c \equiv c_1$ versus the gap $\Delta$ for chosen parameter $\alpha = 0.6$, interlayer separations $D = 25$ nm for chosen concentrations $n_A = 50 \times 10^{11}$ cm$^{-2}$ and $n_B = 50 \times 10^{11}$ cm$^{-2}$ of A and B excitons, respectively. According to Figure 8, the sound velocity $c$ is a decreasing function of $\Delta$.

In Figure 9, we show the sound velocity $c \equiv c_1$ as a function of the interlayer separation $D$ for hopping parameter $\alpha = 0.6$ and gap $\Delta = 0.5\,\hbar v_F/a$, for fixed concentrations $n_A = 50 \times 10^{11}$ cm$^{-2}$ and $n_B = 50 \times 10^{11}$ cm$^{-2}$ of A and B excitons, respectively. According to Figure 9, the sound velocity $c$ is an increasing function of $D$.

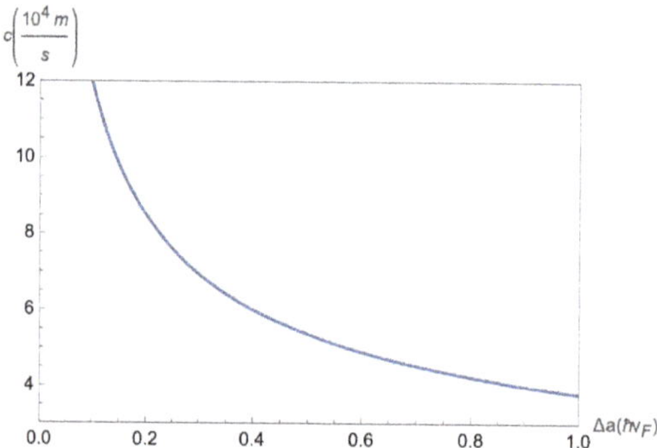

**Figure 8.** The sound velocity $c \equiv c_1$ versus the gap $\Delta$ for chosen parameter $\alpha = 0.6$, interlayer separations $D = 25$ nm at the fixed concentrations $n_A = 50 \times 10^{11}$ cm$^{-2}$ and $n_B = 50 \times 10^{11}$ cm$^{-2}$ of A and B excitons, respectively.

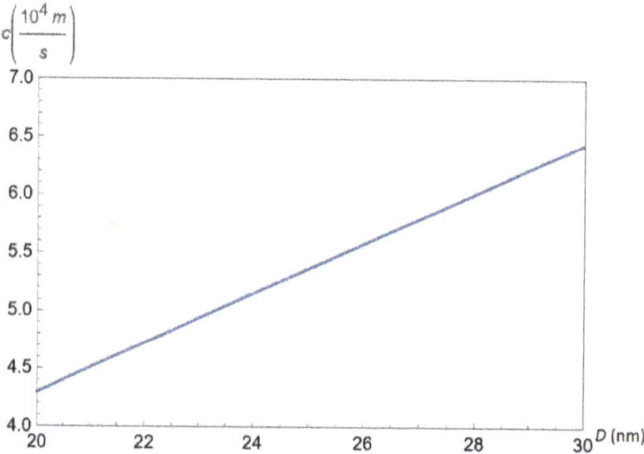

**Figure 9.** The sound velocity $c \equiv c_1$ versus the interlayer separation $D$ for chosen parameter $\alpha = 0.6$ and gap $\Delta = 0.5\, \hbar v_F / a$, at fixed concentrations $n_A = 50 \times 10^{11}$ cm$^{-2}$ and $n_B = 50 \times 10^{11}$ cm$^{-2}$ of A and B excitons, respectively.

In Figure 10, we illustrate the dependence of the sound velocity $c \equiv c_1$ on the concentrations $n_A$ and $n_B$ of A and B excitons, respectively for chosen hopping parameter $\alpha = 0.6$ and gap $\Delta = 0.5\, \hbar v_F / a$, at fixed interlayer separation $D = 25$ nm. According to Figure 10, the sound velocity $c$ is an increasing function of both concentrations $n_A$ and $n_B$.

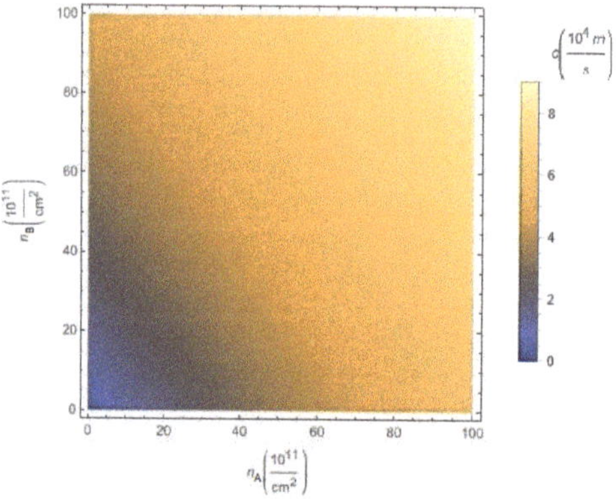

**Figure 10.** The sound velocity $c \equiv c_1$ versus the concentrations $n_A$ and $n_B$ of A and B excitons, respectively, for chosen parameter $\alpha = 0.6$ and gap $\Delta = 0.5\, \hbar v_F / a$, at the fixed interlayer separation $D = 25$ nm.

In Figure 11, we present the mean-field phase transition critical temperature $T_c(n_A, n_B, \alpha, \Delta, D)$ as a function of the parameter $\alpha$ for chosen gap $\Delta = 0.5\ \hbar v_F/a$, interlayer separations $D = 25$ nm at the fixed concentrations $n_A = 50 \times 10^{11}$ cm$^{-2}$ and $n_B = 50 \times 10^{11}$ cm$^{-2}$ of A and B excitons, respectively. According to Figure 11, $T_c$ is a decreasing function of $\alpha$ at $\alpha \lesssim 0.9$, while at $\alpha \gtrsim 0.9$ the critical temperature $T_c$ is an increasing function of $\alpha$.

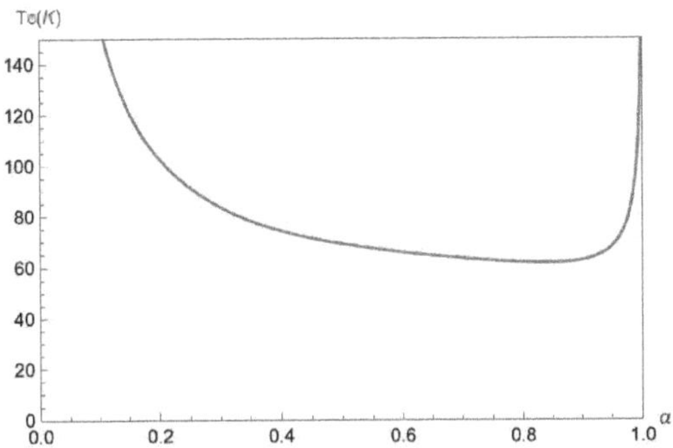

**Figure 11.** The mean-field phase transition critical temperature $T_c(n_A, n_B, \alpha, \Delta, D)$ versus the parameter $\alpha$ for chosen gap $\Delta = 0.5\ \hbar v_F/a$, interlayer separations $D = 25$ nm at the fixed concentrations $n_A = 50 \times 10^{11}$ cm$^{-2}$ and $n_B = 50 \times 10^{11}$ cm$^{-2}$ of A and B excitons, respectively.

In Figure 12, we present the mean-field phase transition critical temperature $T_c(n_A, n_B, \alpha, \Delta, D)$ as a function of the gap $\Delta$ for chosen parameter $\alpha = 0.6$, interlayer separations $D = 25$ nm at the fixed concentrations $n_A = 50 \times 10^{11}$ cm$^{-2}$ and $n_B = 50 \times 10^{11}$ cm$^{-2}$ of A and B excitons, respectively. According to Figure 12, the criticaltemperature $T_c$ is a decreasing function of $\Delta$.

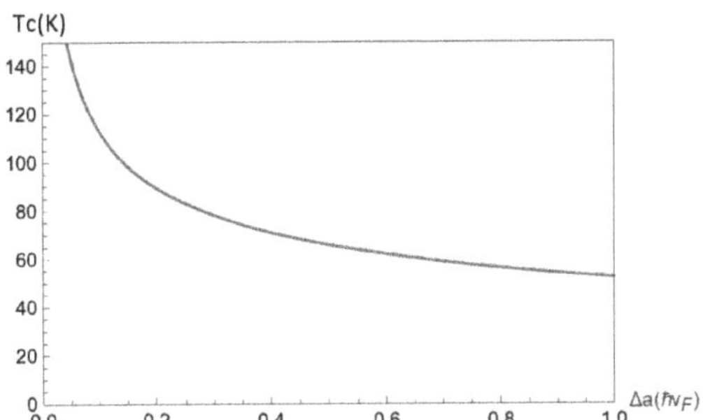

**Figure 12.** The mean-field phase transition critical temperature $T_c(n_A, n_B, \alpha, \Delta, D)$ versus the gap $\Delta$ for chosen parameter $\alpha = 0.6$, interlayer separations $D = 25$ nm at the fixed concentrations $n_A = 50 \times 10^{11}$ cm$^{-2}$ and $n_B = 50 \times 10^{11}$ cm$^{-2}$ of A and B excitons, respectively.

In Figure 13, we demonstrate the mean-field phase transition critical temperature $T_c(n_A, n_B, \alpha, \Delta, D)$ as a function of the interlayer separation $D$ for chosen parameter $\alpha = 0.6$ and gap $\Delta = 0.5\ \hbar v_F/a$, at the fixed concentrations $n_A = 50 \times 10^{11}$ cm$^{-2}$ and $n_B = 50 \times 10^{11}$ cm$^{-2}$ of A and B excitons, respectively. According to Figure 13, the critical temperature $T_c$ is an increasing function of $D$.

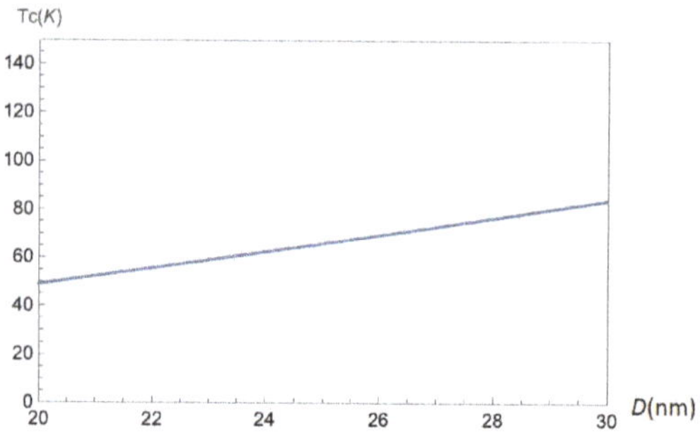

**Figure 13.** The mean-field phase transition critical temperature $T_c(n_A, n_B, \alpha, \Delta, D)$ versus the inter-layer separation $D$ for chosen parameter $\alpha = 0.6$ and gap $\Delta = 0.5\ \hbar v_F/a$, at fixed concentrations $n_A = 50 \times 10^{11}$ cm$^{-2}$ and $n_B = 50 \times 10^{11}$ cm$^{-2}$ of A and B excitons, respectively.

In Figure 14, we present density plots for the mean-field phase transition critical temperature $T_c(n_A, n_B, \alpha, \Delta, D)$ as a function of the concentrations $n_A$ and $n_B$ of A and B excitons, respectively for chosen parameter $\alpha = 0.6$ and gap $\Delta = 0.5\ \hbar v_F/a$, at the fixed interlayer separation $D = 25$ nm. According to Figure 14, the the critical temperature $T_c$ is an increasing function of both the concentrations $n_A$ and $n_B$.

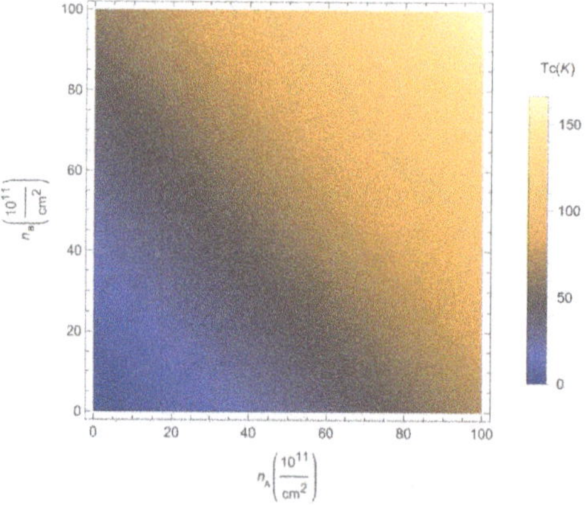

**Figure 14.** Density plot for the mean-field phase transition critical temperature $T_c(n_A, n_B, \alpha, \Delta, D)$ versus the concentrations $n_A$ and $n_B$ of A and B excitons, respectively, for chosen parameter $\alpha = 0.6$ and gap $\Delta = 0.5\ \hbar v_F/a$, at the fixed interlayer separation $D = 25$ nm.

At a formal level, the weakly interacting Bose gas of A and B dipolar excitons in a GHAT3 double layer are similar to the two-component weakly interacting Bose gas of trapped cold atoms in a planar harmonic trap. The spectrum of collective excitations in the Bogoliubov approximation for dipolar excitons in a GHAT3 double layer is similar to one for a two-component BEC of trapped cold atoms, studied in [53,54].

The gap parameter $\Delta$, has a dual role, since it appears as chemical potential in the Hamiltonian, as also in the mass of the excitons through the band curvature. According to Figures 2 and 12, the dipolar exciton binding energy is an increasing function of the gap $\Delta$, while is the mean-field phase transition temperature $T_c$ is a decreasing function of the gap $\Delta$. Therefore, there should be an optimal value for $\Delta$, which would correspond to relatively high $T_c$ at the relatively high dipolar exciton binding energy. The latter condition provides the formation of the superfluid phase by the relatively stable dipolar excitons.

Note that electron-hole superfluids can be formed not only in the BEC regime but also in the BCS-BEC crossover regime [7]. Quantum Monte Carlo simulations analyzing the BCS-BEC crossover regime for electron-hole systems have been performed [57]. In this paper, we concentrate on the dilute electron-hole system, which corresponds to the BEC, which matches experimentally achievable densities in the electron-hole systems in 2D materials. BCS regime requires higher concentrations beyond the model of weakly interacting Bose gas. The studies of the BCS regime, and BEC-BCS crossover for an electron-hole superfluid in a GHAT3 double layer, seem to be a promising direction for future studies.

The considered system of dipolar excitons in a GHAT3 double layer has also a strong similarity, with photon condensation in a cavity. The collective modes and possibility of the Kosterlitz-Thouless phase transition to the superfluid phase [58,59] has been studied for a photon condensation in a cavity in [60]. If we consider only one type of excitons in a GHAT3 double layer, assuming the concentration of the excitons of another type to be zero, the expressions for the spectrum of collective excitations reported in this paper can be reduced to the expressions similar to [60].

The Kosterlitz-Thouless phase transition to the superfluid phase [58,59] can be inferred from the variation of the superfluid density, which has been computed in this paper.

Note that, in this paper, we did not consider vortices, as within the mean-field approximation it was assumed that the number of quasiparticles are relatively low. However, beyond the mean-field approximation, it is possible to consider the properties of vortices in the system of dipolar excitons. Thus, the dynamical creation of fractionalized vortices and vortex lattices can be considered by applying the approach, developed for the BEC of cold atoms in [61].

The Josephson phenomena for two trapped condensates of dipolar excitons can be studied by applying an approach similar to the one developed for non-Abelian Josephson effect between two $F = 2$ spinor Bose-Einstein condensates of cold atoms in double optical traps [62].

## 6. Conclusions

This paper is devoted to an investigation of the existence of BEC and superfluidity of dipolar excitons in double layers of GHAT3 which was proposed and analyzed. We have derived the solution of a two-body problem for an electron and a hole for the model Hamiltonian representing double-layer GHAT3. We predict the formation of two types of dipolar excitons, characterized by different binding energies and effective masses, in the double layer of GHAT3. We have calculated the binding energy, effective mass, spectrum of collective excitations, superfluid density, and the mean-field critical temperature of the phase transition to the superfluid state for the two-component weakly interacting Bose gas of A and B dipolar excitons in double-layer GHAT3. We have demonstrated that at fixed exciton density, the mean-field critical temperature for superfluidity of dipolar excitons is decreased as a function of the gap $\Delta$. Our results show that $T_c$ is increased as a function of the density $n$ and is decreased as a function of the gap $\Delta$ and the interlayer separation $D$.

The occupancy of the superfluid state at $T < T_c$ can result in the existence of persistent dissipationless superconducting oppositely directed electric currents in each GHAT3 layer, forming a double layer. According to the presented results of our calculations, while the external weak magnetic field, responsible for the formation of the gap $\Delta$ in the double layer of $\alpha - T_3$ increases the exciton binding energy, the mean-field transition temperature to the superfluid phase is increased as the weak magnetic field and $\Delta$ are decreased. Therefore, the dipolar exciton system in a double-layer of GHAT3 can be applied to engineer a switch, where transport properties of dipolar excitons can be tuned by an external weak magnetic field, forming the gap $\Delta$. Varying a weak magnetic field may lead to a phase transition between the superfluid and normal phase, which sufficiently changes the transport properties of dipolar excitons.

**Author Contributions:** Conceptualization, O.L.B. and G.G.; computations, G.P.M. and P.F.; writing reviewing and editing, O.L.B. and G.G.; supervision, O.L.B., G.G. and P.F.; project administration, G.G. All authors have read and agreed to the published version of the manuscript.

**Funding:** This research was supported by U.S. ARO grant No. W911NF1810433. G.G. would like to acknowledge the support from the Air Force Research Laboratory (AFRL) through Grant No. FA9453-21-1-0046.

**Institutional Review Board Statement:** Not applicable.

**Informed Consent Statement:** Not applicable.

**Data Availability Statement:** The data presented in this study are available on request from the corresponding author

**Conflicts of Interest:** The authors declare no conflict of interest.

## References


1. Lozovik, Y.E.; Yudson, V.I. Feasibility of superfluidity of paired spatially separated electrons and holes; a new superconductivity mechanism. *Sov. Phys. JETP Lett.* **1975**, *22*, 274.
2. Lozovik, Y.E.; Yudson, V.I. A new mechanism for superconductivity: Pairing between spatially separated electrons and holes. *Sov. Phys. JETP Lett.* **1976**, *44*, 389.
3. Snoke, D.W. Spontaneous Bose coherence of excitons and polaritons. *Science* **2002**, *298*, 1368. [CrossRef] [PubMed]
4. Butov, L.V. Condensation and pattern formation in cold exciton gases in coupled quantum wellsm. *J. Phys. Condens. Matter* **2004**, *16*, R1577. [CrossRef]
5. Eisenstein, J.P.; MacDonald, A.H. Bose–Einstein condensation of excitons in bilayer electron systems. *Nature* **2004**, *432*, 691. [CrossRef]
6. Snoke, D.W. *Quantum Gases: Finite Temperature and Non-equilibrium Dynamics*; Cold Atom Series; Proukakis, N.P., Gardiner, S.A., Davis, M.J., Szymanska, M.H., Eds.; Imperial College Press: London, UK, 2013; Volume 1, p. 419.
7. Saberi-Pouya, S.; Conti, S.; Perali, A.; Croxall, A.F.; Hamilton, A.R.; Peeters, F.M.; Neilson, D. Experimental conditions for the observation of electron-hole superfluidity in GaAs heterostructures. *Phys. Rev. B* **2020**, *101*, 140501(R). [CrossRef]
8. Berman, O.L.; Lozovik, Y.E.; Gumbs, G. Dynamical equation for an electron-hole pair condensate in a system of two graphene layers. *Phys. Rev. B* **2008**, *77*, 155433. [CrossRef]
9. Lozovik, Y.E.; Sokolik, A.A. Electron-hole pair condensation in a graphene bilayer. *JETP Lett.* **2008**, *87*, 55. [CrossRef]
10. Lozovik, Y.E.; Sokolik, A.A. Multi-band pairing of ultrarelativistic electrons and holes in graphene bilayer. *Phys. Lett. A* **2009**, *374*, 326. [CrossRef]
11. Bistritzer, R.; MacDonald, A.H. Influence of disorder on electron-hole pair condensation in graphene bilayers. *Phys. Rev. Lett.* **2008**, *101*, 256406. [CrossRef]
12. Berman, O.L.; Kezerashvili, R.Y.; Ziegler, K. Superfluidity of dipole excitons in the presence of band gaps in two-layer graphene. *Phys. Rev. B* **2012**, *85*, 035418. [CrossRef]
13. Perali, A.; Neilson, D.; Hamilton, A.R. High-temperature superfluidity in double-bilayer graphene. *Phys. Rev. Lett.* **2013**, *110*, 146803. [CrossRef]
14. Fogler, M.M.; Butov, L.V.; Novoselov, K.S. High-temperature superfluidity with indirect excitons in van der Waals heterostructures. *Nat. Commun.* **2014**, *5*, 4555. [CrossRef]
15. Wu, F.-C.; Xue, F.; MacDonald, A.H. Theory of two-dimensional spatially indirect equilibrium exciton condensates. *Phys. Rev. B* **2015**, *92*, 165121. [CrossRef]
16. Berman, O.L.; Kezerashvili, R.Y. High-temperature superfluidity of the two-component Bose gas in a transition metal dichalcogenide bilayer. *Phys. Rev. B* **2016**, *93*, 245410. [CrossRef]
17. Berman, O.L.; Kezerashvili, R.Y. Superfluidity of dipolar excitons in a transition metal dichalcogenide double layer. *Phys. Rev. B* **2017**, *96*, 094502. [CrossRef]
18. Conti, S.; Van der Donck, M.; Perali, A.; Peeters, F.M.; Neilson, D. Doping-dependent switch from one- to two-component superfluidity in coupled electron-hole van der Waals heterostructures. *Phys. Rev. B* **2020**, *101*, 220504(R). [CrossRef]
19. Berman, O.L.; Gumbs, G.; Kezerashvili, R.Y. Bose-Einstein condensation and superfluidity of dipolar excitons in a phosphorene double layer. *Phys. Rev. B* **2017**, *96*, 014505. [CrossRef]
20. Saberi-Pouya, S.; Zarenia, M.; Perali, A.; Vazifehshenas, T.; Peeters, F.M. High-temperature electron-hole superfluidity with strong anisotropic gaps in double phosphorene monolayers. *Phys. Rev. B* **2018**, *97*, 174503. [CrossRef]
21. Mazuz-Harpaz, Y.; Cohen, K.; Leveson, M.; West, K.; Pfeiffer, L.; Khodas, M.; Rapaport, R. Dynamical formation of a strongly correlated dark condensate of dipolar excitons. *Proc. Nat. Acad. Sci. USA* **2019**, *116*, 18328. [CrossRef]
22. Raoux, A.; Morigi, M.; Fuchs, J.-N.; Piéchon, F.; Montambaux, G. From dia- to paramagnetic orbital susceptibility of massless Fermions. *Phys. Rev. Lett.* **2014**, *112*, 026402. [CrossRef]
23. Sutherland, B. Localization of electronic wave functions due to local topology. *Phys. Rev. B* **1986**, *34*, 5208. [CrossRef]
24. Illes, E.; Carbotte, J.P.; Nicol, E.J. Hall quantization and optical conductivity evolution with variable Berry phase in the $\alpha - T_3$ model. *Phys. Rev. B* **2015**, *92*, 245410. [CrossRef]
25. Islam, S.K.F.; Dutta, P. Valley-polarized magnetoconductivity and particle-hole symmetry breaking in a periodically modulated $\alpha - T_3$ lattice. *Phys. Rev. B* **2017**, *96*, 045418. [CrossRef]

26. Illes, E.; Nicol, E.J. Magnetic properties of the $\alpha - T_3$ model: Magneto-optical conductivity and the Hofstadter butterfly. *Phys. Rev. B* **2016**, *94*, 125435. [CrossRef]

27. Dey, B.; Ghosh, T.K. Photoinduced valley and electron-hole symmetry breaking in $\alpha - T_3$ lattice: The role of a variable Berry phase. *Phys. Rev. B* **2018**, *98*, 075422. [CrossRef]

28. Dey, B.; Ghosh, T.K. Floquet topological phase transition in the $\alpha - T_3$ lattice. *Phys. Rev. B* **2019**, *99*, 205429. [CrossRef]

29. Biswas, T.; Ghosh, T.K. Dynamics of a quasiparticle in the $\alpha - T_3$ model: Role of pseudospin polarization and transverse magnetic field on zitterbewegung. *J. Phys. Condens. Matter* **2018**, *30*, 075301. [CrossRef]

30. Kovács, A.D.; David, G.; Dora, B.; Coorti, J. Frequency-dependent magneto-optical conductivity in the generalized $\alpha - T_3$ model. *Phys. Rev. B* **2017**, *95*, 035414. [CrossRef]

31. Biswas, T.; Ghosh, T.K. Magnetotransport properties of the $\alpha - T_3$ model. *J. Phys. Condens. Matter* **2016**, *28*, 495302. [CrossRef]

32. Oriekhov, D.O.; Gusynin, V.P. RKKY interaction in a doped pseudospin-1 fermion system at finite temperature. *arXiv* **2020**, arXiv:2001.00272.

33. Huang, D.; Iurov, A.; Xu, H.-Y.; Lai, Y.-C.; Gumbs, G. Interplay of Lorentz-Berry forces in position-momentum spaces for valley-dependent impurity scattering in $\alpha - T_3$ lattices. *Phys. Rev. B* **2019**, *99*, 245412. [CrossRef]

34. Li, Y.; Kita, S.; Munoz, P.; Reshef, O.; Vulis, D.I.; Yin, M.; Loncar, M.; Mazur, E. On-chip zero-index metamaterials. *Nat. Photon* **2015**, *9*, 738. [CrossRef]

35. Xu, H.-Y.; Huang, L.; Huang, D.H.; Lai, Y.-C. Geometric valley Hall effect and valley filtering through a singular Berry flux. *Phys. Rev. B* **2017**, *96*, 045412. [CrossRef]

36. Leykam, D.; Andreanov, A.; Flach, S. Artificial flat band systems: From lattice models to experiments. *Adv. Phys. X* **2018**, *3*, 677. [CrossRef]

37. Sherafati, M.; Satpathy, S. Analytical expression for the RKKY interaction in doped graphene. *Phys. Rev. B* **2011**, *84*, 125416. [CrossRef]

38. Iurov, A.; Gumbs, G.; Huang, D. Peculiar electronic states, symmetries, and Berry phases in irradiated $\alpha - T_3$ materials. *Phys. Rev. B* **2019**, *99*, 205135. [CrossRef]

39. Iurov, A.; Zhemchuzhna, L.; Dahal, D.; Gumbs, G.; Huang, D. Quantum-statistical theory for laser-tuned transport and optical conductivities of dressed electrons in $\alpha - T_3$ materials. *Phys. Rev. B* **2020**, *101*, 035129. [CrossRef]

40. Weekes, N.; Iurov, A.; Zhemchuzhna, L.; Gumbs, G.; Huang, D. Generalized WKB theory for electron tunneling in gapped $\alpha - T_3$ lattices. *Phys. Rev. B* **2021**, *103*, 165429. [CrossRef]

41. Iurov, A.; Zhemchuzhna, L.; Gumbs, G.; Huang, D.; Fekete, P.; Anwar, F.; Dahal, D.; Weekes, N. Tailoring plasmon excitations in $\alpha - T_3$ armchair nanoribbons. *Sci. Rep.* **2021**, *11*, 20577. [CrossRef]

42. Abranyos, Y.; Berman, O.L.; Gumbs, G. Superfluidity of dipolar excitons in doped double-layered hexagonal lattice in a strong magnetic field. *Phys. Rev. B* **2020**, *102*, 155408. [CrossRef]

43. Balassis, A.; Gumbs, G.; Roslyak, O. Temperature-Induced Plasmon Excitations for the $\alpha - T_3$ Lattice in Perpendicular Magnetic Field. *Nanomaterials* **2021**, *11*, 1720. [CrossRef]

44. Balassis, A.; Dahal, D.; Gumbs, G.; Iurov, A.; Huang, D.; Roslyak, O. Magnetoplasmons for the $\alpha - T_3$ model with filled Landau levels. *J. Phys. Condens. Matter* **2020**, *32*, 485301. [CrossRef]

45. Malcolm, J.D.; Nicol, E.J. Frequency-dependent polarizability, plasmons, and screening in the two-dimensional pseudospin-1 dice lattice. *Phys. Rev. B* **2016**, *93*, 165433. [CrossRef]

46. Illes, E. Properties of the $\alpha - T_3$ Model. Ph.D. Thesis, University of Guelph, Guelph, ON, Canada, 2017.

47. Wunsch, B.; Stauber, T.; Sols, F.; Guinea, F. Dynamical polarization of graphene at finite doping. *New J. Phys.* **2006**, *8*, 318. [CrossRef]

48. Cai, Y.; Zhang, L.; Zeng, Q.; Cheng, L.; Xu, Y. Infrared reflectance spectrum of BN calculated from first principles. *Solid State Commun.* **2007**, *141*, 262. [CrossRef]

49. Maksym, P.A.; Chakraborty, T. Quantum dots in a magnetic field: Role of electron-electron interactions. *Phys. Rev. Lett.* **1990**, *65*, 108. [CrossRef]

50. Iyengar, A.; Wang, J.; Fertig, H.A.; Brey, L. Excitations from filled Landau levels in graphene. *Phys. Rev. B* **2007**, *75*, 125430. [CrossRef]

51. Berman, O.L.; Kezerashvili, R.Y.; Kolmakov, G.V.; Lozovik, Y.E. Turbulence in a Bose-Einstein condensate of dipolar excitons in coupled quantum wells. *Phys. Rev. B* **2012**, *86*, 045108. [CrossRef]

52. Lifshitz, E.M.; Pitaevskii, L.P. *Statistical Physics, Part 2*; Pergamon Press: Oxford, UK, 1980.

53. Tommasini, P.; de Passos, E.J.V.; de Toledo Piza, A.F.R.; Hussein, M.S.; Timmermans, E. Bogoliubov theory for mutually coherent condensates. *Phys. Rev. A* **2003**, *67*, 023606. [CrossRef]

54. Sun, B.; Pindzola, M.S. Bogoliubov modes and the static structure factor for a two-species Bose–Einstein condensate. *J. Phys. B* **2010**, *43*, 055301. [CrossRef]

55. Abrikosov, A.A.; Gorkov, L.P.; Dzyaloshinskii, I.E. *Methods of Quantum Field Theory in Statistical Physics*; Prentice-Hall: Englewood Cliffs, NJ, USA, 1963.

56. Pitaevskii, L.; Stringari, S. *Bose-Einstein Condensation*; Clarendon Press: Oxford, UK, 2003.

57. López Ríos, P.; Perali, A.; Needs, R.J.; Neilson, D. Evidence from quantum Monte Carlo simulations of large-gap superfluidity and BCS-BEC crossover in double electron-hole layers. *Phys. Rev. Lett.* **2018**, *120*, 177701. [CrossRef]

58. Kosterlitz, J.M.; Thouless, D.J. Ordering, metastability and phase transitions in two-dimensional systems. *J. Phys. C* **1973**, *6*, 1181. [CrossRef]

59. Nelson, D.R.; Kosterlitz, J.M. Universal jump in the superfluid density of two-dimensional superfluids. *Phys. Rev. Lett.* **1977**, *39*, 1201. [CrossRef]

60. Vyas, V.M.; Panigrahia, P.K.; Banerji, J. A scheme to observe universal breathing mode and Berezinskii–Kosterlitz–Thouless phase transition in a two-dimensional photon gas. *Phys. Lett. A* **2014**, *378*, 1434. [CrossRef]

61. Ji, A.-C.; Liu, W.M.; Song, J.L.; Zhou, F. Dynamical creation of fractionalized vortices and vortex lattices. *Phys. Rev. Lett.* **2008**, *101*, 010402. [CrossRef]

62. Qi, R.; Yu, X.-L.; Li, Z.B.; Liu, W.M. Non-Abelian Josephson effect between two spinor Bose-Einstein condensates in double optical traps. *Phys. Rev. Lett.* **2009**, *102*, 185301. [CrossRef]

*nanomaterials*

*Article*

# Simultaneous Extraction of the Grain Size, Single-Crystalline Grain Sheet Resistance, and Grain Boundary Resistivity of Polycrystalline Monolayer Graphene

Honghwi Park [1], Junyeong Lee [1], Chang-Ju Lee [1], Jaewoon Kang [1], Jiyeong Yun [1], Hyowoong Noh [1], Minsu Park [1], Jonghyung Lee [1], Youngjin Park [1], Jonghoo Park [1], Muhan Choi [1,2], Sunghwan Lee [3] and Hongsik Park [1,2,*]

1   School of Electronic and Electrical Engineering, Kyungpook National University, Daegu 41566, Korea; hoepark@ee.knu.ac.kr (H.P.); jyl2015@ee.knu.ac.kr (J.L.); chjlee@knu.ac.kr (C.-J.L.); jwnkang@knu.ac.kr (J.K.); yeong112@knu.ac.kr (J.Y.); hwn1327@knu.ac.kr (H.N.); msp7352@knu.ac.kr (M.P.); jongh1019@knu.ac.kr (J.L.); dudwls1218@knu.ac.kr (Y.P.); jonghoopark@knu.ac.kr (J.P.); mhchoi@ee.knu.ac.kr (M.C.)
2   School of Electronics Engineering, Kyungpook National University, Daegu 41566, Korea
3   School of Engineering Technology, Purdue University, West Lafayette, IN 47907, USA; sunghlee@purdue.edu
*   Correspondence: hpark@ee.knu.ac.kr

**Abstract:** The electrical properties of polycrystalline graphene grown by chemical vapor deposition (CVD) are determined by grain-related parameters—average grain size, single-crystalline grain sheet resistance, and grain boundary (GB) resistivity. However, extracting these parameters still remains challenging because of the difficulty in observing graphene GBs and decoupling the grain sheet resistance and GB resistivity. In this work, we developed an electrical characterization method that can extract the average grain size, single-crystalline grain sheet resistance, and GB resistivity simultaneously. We observed that the material property, graphene sheet resistance, could depend on the device dimension and developed an analytical resistance model based on the cumulative distribution function of the gamma distribution, explaining the effect of the GB density and distribution in the graphene channel. We applied this model to CVD-grown monolayer graphene by characterizing transmission-line model patterns and simultaneously extracted the average grain size (~5.95 μm), single-crystalline grain sheet resistance (~321 Ω/sq), and GB resistivity (~18.16 kΩ-μm) of the CVD-graphene layer. The extracted values agreed well with those obtained from scanning electron microscopy images of ultraviolet/ozone-treated GBs and the electrical characterization of graphene devices with sub-micrometer channel lengths.

**Keywords:** CVD graphene; polycrystalline; grain size; single-crystalline grain; grain boundary (GB); GB distribution; sheet resistance; transmission-line model measurement

**Citation:** Park, H.; Lee, J.; Lee, C.-J.; Kang, J.; Yun, J.; Noh, H.; Park, M.; Lee, J.; Park, Y.; Park, J.; et al. Simultaneous Extraction of the Grain Size, Single-Crystalline Grain Sheet Resistance, and Grain Boundary Resistivity of Polycrystalline Monolayer Graphene. *Nanomaterials* **2022**, *12*, 206. https://doi.org/10.3390/nano12020206

Academic Editor: Eugene Kogan

Received: 30 November 2021
Accepted: 5 January 2022
Published: 9 January 2022

**Publisher's Note:** MDPI stays neutral with regard to jurisdictional claims in published maps and institutional affiliations.

## 1. Introduction

Chemical vapor deposition (CVD) is the most effective method for uniformly growing monolayer graphene on a wafer scale in a reproducible way [1]. However, CVD graphene can typically be grown as a polycrystalline structure composed of multiple single-crystalline grains connected by disordered grain boundaries (GBs) [2–6]. Scattering at the GB (i.e., structural line defect) affects carrier transport, as does scattering within a single-crystalline grain; thus, both the GB and grain act as major resistive sources in polycrystalline graphene [4–6]. Because the electrical properties of graphene are determined by the competition between these two resistive sources—i.e., relatively high-resistive GB and low-resistive grain—the average grain size has a significant impact on the electrical properties [7–10]. Therefore, unlike single-crystalline graphene, whose electrical performance can be explained only by the sheet resistance of the layer, the performance of polycrystalline CVD graphene should be explained by a combination of various grain-related parameters, such as the average grain size, single-crystalline grains sheet resistance, and GB resistivity. For this reason, rigorous evaluation of these grain parameters is crucial for the design and fabrication of CVD-graphene devices.

Accordingly, various techniques for characterizing grain parameters have been actively studied over the last decade. In the case of grain size, it can be evaluated through structural characterization and identification of GBs using spectroscopic or microscopic measurements. For instance, spatial mapping of the Raman peak intensities for graphene (D peak at ~1350 cm$^{-1}$, G peak at ~1580 cm$^{-1}$, and 2D peak at ~2690 cm$^{-1}$) enables the location and shape of GBs and grains to be identified [11]. This is an effective method for evaluating the size of individual grains, but estimating the average grain size of CVD graphene across the entire grown region is difficult because of the limited inspection area and the extremely slow mapping speed. Instead of mapping the Raman peaks, the GBs with an angstrom-scale width can be imaged directly by performing an ultraviolet (UV)/ozone treatment after growing graphene on a Cu substrate. The UV/ozone treatment selectively oxidizes Cu beneath the GBs through strong chemical reactions with O and OH radicals, allowing the GBs to be visualized and examined under an optical microscope (OM) or a scanning electron microscope (SEM) [12,13]. Although this process provides a convenient way to observe multiple grains and to evaluate their sizes, for a global estimation of the average grain size, a time-consuming manual process that evaluates the sizes of individual grains from a large amount of microscopy images covering a wide area of CVD-grown graphene should accompany it [14].

For the electrical properties of GBs and single-crystalline grains, four-terminal measurement-based evaluation techniques have been widely used. To extract the resistances of a GB and grain separately using these techniques, the location of the GB should first be identified with non-destructive transmission electron microscopy (TEM) [6]. An electron-beam lithography system is then used to fabricate a Hall-bar pattern across two grains joined by the GB. The sheet resistance of the single-crystalline grain and the resistivity of the GB can be extracted by performing a series of four-terminal measurements with this pattern [6]. This technique is significantly useful for understanding the electrical properties of individual grains and GBs; however, high-level technical skills are required to accurately estimate the location of the GB and fabricate the Hall-bar pattern aligned well with the GB location [4]. Furthermore, because of the limited TEM resolution capable of identifying the GBs, it is difficult to extend this technique to a global evaluation method that can extract the average grain sheet resistance and GB resistivity from multiple single-crystalline grains and GBs. As an alternative approach, an electrical characterization technique based on ohmic scaling law was developed for the global evaluation of these two electrical parameters [4]. In this technique, the average grain sheet resistance and GB resistivity can be extracted on a large scale by measuring the channel sheet resistance of each CVD-graphene sample as a function of the average grain size and then fitting it to the ohmic scaling law (i.e., a simple 1D series-resistance model) [4,5]. However, to apply this technique effectively, it is necessarily required to investigate the exact average grain size of each graphene sample, and further, prepare multiple CVD graphene samples with different average grain sizes but identical average grain sheet resistance and GB resistivity. These requirements may limit the practical application and an accurate extraction of the average grain sheet resistance and GB resistivity.

In this paper, we propose for the first time an electrical characterization method for extracting the average grain size, grain sheet resistance, and GB resistivity of monolayer CVD graphene simultaneously. For this purpose, we investigate the probability distribution of the number of GBs depending on the graphene–channel dimension, from which we develop an analytical resistance model that can explain the relationship between the electrical properties of polycrystalline graphene and its grain parameters. With this resistance model, we show that the three-grain parameters can be extracted simultaneously with an accuracy greater than 99% from the dependence of the material's electrical property (i.e., sheet resistance) on the channel dimension. To validate the developed resistance model and its applicability for parameter extraction, we fabricate a transmission-line model (TLM) pattern on monolayer CVD graphene and characterize the channel sheet resistance ($R_{sh}$) as a function of channel length ($L_{ch}$). We show that the average grain size, grain sheet resistance, and GB resistivity of CVD graphene can be extracted simultaneously from the analytical resistance model that is fitted to the measured $R_{sh}$–$L_{ch}$ curve. The three extracted values are then compared with those obtained using conventional methods and those reported in the literature.

## 2. Materials and Methods

### 2.1. Graphene Growth and Transfer

After the development of an analytical resistance model, to verify its applicability for grain-parameter extraction, we synthesized monolayer graphene on a 25-μm-thick Cu foil (99.8% purity, Alfa-Aesar Inc., Ward Hill, MA, USA) by a thermal CVD system (TCVD 100, Scientec Inc., Suwon, South Korea). To grow the polycrystalline graphene layer with uniform-size grains, electropolishing of the Cu surface was first performed in an 85% phosphoric acid bath using a constant voltage of 1.2 V for 20 min [15]. The polished Cu foil was then loaded into the CVD chamber and annealed at 1050 °C for 1 h with a flow of Ar (570 sccm) and $H_2$ (100 sccm) for surface treatment. Following the annealing step, monolayer graphene was grown at 1050 °C for 1 h under a chamber pressure of 2 Torr with a flow of Ar (570 sccm), $H_2$ (100 sccm), and $CH_4$ (2 sccm). The CVD chamber was then cooled down to room temperature in an Ar environment before the as-grown graphene sample was finally removed from the chamber.

For device fabrication, CVD-grown monolayer graphene was transferred onto thermally oxidized Si substrates (90-nm-thick $SiO_2$) by the optimized poly (methyl methacrylate) (PMMA)-mediated transfer method [16]. First, a PMMA solution (495 A4, MicroChem Inc., Westborough, MA, USA) was spin-coated onto the top side of the graphene-grown Cu foil at 1000 rpm for 60 s, followed by drying at room temperature for 24 h. The graphene layer grown on the back side of Cu was etched using $O_2$ plasma ashing (RF power = 30 W, working pressure = 30 m Torr) for 3 min. The Cu foil was then etched using diluted ammonium persulfate solution (0.02 M, Sigma Aldrich Inc., St. Louis, MO, USA) at room temperature for 24 h, after which the remaining PMMA/graphene stack was rinsed repeatedly with deionized water. The rinsed PMMA/graphene stack was finally transferred onto the oxidized Si substrate and the transferred sample was baked at 160 °C in a vacuum for 30 min to remove residual water and improve the adhesion between the graphene layer and the substrate. To minimize the PMMA residue and wrinkles, the PMMA layer was removed using acetic acid at room temperature for 3 h, followed by annealing at 300 °C for 3 h under an ultrahigh vacuum ($\sim 10^{-8}$ Torr) [16].

### 2.2. Device Fabrication

To extract the grain parameters of CVD graphene using the developed analytical resistance model, TLM patterns composed of back-gated field-effect transistors (FETs) with varying channel lengths (2–100 μm) were fabricated on CVD graphene transferred onto 90-nm-thick oxidized Si substrates. First, channel regions of the FETs were patterned by *i*-line mask-aligner lithography (MA6/BA6, Karl Suss Inc., Munich, Germany) with a positive-tone photoresist (AZ GXR-601, AZ Electronic Materials Inc., Branchburg, NJ, USA), and graphene channels were defined using $O_2$ plasma etching (30 W, 30 m Torr) for 2 min. Following the channel definition, the second photolithography process was performed using an image reversal photoresist (AZ 5214-E, AZ Electronic Materials Inc.) for source/drain electrode patterning. Then, a 20-nm-thick Pd and 50-nm-thick Au were sequentially deposited by an electron-beam evaporation system (SRN-200, Sorona Inc., Anseong, South Korea) for contact formation, after which the residual photoresist layer and the metal deposited on top of the photoresist were removed by the lift-off process in a warm acetone bath. CVD-graphene FETs with sub-micrometer channel lengths (0.18–0.75 μm) were fabricated using electron-beam lithography (Raith 150-TWO, Raith Inc., Dortmund,

Germany) and a positive-tone electron-beam resist (AR-P 672.045, Allresist Inc., Strausberg, Germany). For patterning the graphene channels and source/drain electrodes, electron-beam conditions of an area dose of 200 µC/cm$^2$, a step size of 5 nm, and an acceleration voltage of 30 kV were used.

*2.3. Characterizations*

The electrical characteristics of the fabricated graphene FETs were measured using a semiconductor parameter analyzer (4156C, Agilent Inc., Palo Alto, CA, USA) in a probe station under a high vacuum (~$10^{-7}$ Torr). Before the measurements, the graphene surface was annealed at 120 °C for 3 h in the vacuum probe station to remove any moisture, oxygen, and photoresist residue, which act as p-type dopants [16–18]. A Raman spectroscope (inVia reflex, Renishaw Inc., Wotton-under-Edge, UK) with 532-nm excitation was used to evaluate the material quality of monolayer CVD graphene. The top-view images of the CVD-graphene surface and fabricated graphene FETs were obtained using a field-emission SEM (SU8220, Hitachi Inc., Tokyo, Japan) and an OM (BX51, Olympus Inc., Tokyo, Japan) system. The GB visualization for estimating the average grain size was performed in a UV/ozone chamber (PSD-Pro, Novascan Technologies Inc., Boone, IA, USA) by irradiating a 254-nm UV light with an output power of 20 mW/cm$^2$ under ambient conditions. The grain sizes of CVD graphene observed in the top-view SEM images after the UV/ozone treatment were measured using the ImageJ software provided by the National Institutes of Health, USA. Note that the grain-size estimation from the SEM images was performed to check the accuracy of the average grain size extracted by the proposed electrical characterization method.

## 3. Results and Discussion

To develop a characterization method for extracting the average grain size, grain sheet resistance, and GB resistivity, it is necessary to investigate the effects of these parameters on the electrical characteristics of polycrystalline-graphene devices. Thus, we first theoretically calculated the channel sheet resistance of polycrystalline graphene as a function of channel length using a parallel-resistance model [14], and investigated its dependence on the channel length. For the sheet-resistance computation, the Voronoi tessellation (VT) method was used to generate 2D polycrystalline structures (Figure 1a), which can depict a real polycrystalline morphology with non-uniform sizes and shapes of grains [14,19–22]. Because the sizes and shapes of polycrystalline graphene are not uniform, the number of GBs impeding carrier transport between two electrodes varies with the location in the poly-graphene channel. This indicates that the sheet resistance of the graphene channel can vary locally—i.e., the sheet resistance would not be uniform within the polycrystalline channel. Thus, for rigorous resistance modeling that takes into account the local non-uniformity of the GB number, the channel width was divided into extremely narrow elements, and the number of GBs within each width element was counted separately. The resistance of each width element, which is the sum of the resistances of numerous grains and GBs connected in series (Figure 1a), was then calculated from the following 1D ohmic scaling law [4]:

$$\Delta R_i = R_{sh}^G \frac{L_{ch}}{\Delta W_{ch}} + \rho_{GB} \frac{n_i}{\Delta W_{ch}} \tag{1}$$

where $\Delta R_i$ is the resistance of each width element, $R_{sh}^G$ is the average sheet resistance of single-crystalline grains, $\rho_{GB}$ is the average resistivity of GBs, $L_{ch}$ is the channel length, $W_{ch}$ is the channel width, and $n_i$ is the number of GBs within each width element. Subsequently, the total channel resistance of polycrystalline graphene was calculated from the sum of the resistances of every width element connected in parallel (Figure 1a). The equation for calculating the channel resistance as a function of the channel length is as follows:

$$\frac{1}{R_{ch}(L_{ch})} = \sum_{i=1}^{m} \frac{1}{\Delta R_i} = \sum_{i=1}^{m} \frac{1}{R_{sh}^G \frac{L_{ch}}{\Delta W_{ch}} + \rho_{GB} \frac{n_i}{\Delta W_{ch}}} \tag{2}$$

where $R_{ch}$ is the channel resistance and $m$ is the number of divided width elements ($=W_{ch}/\Delta W_{ch}$). In this study, the channel width was divided into $10^5$ elements for the parallel-resistance modeling. Furthermore, to estimate the average channel resistance as a function of the channel length, we randomly generated 1000 polycrystalline structures with an average grain size of 5 µm using the VT method and repeated this calculation process (see Figure S1 for the calculation results of poly-graphene channels with different average grain sizes of 2.5 and 15 µm). The $W_{ch}$ used in the calculation was 20 µm, and the $R_{sh}^G$ and $\rho_{GB}$ values were 300 Ω/sq and 10.6 kΩ-µm, respectively, which were selected within the range of those of CVD graphene reported in the literature [4,5]. (The results calculated using different $W_{ch}$, $R_{sh}^G$, and $\rho_{GB}$ values are shown in Figure S2). Note also that the source/drain contact resistance was not taken into account in the resistance modeling and computation processes. The theoretically calculated average resistance of the poly-graphene channel (average grain size of 5 µm) simulated using the VT method is shown in Figure 1b. The channel resistance is directly proportional to the channel length in the long-channel region (i.e., constant $dR_{ch}/dL_{ch}$), whereas the slope ($dR_{ch}/dL_{ch}$) varies with the channel length in the relatively short-channel region (particularly the $L_{ch}$ around the average grain size). This indicates that the sheet resistance of the polycrystalline channel depends on the channel dimension, unlike the single-crystalline graphene or other single-crystalline semiconductors. The average channel sheet resistance ($R_{sh}$) as a function of channel length is shown in Figure 1c. The sheet resistance was calculated from $R_{ch} \times W_{ch}/L_{ch}$. The channel sheet resistance is constant for long channels; however, it decreases sharply as the channel length is reduced below approximately 25 µm. The decrease in channel sheet resistance is most prominent at channel lengths around the average grain size, which is due to the significantly lowered probability of the existence of GBs at those channel lengths [14]. This implies that the GB density and distribution within the channel region play a critical role in determining the dependence of the channel sheet resistance on the channel length.

As shown in Figure 1, it is critical to estimate the GB density and distribution within the graphene channel to understand the dependence of the channel sheet resistance on the channel length. Thus, we investigated the proportion distribution of the number of GBs depending on the channel length by counting the GB number within narrow width elements divided into $10^5$. For statistical evaluation, we repeated the process with 1,000 polycrystalline-graphene structures (average grain size of 5 µm) generated randomly using the VT method, as in the channel resistance calculation. The average histogram distributions of the proportion of the GB number within the channel region at three different channel lengths are shown in Figure 2a–c (see Figure S3 for the results at various channel lengths). The number of GBs within the channel region is limited to 0–3 at channel lengths less than or equal to the average grain size (i.e., $L_{ch} \leq 5$ µm), whereas it is evenly distributed at the channel length greater than the average grain size. This explains why there is a significant decrease in sheet resistance at channel lengths near the average grain size (Figure 1c). We investigated several distribution functions to find a way to estimate such a proportion distribution without counting the number of GBs and found that the envelope of the proportional distribution of the GB number follows the continuous probability density function of the gamma distribution (i.e., gamma PDF) [23]:

$$GamPDF(x) = \frac{1}{\beta^\alpha \int_0^\infty u^{\alpha-1}e^{-u}du}x^{\alpha-1}e^{-\frac{x}{\beta}} \; for \; x, \, \alpha, \, \beta > 0 \tag{3}$$

where $\alpha \cdot \beta$ is the mean and $\alpha \cdot \beta^2$ is the variance. When $\alpha = 3.85 \times L_{ch}/l_G$ and $\beta = 0.33$ (where $l_G$ is the average grain size), it was empirically confirmed that the gamma PDF agrees well with the envelope of the proportional distribution of the GB number for all channel lengths (Figure 2 and Figure S3). Furthermore, this observation was also valid for poly-graphene channels with different average grain sizes (see Figure S4 and Figure S5 for $l_G = 2.5$ and 15 μm, respectively). Following this, to obtain the discrete distribution of the GB number from the continuous gamma PDF, we used the cumulative distribution function of the gamma distribution (i.e., gamma CDF), which is an integral form of the gamma PDF as follows:

$$GamCDF(n) = \int_0^n GamPDF(x)dx$$

$$= \int_0^n \frac{1}{0.33^{\frac{3.85L_{ch}}{l_G}} \int_0^\infty u^{\left(\frac{3.85L_{ch}}{l_G}-1\right)} e^{-u}du} x^{\left(\frac{3.85L_{ch}}{l_G}-1\right)} e^{-\frac{x}{0.33}} dx \tag{4}$$

where $n$ is the number of GBs. The proportion distribution of the GB number can be estimated from $GamCDF\,(n + 0.5) - GamCDF\,(n - 0.5)$ with an accuracy greater than 98% for all channel lengths, as shown in the red symbols of Figures 2 and S3.

Because the proportion distribution of the GB number within the channel region can now be accurately estimated without counting the number of GBs, we can develop an analytical resistance model that is more generalized for explaining the dependence between the sheet resistance and channel length. For this purpose, the divided narrow width elements ($\Delta W_{ch}$) were grouped and rearranged by the number of GBs considering its proportion within the channel region estimated from the gamma CDF (Figure 3a). The rearranged width element by the number of GBs ($W_{ch,n}$) can be expressed as follows:

$$W_{ch,n} = W_{ch} \times \{GamCDF(n+0.5) - GamCDF(n-0.5)\}, \quad W_{ch} = \sum_{n=0}^{k} W_{ch,n} \tag{5}$$

where $k$ is the maximum number of GBs existing within the divided width elements. Based on the rearranged width elements, Equation (2) can then be generalized as:

$$\frac{1}{R_{ch}(L_{ch})} = \sum_{n=0}^{k} \frac{W_{ch,n}}{R_{sh}^G L_{ch} + \rho_{GB} n} \tag{6}$$

Following that, the analytical model for the channel sheet resistance, composed of the three-grain parameters ($l_G$, $R_{sh}^G$, and $\rho_{GB}$), can be finally induced as a function of the channel length:

$$R_{sh}(L_{ch}) = R_{ch}(L_{ch}) \times \frac{W_{ch}}{L_{ch}}$$

$$= \left\{ \sum_{n=0}^{k} \frac{\left( \int_{n-0.5}^{n+0.5} \frac{1}{0.33^{\frac{3.85L_{ch}}{l_G}} \int_0^\infty u^{\left(\frac{3.85L_{ch}}{l_G}-1\right)} e^{-u}du} x^{\left(\frac{3.85L_{ch}}{l_G}-1\right)} e^{-\frac{x}{0.33}} dx \right) L_{ch}}{R_{sh}^G L_{ch} + \rho_{GB} n} \right\}^{-1} \tag{7}$$

Using the developed analytical resistance model as a fitting function, we estimated the dependence of the channel sheet resistance on the channel length observed in the theoretical calculation result, Figure 1c, by adjusting the three unknown fit parameters (i.e., $l_G$, $R_{sh}^G$, and $\rho_{GB}$). As a result, the analytical model was fitted with the calculation result well, with a fitting accuracy greater than 99.98% (Figure 3b). Note that any predetermined grain-parameter information is not required in the fitting process. The important aspect of this result is that the three-grain parameters can be extracted from the analytical resistance model that is fitted to the $R_{sh}$–$L_{ch}$ curve. The three-parameter values provided for the theoretical sheet-resistance calculation (Figure 1) and those extracted from the best-fitted analytical model are summarized in Table 1. The result shows that the average grain size, the sheet resistance of single-crystalline grains, and the resistivity of GBs, can be extracted simultaneously with a high accuracy (>99%). Likewise, in the theoretical sheet-resistance results simulated with different average grain sizes or different grain sheet resistance and GB resistivity values, it was also confirmed that the three grain parameters can be extracted with a high accuracy using the analytical model (see Figure S1 and Figure S2). These results indicate that the proposed method can simultaneously extract the average grain size, grain sheet resistance, and GB resistivity of polycrystalline graphene from the electrical characteristics of graphene devices without using any predetermined grain parameter values, and furthermore, it can be applied to various polycrystalline graphene layers with different grain sizes, as well as the electrical properties of GBs and single-crystalline grains.

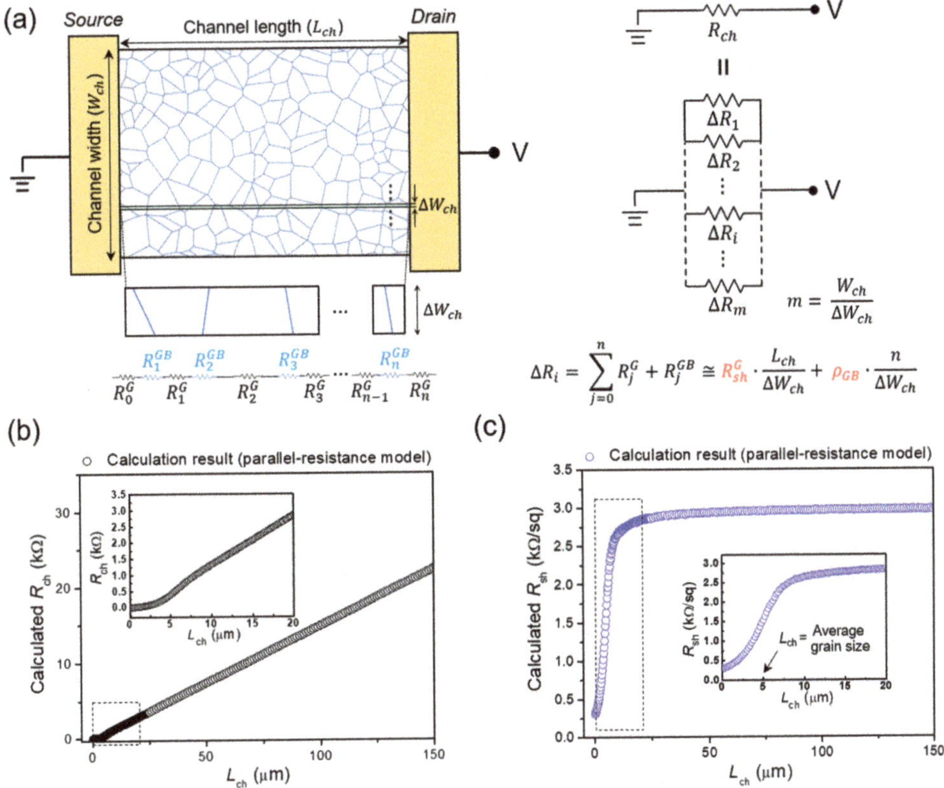

**Figure 1.** Parallel-resistance model for theoretical computation of the sheet resistance depending on the channel dimension. (**a**) A schematic of the electrical device with a poly-graphene channel simulated using the VT method. For poly-graphene resistance modeling, the channel width is divided into extremely narrow elements ($\Delta W_{ch} = W_{ch}/10^5$). The resistance of the poly-graphene channel is calculated from the parallel connection of the divided elements. (**b**) The calculated channel resistance as a function of channel length for polycrystalline graphene (with an average grain size of 5 μm) simulated using the VT method. Inset: the calculated channel resistances in relatively short channels (denoted by the dashed box), which shows that the slope ($dR_{ch}/dL_{ch}$) varies with the channel length. (**c**) The calculated channel sheet resistance as a function of the channel length. Inset: the calculated channel sheet resistances in relatively short channels (denoted by the dashed box), which shows that the sheet resistance decreases significantly as the channel length approaches the average grain size.

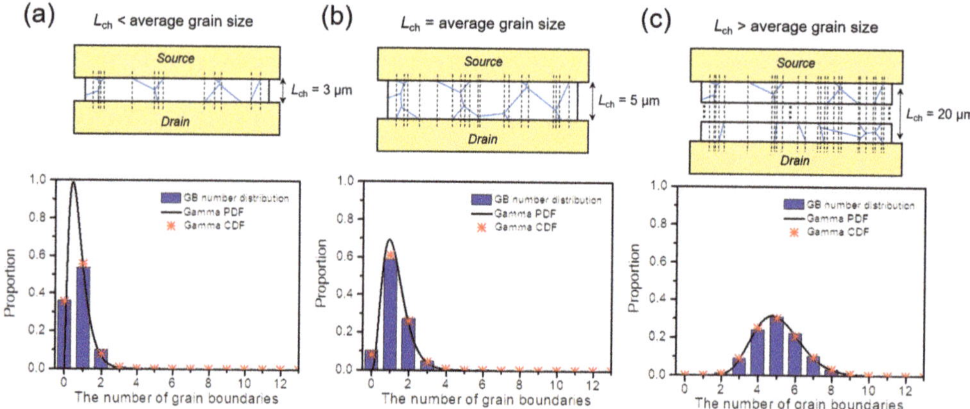



**Figure 2.** Histogram distributions of the proportion of the GB number within the channel region when the channel length is (**a**) less than the average grain size, (**b**) equal to the average grain size, and (**c**) greater than the average grain size. Each distribution was evaluated by counting the GB number within narrow width elements divided into $10^5$. The black dashed lines in (**a**–**c**) indicate the locations where the number of GBs changes. For all channel lengths, the envelope of the proportional distribution of the number of GBs follows the continuous probability density function of the gamma distribution (gamma PDF), and the discrete proportion distribution of the GB number can be estimated from the cumulative distribution function of the gamma distribution (gamma CDF) with an accuracy greater than 98%.

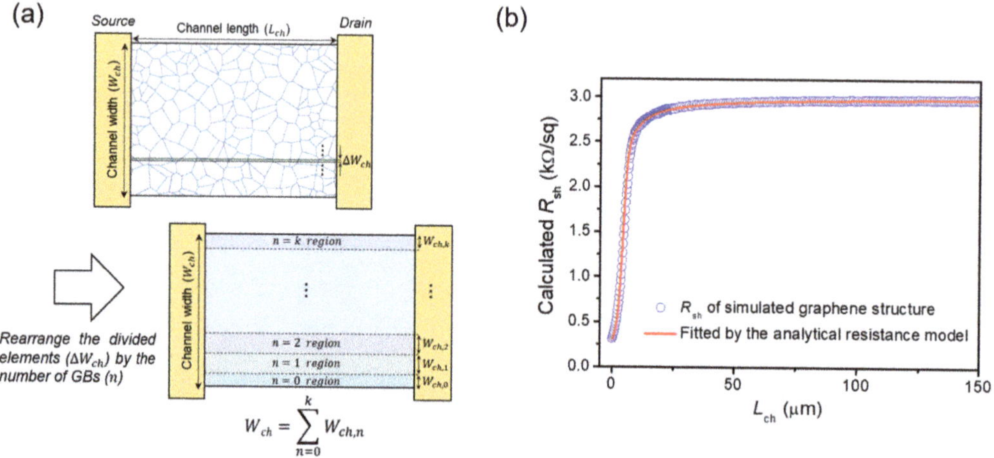



**Figure 3.** GB distribution-based analytical resistance model for the simultaneous extraction of the grain parameters (the average grain size, grain sheet resistance, GB resistivity) of polycrystalline graphene. (**a**) Considering the proportion distribution of the GB number estimated from the gamma CDF, the divided width elements ($\Delta W_{ch}$) can be grouped and rearranged by the number of GBs. Based on the rearranged width elements ($W_{ch,n}$), a sheet-resistance model composed of the three-grain parameters can be induced as a function of the channel length. (**b**) The developed analytical resistance model was used to fit the calculated channel sheet resistance as a function of the channel length. The sheet-resistance dependence on the channel length can be estimated with a high fitting accuracy greater than 99.98% by adjusting three fit (i.e., grain) parameters, from which the three-grain parameters can be extracted from the analytical resistance model fitted to the $R_{sh}$–$L_{ch}$ curve.

**Table 1.** Comparison of three-grain parameters given for the theoretical calculation with those extracted by the developed analytical resistance model.

|  | $l_G$ (µm) | $R_{sh}^G$ (Ω/sq) | $\rho_{GB}$ (kΩ-µm) |
| --- | --- | --- | --- |
| Given parameter | 5.0 | 300 | 10.6 |
| Extracted parameter | 5.02 | 299.7 | 10.59 |

To verify whether the proposed parameter extraction method is practical, we fabricated TLM patterns on CVD graphene and extracted three-grain parameters by characterizing the dependence of the channel sheet resistance on the channel length using the analytical resistance model. For this purpose, we synthesized monolayer graphene through CVD and transferred it onto thermally oxidized Si substrates (90-nm-thick $SiO_2$) for the device fabrication. The material and layer quality of as-transferred CVD graphene was then evaluated using Raman spectroscopy measurements, confirming the defect-free monolayer graphene with an intensity ratio of 2D to G peaks of ~2.8 (see Figure S6). To characterize the dependence between the channel sheet resistance and the channel length, the back-gated graphene FETs with channel lengths of 2–100 µm and a channel width of 20 µm were fabricated on transferred CVD graphene (Figure 4a) and the total device resistance (at the charge neutrality point) of the graphene FETs was then measured as a function of the channel length. To obtain the channel resistance, the source/drain contact resistance ($R_C$) was separated from the measured total device resistance ($R_{tot}$) by using the TLM method [24]—i.e., $R_{ch} = R_{tot} - 2R_C$ (Figure S7). Thereafter, the channel sheet resistance was calculated from $R_{ch} \times W_{ch}/L_{ch}$. The dots in Figure 4b show the channel sheet resistance as a function of the channel length, obtained from the measurement results of 5–15 FETs per channel length using the three identical graphene samples grown by the same CVD run. Note that the dependence between $R_{sh}$–$L_{ch}$ is similar to that observed in the theoretical calculation result (Figure 1c). From such dependence, the three-grain parameters were extracted simultaneously by the analytical resistance model fitted to the $R_{sh}$–$L_{ch}$ curve as shown by the red line in Figure 4b. The extracted average grain size, the sheet resistance of single-crystalline grains, and the resistivity of GBs were determined to be ~5.95 µm, ~321 Ω/sq, and ~18.16 kΩ-µm, respectively. The detailed procedure and flow chart for extracting the three grain parameters from the electrical measurement results using the analytical resistance model are summarized in Figure S7 and Figure S8, respectively.

To confirm whether the extracted three-parameter values are rational, we compared the extracted values with those evaluated using conventional methods and those reported in the literature. First, the GB visualization technique based on UV/ozone treatment was used to estimate the average grain size of CVD graphene. Figure 4c shows a representative SEM image of CVD graphene grown on a Cu foil after the UV/ozone treatment. Note that the UV/ozone-treated GBs were highlighted in yellow to make them more visible (see Figure S9 for the original image). We evaluated the sizes of 376 grains observed in multiple SEM images (Figure S9), from which the average grain size was estimated to be 5.88 ± 1.5 µm. This value agrees well with the extracted average grain size from the fitted analytical resistance model (~5.95 µm). The sheet resistance of single-crystalline grains was estimated by characterizing TLM patterns composed of short-channel graphene FETs with $L_{ch}$ of 0.18–0.75 µm (Figure 4d). Because the probability of the GBs existing in these channel lengths is significantly low, the channel resistance increased in direct proportion to the channel length, just like in a single-crystalline material. Thus, the sheet resistance of single-crystalline grains could be estimated using the conventional TLM method [24]—i.e., from the slope of the measured width-normalized channel resistance as a function of the channel length (Figure 4e). The estimated grain sheet resistance from this approach was determined as 362 Ω/sq, which is comparable to the average $R_{sh}^G$ extracted from the fitted analytical model (~321 Ω/sq). The slight difference between the two values may be due to the influence of one GB that remains within the channel region, even as the channel length decreases, as shown in the inset of Figure 4e. Accordingly, the grain sheet resistance evaluated using the conventional TLM method can be slightly overestimated because of the carrier scattering at the GB. The extracted GB resistivity from the fitted analytical resistance model (~18.16 k$\Omega$-$\mu$m) was verified by comparing the value with those reported in the literature [4–6,25–28]. A summary of experimental results for the GB resistivity of CVD graphene that has been reported in the literature is shown in Figure 4f. Note that all resistivity values shown in the summary plot were extracted at the charge neutrality point. The average GB resistivity value in this study is in the range of those reported in the literature to date and is similar to their average value. Consequently, these validation results support that the extracted three-grain parameters are within the rational range. Therefore, we can conclude that the proposed electrical characterization method can extract the average grain size, single-crystalline grain sheet resistance, and GB resistivity simultaneously using the GB distribution-based analytical resistance model. This method of simultaneous extraction of the grain-related parameters from the dependence between $R_{sh}$–$L_{ch}$ obtained from simple TLM measurements will provide a convenient way for the electrical characterization of CVD graphene and its efficient device applications. Furthermore, it is expected that the proposed method can be extended to various 2D materials with polycrystalline structures.

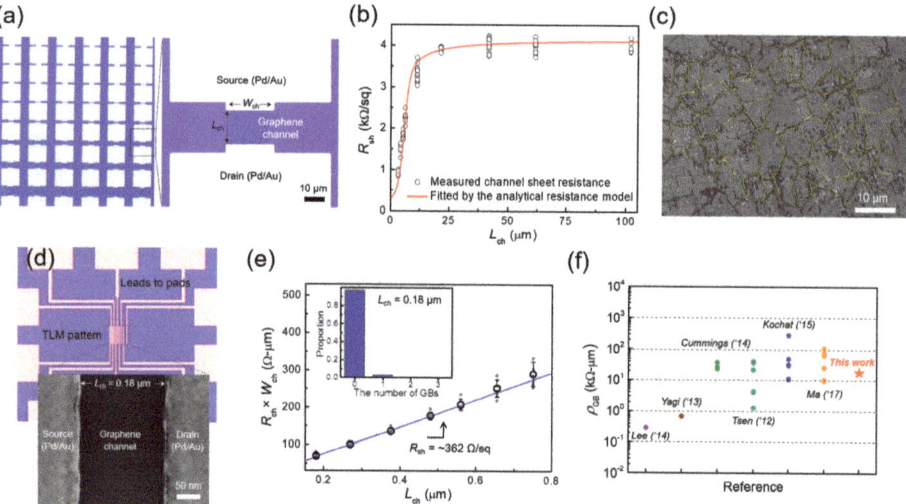


**Figure 4.** Experimental verification of the GB distribution-based analytical resistance model and parameter extraction method. (**a**) OM images of fabricated TLM patterns comprising the CVD-graphene FETs with varying channel lengths ($L_{ch}$ of 2–100 $\mu$m). (**b**) Measured channel sheet resistance as a function of channel length and fitting result using the analytical resistance model. The three-grain parameters extracted from the fitted model are ~5.95 $\mu$m (for the average grain size), ~321 $\Omega$/sq (for the average grain sheet resistance), and ~18.16 k$\Omega$·$\mu$m (for the average GB resistivity). (**c**) The representative SEM image of CVD graphene grown on a Cu foil, with UV/ozone-treated GBs highlighted in yellow. The average grain size estimated from 376 grains is 5.88 $\pm$ 1.5 $\mu$m. (**d**) OM and SEM image of a fabricated TLM pattern comprising the graphene FETs with sub-micrometer channel lengths ($L_{ch}$ of 0.18–0.75 $\mu$m). (**e**) The measured width-normalized channel resistance as a function of the channel length, in which the linear slope indicates the sheet resistance of single-crystalline grains due to the extremely low probability of the presence of GBs within the short-channel regions. Inset: the histogram distribution of the number of GBs when the channel length is significantly smaller than the average grain size ($L_{ch}$ = 0.18 $\mu$m and $l_G$ = 5 $\mu$m). (**f**) Summary of experimental results for the GB resistivity reported in the literature [4–6,25–28], where all represented resistivity values were extracted at the charge neutrality point. This summary plot shows that the GB resistivity extracted in this study falls within the range of the reported resistivity values.

## 4. Conclusions

In summary, we demonstrated an electrical characterization method for extracting the average grain size, grain sheet resistance, and GB resistivity of monolayer CVD graphene simultaneously. We developed an analytical resistance model to explain the relationship between the electrical properties of polycrystalline graphene and the channel dimension by precisely estimating the proportion distribution of the number of GBs within graphene channels. With the developed analytical resistance model, we showed that the three-grain parameters can be extracted simultaneously from the dependence of the graphene sheet resistance on the channel dimension with an accuracy greater than 99%. The proposed parameter extraction method using the GB distribution-based analytical resistance model was experimentally verified by characterizing TLM patterns fabricated on monolayer CVD graphene. The result showed that the average grain size, grain sheet resistance, and GB resistivity of CVD graphene can be extracted simultaneously with high accuracy from the analytical resistance model fitted to the measured $R_{sh}$–$L_{ch}$ curve. We believe that this method will be a useful tool for the electrical characterization of CVD graphene and other polycrystalline 2D materials.

**Supplementary Materials:** The following are available online at https://www.mdpi.com/article/10.3390/nano12020206/s1, MATLAB ® algorithm/code file used for the grain-parameter extraction, Figure S1: Theoretically-calculated $R_{ch}$–$L_{ch}$ and $R_{sh}$–$L_{ch}$ results for polycrystalline graphene with different average grain sizes and the grain-parameter extraction results using the analytical resistance model, Figure S2: Theoretically-calculated $R_{ch}$–$L_{ch}$ and $R_{sh}$–$L_{ch}$ result using different $W_{ch}$, $R_{sh}^{G}$, and $\rho_{GB}$ values and the grain-parameter extraction results using the analytical resistance model, Figure S3: Average histogram distributions of the proportion of the number of GBs within the channel region at various channel lengths, Figure S4: Average histogram distributions of the proportion of the number of GBs within the channel region for polycrystalline graphene with an average grain size of 2.5 μm, Figure S5: Average histogram distributions of the proportion of the number of GBs within the channel region for polycrystalline graphene with an average grain size of 15 μm, Figure S6: Raman spectrum of the graphene layer grown by CVD, Figure S7: A procedure for measuring the graphene sheet resistance as a function of the channel length and extracting the three grain parameters using the analytical resistance model, Figure S8: A flow chart explaining the algorithm for the process of extracting grain parameters using the analytical resistance model, Figure S9: Top-view SEM images of UV/ozone-treated CVD graphene, used for estimation of the average grain size.

**Author Contributions:** Conceptualization, H.P. (Hongsik Park); methodology, H.P. (Hongsik Park), H.P. (Honghwi Park), J.L. (Junyeong Lee) and C.-J.L.; software, J.L. (Junyeong Lee) and M.P.; validation, J.P., M.C. and S.L.; formal analysis, H.P. (Honghwi Park), J.L. (Junyeong Lee), C.-J.L. and J.K.; investigation, J.Y., H.N., J.L. (Jonghyung Lee) and Y.P.; resources, H.P. (Hongsik Park) and M.C.; data curation, H.P. (Honghwi Park), J.L. (Junyeong Lee), C.-J.L., J.K. and J.Y.; writing—original draft preparation, H.P. (Hongsik Park), H.P. (Honghwi Park) and C.-J.L.; writing—review and editing, H.P. (Hongsik Park) and H.P. (Honghwi Park); visualization, H.P. (Honghwi Park) and J.L. (Junyeong Lee); supervision, H.P. (Hongsik Park), J.P., M.C. and S.L.; project administration, H.P. (Hongsik Park), J.P., M.C. and S.L.; funding acquisition, H.P. (Hongsik Park) and M.C. All authors have read and agreed to the published version of the manuscript.

**Funding:** This work was supported by the National Research Foundation of Korea (NRF) grant funded by the Korea government (MSIT) (2019R1A2C1088324), the NRF grant funded by the Korea government (MSIT) (2020R1A4A1019518), and the Bio & Medical Technology Development Program of the NRF funded by the Ministry of Science & ICT (2017M3A9G8083382). This work was also supported by the BK21 FOUR Project funded by the Ministry of Education, Korea (4199990113966) and the Samsung Electronics' University R&D program.

**Institutional Review Board Statement:** Not applicable.

**Informed Consent Statement:** Not applicable.

**Data Availability Statement:** Data supporting the results are presented in the article and Supplementary Materials in the form of graphs, tables, and schematic diagrams. Raw and/or additional data are available on request from the corresponding author.

**Conflicts of Interest:** The authors declare no conflict of interest.

## References


1. Li, X.; Cai, W.; An, J.; Kim, S.; Nah, J.; Yang, D.; Piner, R.; Velamakanni, A.; Jung, I.; Tutuc, E.; et al. Large-area synthesis of high-quality and uniform graphene films on copper foils. *Science* **2009**, *324*, 1312–1314. [CrossRef]
2. Yazyev, O.V.; Chen, Y.P. Polycrystalline graphene and other two-dimensional materials. *Nat. Nanotechnol.* **2014**, *9*, 755–767. [CrossRef] [PubMed]
3. Yao, W.; Wu, B.; Liu, Y. Growth and grain boundaries in 2D materials. *ACS Nano* **2020**, *14*, 9320–9346. [CrossRef] [PubMed]
4. Cummings, A.W.; Duong, D.L.; Nguyen, V.L.; Tuan, D.V.; Kotakoski, J.; Vargas, J.E.B.; Lee, Y.H.; Roche, S. Charge transport in polycrystalline graphene: Challenges and opportunities. *Adv. Mater.* **2014**, *26*, 5079–5094. [CrossRef] [PubMed]
5. Isacsson, A.; Cummings, A.W.; Colombo, L.; Colombo, L.; Kinaret, J.M.; Roche, S. Scaling properties of polycrystalline graphene: A review. *2D Mater.* **2017**, *4*, 012002. [CrossRef]
6. Tsen, A.W.; Brown, L.; Levendorf, M.P.; Ghahari, F.; Huang, P.Y.; Havener, R.W.; Ruiz-Vargas, C.S.; Muller, D.A.; Kim, P.; Park, J. Tailoring electrical transport across grain boundaries in polycrystalline graphene. *Science* **2012**, *336*, 1143–1146. [CrossRef] [PubMed]
7. Barrios-Vargas, J.E.; Mortazavi, B.; Cummings, A.W.; Martinez-Gordillo, R.; Pruneda, M.; Colombo, L.; Rabczuk, T.; Roche, S. Electrical and thermal transport in coplanar polycrystalline graphene–hBN heterostructures. *Nano Lett.* **2017**, *17*, 1660–1664. [CrossRef] [PubMed]
8. Ma, T.; Liu, Z.; Wen, J.; Gao, Y.; Ren, X.; Chen, H.; Jin, C.; Ma, X.-L.; Xu, N.; Cheng, H.-M.; et al. Tailoring the thermal and electrical transport properties of graphene films by grain size engineering. *Nat. Commun.* **2017**, *8*, 14486. [CrossRef] [PubMed]
9. Balasubramanian, K.; Biswas, T.; Ghosh, P.; Suran, S.; Mishra, A.; Mishra, R.; Sachan, R.; Jain, M.; Varma, M.; Pratap, R.; et al. Reversible defect engineering in graphene grain boundaries. *Nat. Commun.* **2019**, *10*, 1090. [CrossRef] [PubMed]
10. Zhao, T.; Xu, C.; Ma, W.; Liu, Z.; Zhou, T.; Liu, Z.; Feng, S.; Zhu, M.; Kang, N.; Sun, D.-M.; et al. Ultrafast growth of nanocrystalline graphene films by quenching and grain-size-dependent strength and bandgap opening. *Nat. Commun.* **2019**, *10*, 4854. [CrossRef]
11. Yu, Q.; Jauregui, L.A.; Wu, W.; Colby, R.; Tian, J.; Su, Z.; Cao, H.; Liu, Z.; Pandey, D.; Wei, D.; et al. Control and characterization of individual grains and grain boundaries in graphene grown by chemical vapour deposition. *Nat. Mater.* **2011**, *10*, 443–449. [CrossRef] [PubMed]
12. Duong, D.L.; Han, G.H.; Lee, S.M.; Gunes, F.; Kim, E.S.; Kim, S.T.; Kim, H.; Ta, Q.H.; So, K.P.; Yoon, S.J.; et al. Probing graphene grain boundaries with optical microscopy. *Nature* **2012**, *490*, 235–239. [CrossRef]
13. Cheng, Y.; Song, Y.; Zhao, D.; Zhang, X.; Yin, S.; Wang, P.; Wang, M.; Xia, Y.; Maruyama, S.; Zhao, P.; et al. Direct identification of multilayer graphene stacks on copper by optical microscopy. *Chem. Mater.* **2016**, *28*, 2165–2171. [CrossRef]
14. Park, H.; Lee, J.; Lee, C.-J.; Kim, J.; Kang, J.; Noh, H.; Lee, J.; Park, Y.; Park, J.; Choi, M.; et al. Evaluation of the average grain size of polycrystalline graphene using an electrical characterization method. *Solid-State Electron.* **2021**, *186*, 108172. [CrossRef]
15. Kang, J.; Lee, C.-J.; Kim, J.; Park, H.; Lim, C.; Lee, J.; Choi, M.; Park, H. Effect of copper surface morphology on grain size uniformity of graphene grown by chemical vapor deposition. *Curr. Appl. Phys.* **2019**, *19*, 1414–1420. [CrossRef]
16. Park, H.; Lim, C.; Lee, C.-J.; Kang, J.; Kim, J.; Choi, M.; Park, H. Optimized poly(methyl methacrylate)-mediated graphene-transfer process for fabrication of high-quality graphene layer. *Nanotechnology* **2018**, *29*, 415303. [CrossRef]
17. Lee, C.-J.; Park, H.; Kang, J.; Lee, J.; Choi, M.; Park, H. Extraction of intrinsic field-effect mobility of graphene considering effects of gate-bias-induced contact modulation. *J. Appl. Phys.* **2020**, *127*, 185105. [CrossRef]
18. Jung, J.; Park, H.; Won, H.; Choi, M.; Lee, C.-J.; Park, H. Effect of graphene doping level near the metal contact region on electrical and photoresponse characteristics of graphene photodetector. *Sensors* **2020**, *20*, 4661. [CrossRef]
19. Darling, K.A.; Rajagopalan, M.; Komarasamy, M.; Bhatia, M.A.; Hornbuckle, B.C.; Mishra, R.S.; Solanki, K.N. Extreme creep resistance in a microstructurally stable nanocrystalline alloy. *Nature* **2016**, *537*, 378–381. [CrossRef] [PubMed]
20. Bhattacharya, D.; Razavi, S.A.; Wu, H.; Dai, B.; Wang, K.L.; Atulasimha, J. Creation and annihilation of non-volatile fixed magnetic skyrmions using voltage control of magnetic anisotropy. *Nat. Electron.* **2020**, *3*, 539–545. [CrossRef]
21. Zhu, Y.; Ding, W.; Yu, T.; Xu, J.; Fu, Y.; Su, H. Investigation on stress distribution and wear behavior of brazed polycrystalline cubic boron nitride superabrasive grains: Numerical simulation and experimental study. *Wear* **2017**, *376–377*, 1234–1244. [CrossRef]
22. Jeong, C.; Nair, P.; Khan, M.; Lundstrom, M.; Alam, M.A. Prospects for nanowire-doped polycrystalline graphene films for ultratransparent, highly conductive electrodes. *Nano Lett.* **2011**, *11*, 5020–5025. [CrossRef]
23. Pineda, E.; Bruna, P.; Crespo, D. Cell size distribution in random tessellations of space. *Phys. Rev. E* **2004**, *70*, 066119. [CrossRef] [PubMed]
24. Berger, H.H. Models for contacts to planar devices. *Solid-State Electron.* **1972**, *15*, 145–158. [CrossRef]
25. Lee, D.; Kwon, G.D.; Kim, J.H.; Moyen, E.; Lee, Y.H.; Baik, S.; Pribat, D. Significant enhancement of the electrical transport properties of graphene films by controlling the surface roughness of Cu foils before and during chemical vapor deposition. *Nanoscale* **2014**, *6*, 12943–12951. [CrossRef] [PubMed]

26. Yagi, K.; Yamada, A.; Hayashi, K.; Harada, N.; Sato, S.; Yokoyama, N. Dependence of field-effect mobility of graphene grown by thermal chemical vapor deposition on its grain size. *Jpn. J. Appl. Phys.* **2013**, *52*, 110106. [CrossRef]
27. Kochat, V.; Tiwary, C.S.; Biswas, T.; Ramalingam, G.; Hsieh, K.; Chattopadhyay, K.; Raghavan, S.; Jain, M.; Ghosh, A. Magnitude and origin of electrical noise at individual grain boundaries in graphene. *Nano Lett.* **2016**, *16*, 562–567. [CrossRef]
28. Ma, R.; Huan, Q.; Wu, L.; Yan, J.; Guo, W.; Zhang, Y.-Y.; Wang, S.; Bao, L.; Liu, Y.; Du, S.; et al. Direct four-probe measurement of grain-boundary resistivity and mobility in millimeter-sized graphene. *Nano Lett.* **2017**, *17*, 5291–5296. [CrossRef]

 *nanomaterials*

Article

# Improvement of Temperature and Optical Power of an LED by Using Microfluidic Circulating System of Graphene Solution

Yung-Chiang Chung *, Han-Hsuan Chung and Shih-Hao Lin

Department of Mechanical Engineering, Ming Chi University of Technology, New Taipei 24301, Taiwan; gary850311@gmail.com (H.-H.C.); M07118021@mail2.mcut.edu.tw (S.-H.L.)
* Correspondence: ycchung@mail.mcut.edu.tw

**Abstract:** Electric devices have evolved to become smaller, more multifunctional, and increasingly integrated. When the total volume of a device is reduced, insufficient heat dissipation may result in device failure. A microfluidic channel with a graphene solution may replace solid conductors for simultaneously supplying energy and dissipating heat in a light emitting diode (LED). In this study, an automated recycling system using a graphene solution was designed that reduces the necessity of manual operation. The optical power and temperature of an LED using this system was measured for an extended period and compared with the performance of a solid conductor. The temperature difference of the LED bottom using the solid and liquid conductors reached 25 °C. The optical power of the LED using the liquid conductor was higher than that of the solid conductor after 120 min of LED operation. When the flow rate was increased, the temperature difference of the LED bottom between initial and 480 min was lower, and the optical power of the LED was higher. This result was attributable to the higher temperature of the LED with the solid conductor. Moreover, the optical/electric power transfer rate of the liquid conductor was higher than that of the solid conductor after 120 min of LED operation, and the difference increased over time.

**Keywords:** liquid conductor; graphene solution; circulating system; microfluidic channel; temperature; optical power

**Citation:** Chung, Y.-C.; Chung, H.-H.; Lin, S.-H. Improvement of Temperature and Optical Power of an LED by Using Microfluidic Circulating System of Graphene Solution. *Nanomaterials* **2021**, *11*, 1719. https://doi.org/10.3390/nano11071719

Academic Editors: Eugene Kogan and Filippo Giannazzo

Received: 29 April 2021
Accepted: 26 June 2021
Published: 29 June 2021

**Publisher's Note:** MDPI stays neutral with regard to jurisdictional claims in published maps and institutional affiliations.

## 1. Introduction

Heat dissipation and electrical conduction are critical considerations in the operation of an integrated circuit device. Solid metal conductors are typically used for conducting electricity. Several heat dissipation methods, such as the use of microfluidic devices, are employed in electric systems. For example, researchers have considered using nanofluids for cooling electric devices [1,2]. Studies have proposed a thermal contact liquid cooling system [3] as well as a technique for cooling photovoltaic cell systems through the use of rotating magnetic fields and ferrofluids [4]. A related study used a system combining liquid cooling and composite phase change material cooling to dissipate the heat generated in a battery [5]. Liquid cooling systems have also been applied in central processing units and laptops [6,7]. Research has revealed various cooling methods for electric devices [8–10]. Some studies have also employed the recycling of various fluids, such as ammonia [11], $CO_2$ [12], and supercritical water [13] for cooling. Additionally, researchers have introduced a one-section and two-stepwise microchannel for cooling [14] and a recirculating cooling water system to reduce energy consumption [15]. Another study considered the cooling effect of dielectric liquid [16], and yet another reported that a fracture in the neck of the bond between the solder joint and gold wire after 20 thermal shocks would result in metal conductors failing at high temperatures [17].

The dissipation of heat in light emitting diodes (LEDs) has been studied extensively. For example, researchers have investigated the temperature distribution [18] and liquid cooling systems [19,20] of an LED array. Furthermore, composite coatings composed of cupric oxide [21] or various heat pipes and heat sinks [22–28] have been employed to enhance the heat dissipation of LEDs. Other relevant studies have used a dielectric layer with an aluminum nitride insulation plate [29] and a dual synthetic jet actuator for heat dissipation in LEDs [30]. A cooling system with ferrofluid was used in a high-power LED; the results were compared with those of systems employing air and water working fluid, and the effect of ferrofluid was the best [31]. A thermoelectric cooler integrated with a microchannel heat sink was used to control LED temperature [32]. Another study proposed the energy recycling and self-sufficient application of an LED integrated with a thermoelectric generator module and electrical fan [33]. Moreover, graphene has been employed as a novel material in several investigations [34,35]. Some researchers have considered the thermophysical properties and forced convective heat transfer performance of graphene [36,37] and reviewed the applications of graphene [38,39]. Researchers have suggested that the conductivity of water can be improved through the addition of mono and hybrid nano-additives containing graphene and silica [40] and have developed a method that integrates graphene nanocapillaries into a micro heat pipe for enhanced LED cooling [41]. Graphene solutions have been widely applied to increase heat transfer efficiency [42–47], and a higher graphene solution concentration has been demonstrated to result in greater heat transfer [48,49]. Moreover, some researchers have studied thermal conductivity and electrical conductivity of graphene nanoplatelets [50,51]. The aforementioned investigations have concentrated on either dissipating heat or supplying energy but have not considered the combination of energy supply and heat dissipation. Furthermore, numerous studies have examined the energy conversion of artificial light [52–55] and white light LEDs [56–58]. Evidence suggests that the power conversion efficiencies of solar cells and LEDs are lower than 53.6% [59,60] and that the efficiency of LEDs is 42% at 30 °C, dropping to 30% at 50 °C [61]. However, the aforementioned studies focused predominantly on energy conversion from optical power to electric power; energy conversion from electric power to optical power has seldom been discussed. The stability and reliability of an electric apparatus may decrease by 10% when the temperature is increased by 2 °C [62]. In another investigation, a graphene solution was used as a liquid conductor for dissipating heat and transferring energy; the heat dissipation efficiency was excellent, but the optical power of an LED with a liquid conductor was lower than that of an LED with a solid conductor during the first 6 min, and the graphene solution was recycled manually [63]. According to the aforementioned study, the temperature of the LED using the liquid conductor was much lower than that of the LED using the solid conductor. However, the optical power of LEDs using liquid and solid conductors was not compared over longer periods. Therefore, to supply energy and dissipate heat of an LED during longer experiments, the graphene solution can be automatically recycled. Such a system can simultaneously dissipate heat and supply energy as well as improve the LED's optical power. The optical power of the LED using the liquid conductor may be higher than that of the LED using the solid conductor over longer periods. Therefore, in this study, an automated liquid conductor circulating system was developed and the energy supply and heat dissipation of an LED was studied over an extended period.

## 2. Materials and Methods

### 2.1. Principle of Operation

A liquid conductor that can dissipate heat and conduct electricity simultaneously may effectively reduce the temperature of electric products and enhance the photoelectricity transfer efficiency. Related research has indicated that the temperature of an LED using a liquid conductor is much lower than that of an LED using a solid conductor. Thus, the effect of the temperature of the LED on its optical power, especially during extended periods, merits further study.

### 2.2. Chip Design and Fabrication

Microfluidic channels were fabricated using a microelectromechanical process, as displayed in Figure 1a,b. The widths of the inlet and outlet of the channel were 0.5 mm, and the length of the channel was 20 mm; the distance between the channels was 0.5 and 1 mm, respectively. As displayed in Figure 2a, first, the wafer was cleaned using acetone and deionized water. Second, the periphery of wafer was coated with Teflon and baked; the periphery of the wafer was hydrophobic to enhance the photoresist coating. The silicon wafer was coated with SU8 photoresist with a thickness of approximately 500 μm at 200 rpm. Subsequently, the wafer was soft baked and the solvent in the photoresist was removed. The wafer was exposed to define the pattern, and it was baked after exposure to enhance the linking process of the photoresist. After the wafer was developed, the developer removed the undefined region. Subsequently, the polydimethylsiloxane (PDMS) substrate microfluidic channel chip was fabricated. As illustrated in Figure 2b, the Teflon was coated on the wafer and the PDMS was poured into the wafer; the wafer was hydrophobic to facilitate the fabrication of the PDMS substrate. The wafer was then vacuumed to remove the bubbles in the PDMS, and the wafer was heated to solidify the PDMS. Once the wafer had cooled, the PDMS could be separated from the wafer, and the PDMS substrate microfluidic channel chip was complete.

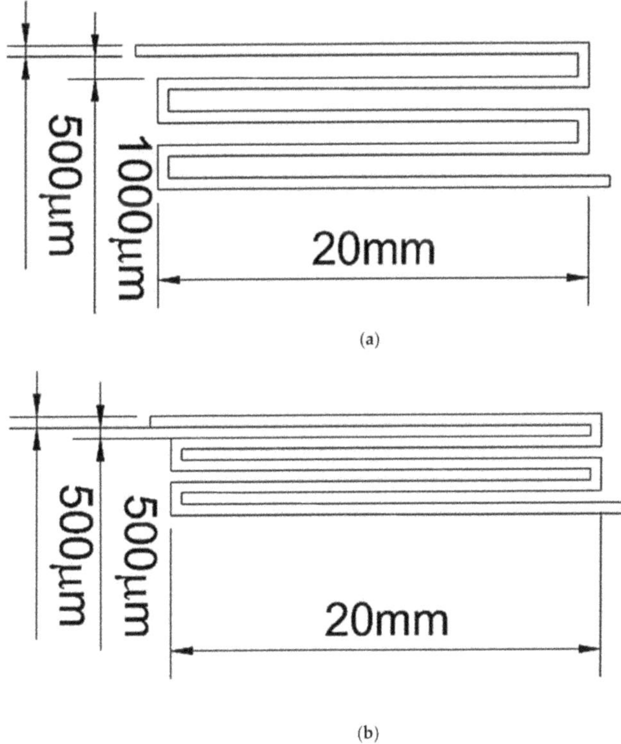


**Figure 1.** Schematics of the LED chip with a microfluidic channel: (**a**) distance between the channel $d = 1$ mm; (**b**) distance between the channel $d = 0.5$ mm.

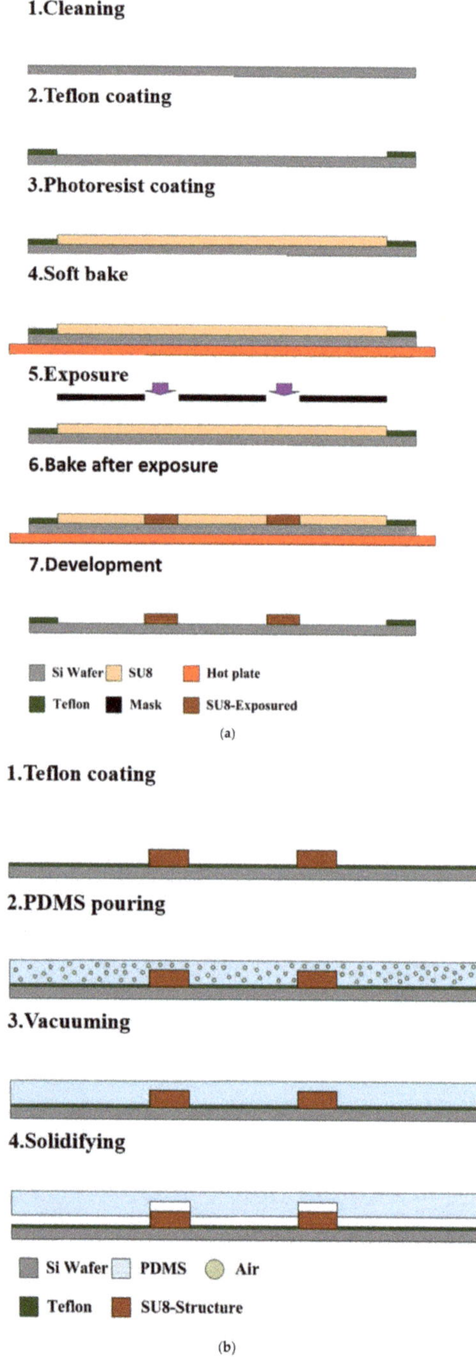



**Figure 2.** Fabrication of the microfluidic channels: (**a**) SU8 mold chip; (**b**) PDMS substrate microfluidic channel chip.

The designated area for a microscope slide ($75 \times 25 \times 1$ mm$^3$) was coated with silver adhesive (OP-901, Double O Technology, Taiwan), as shown in Figure 3a, which was connected to the microfluidic channel and the LED. The PDMS substrate microfluidic channel chip and the microscope slide coated with silver adhesive were bonded together to form the chip. The LED (emission color: white; TY-HNW2-3, TaoYuan Electron Limited, Taiwan) was placed at the center of the chip. The silver adhesive was coated on the electrodes of the LED, connecting it with the silver adhesive of the chip. The size of the liquid recycling reservoir was $75 \times 25 \times 13$ mm$^3$. The area between the needle and the microfluidic channel was sealed with ultraviolet glue to ensure that the liquid would not leak. Finally, the chip, the liquid or solid conductor, the liquid recycling reservoir, and the LED were integrated to form an LED with a liquid circulating system, shown in Figure 3b.

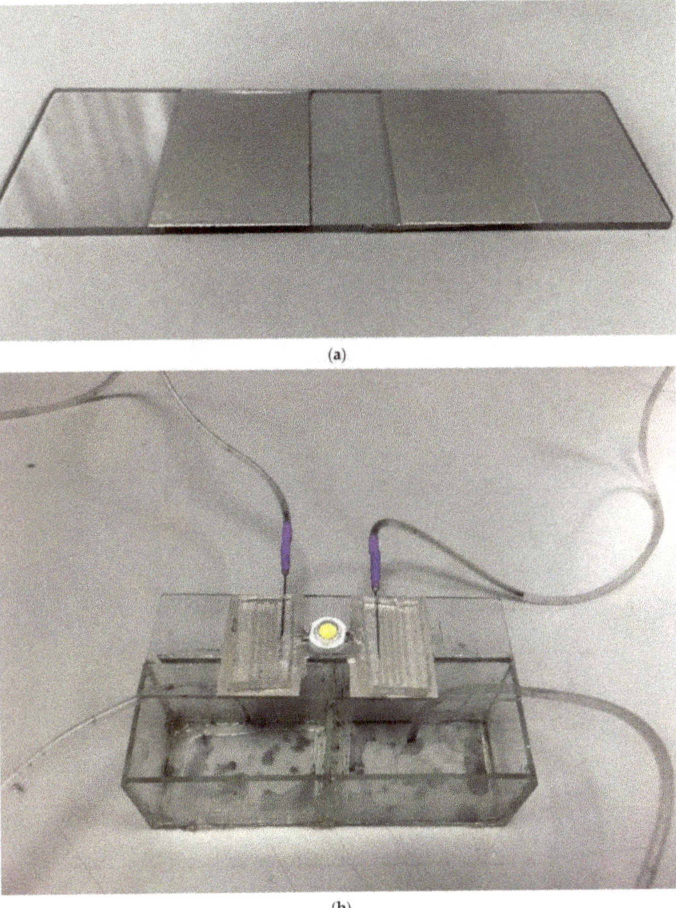

(a)

(b)

**Figure 3.** Schematics of the system: (**a**) image of the microscope slide with silver adhesive; (**b**) image of the LED with a liquid circulating system.

### 2.3. Sample Preparation and Experimental Setup

A graphene solution (Golden Innovation Business Co., Ltd., New Taipei, Taiwan) was used as the liquid conductor in this study. The size of the graphene particles in the solution was 100 nm, the solution was deionized water, and the concentration was 500 ppm. The needle was placed into the microfluidic channel. The graphene solution was fully mixed using an acoustic vibrator, and the solution was injected into the needle and microfluidic channel. The flow rate was regulated using a piezoelectric micropump (CurieJet, PS51I, Microjet Technology, Hsinchu, Taiwan); the flow rate ranged from 0.01 to 10 mL/min. A solid conducting nickel-plated steel wire (diameter: 1 mm) was used as the solid conductor. The power supply system (LPS305, Motech, Tainan, Taiwan) could produce an output voltage of $\pm$ 30 V. The LED system included the power supply, a syringe pump, the chip, and the graphene solution (or solid conductor). The temperature was measured using resistance temperature detectors (RTDs; PT 100 series, OMEGA Engineering Inc., Norwalk, Connecticut, USA), and the optical power of the LED was measured using an optical power meter (Gentec Electro-Optique Inc., Quebec, Quebec, Canada). The integrated system (Figure 4) included the LED system, power supply, syringe pump, temperature data receiver, and optical power sensor. In this study, the temperature and optical power of the LED were measured over an extended period, and the effect of the temperature on the optical power was examined.

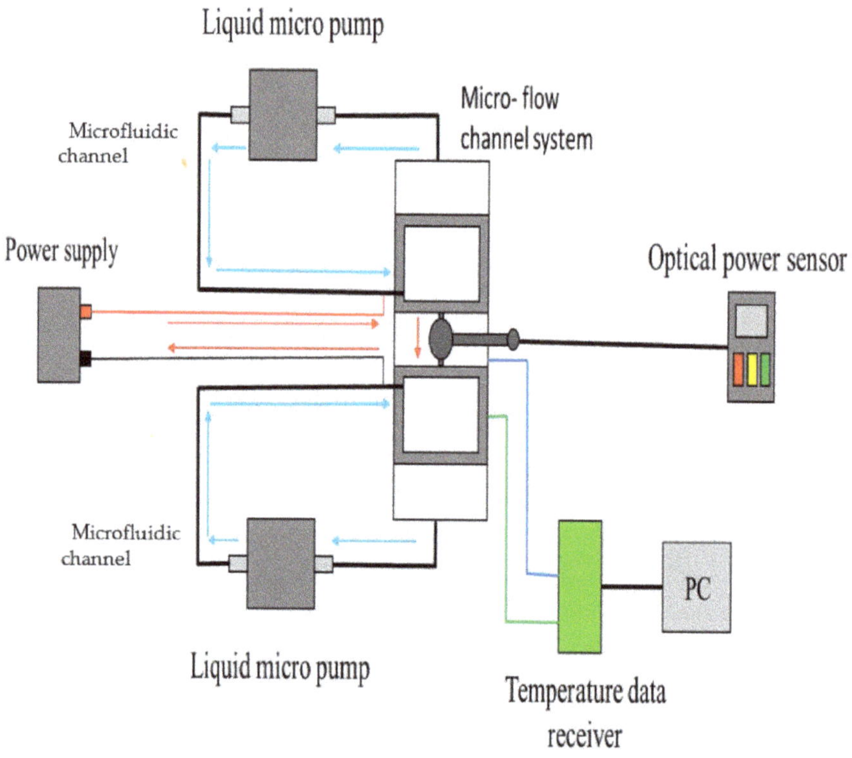

**Figure 4.** Schematic of the integrated system.

## 3. Results and Discussion

In this study, liquid (graphene solution) and solid (nickel-plated steel wire with a diameter of 1 mm, length of 30 mm) conductors were used. The temperatures at the bottom of the LED chip and the microfluidic channel were measured. Moreover, the optical power of the LED under various conditions was measured at a distance of 3 mm from the LED. The uncertainty analysis results of various parameters are listed in Table 1. Experiments were performed five times under each set of experimental conditions, and the average error was less than 20%. The temperatures at the center and four corners were measured, and the temperature variation was lower than 2 °C. Because the temperature at the center of the chip bottom was the highest, it was selected to represent the temperature of the LED chip (Figure 5).

**Table 1.** Uncertainty analysis of various parameters.

| Parameter | Instrument | Uncertainty |
| --- | --- | --- |
| Electric resistance and voltage | Digital multimeters | 0.1% |
| Flow rate | Piezoelectric pump | 0.5% |
| Temperature | Resistance temperature detectors | 0.1 °C |
| Optical power | Optical power meter | 0.5% |

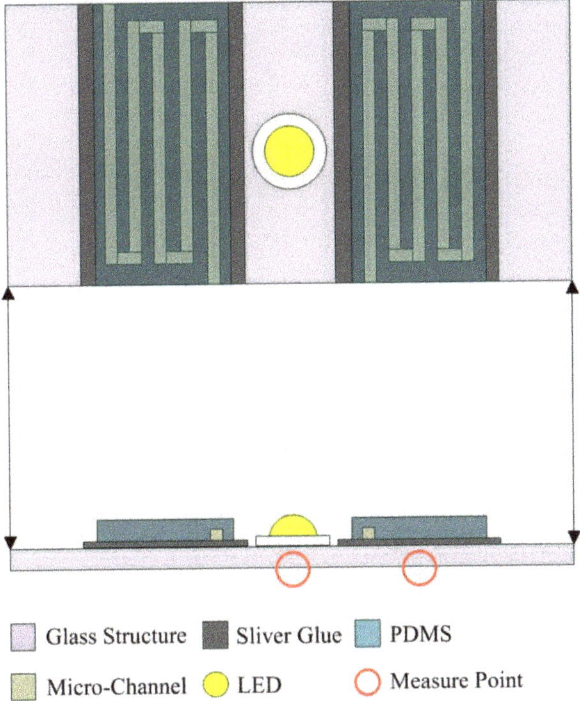


**Figure 5.** Schematic of the temperature measurement points of the LED bottom and microfluidic channel.

### 3.1. Energy Supply

The actual voltage of the LED with the graphene solution conductor differed from the voltage of the power supply. The actual voltages of the LED with the graphene solution were measured at various power supply voltages (Figure 6). The results indicated that the voltage of the graphene solution–conductor LED was 2.8 V when the voltage of the power supply was 5.5 V. The highest operational voltage of the LED was 3 V. Therefore, the actual voltage of the LED with the solid and liquid conductors was selected as 2.8 V in this study.

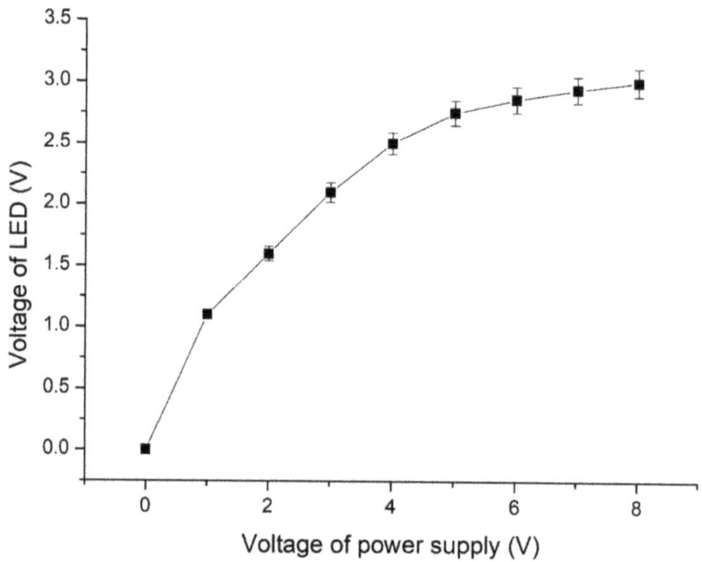

**Figure 6.** Electric voltages of the LED with the liquid conductor at various power supply electric voltages.

### 3.2. Temperature Variation Using Different Conductors

The measured temperatures of the LEDs with solid and liquid conductors are listed in Figure 7. The temperatures of the LED bottom and the microfluidic channel bottom increased over time. For the solid conductor, the temperatures of the LED bottom increased to 58 °C at 50 min, and the temperatures increased gradually to 65.5 °C at 480 min. The temperatures of the microfluidic channel bottom increased to 51.5 °C at 50 min and continued to increase gradually to 59.4 °C at 480 min. The temperature difference between the LED bottom and the microfluidic channel bottom was 6–7 °C. The measured temperatures of the liquid conductors ($d$ = 500 and 1000 μm) also increased over time. When $d$ = 1 mm, the temperatures of the LED bottom and the microfluidic channel bottom increased over time. The temperatures of the LED bottom increased to 37.2 °C at 50 min and continued to increase gradually to 41 °C at 480 min. The temperatures of the microfluidic channel bottom increased to 28.2 °C at 50 min and continued to increase gradually to 30.5 °C at 480 min. The temperature difference between the LED bottom and the microfluidic channel bottom was 9–10.5 °C. When $d$ = 0.5 mm, the temperatures of the LED bottom increased to 36.2 °C at 50 min and increased gradually to 40 °C at 480 min. The temperatures of the microfluidic channel bottom increased to 28 °C at 50 min and increased gradually to 30 °C at 480 min. The temperature difference between the LED bottom and the microfluidic channel bottom was 8.2–10 °C. The temperatures of the LED bottom and microfluidic channel bottom were lowest in the liquid conductor with $d$ = 0.5 mm, but the temperature differences between these two parts did not differ considerably between the liquid conductors with $d$ = 0.5 and 1 mm. Furthermore, the temperature difference of the LED bottom between the solid conductor and the liquid conductor reached 25 °C. This result emphasizes the temperature reduction effect of the LED using the liquid conductor.

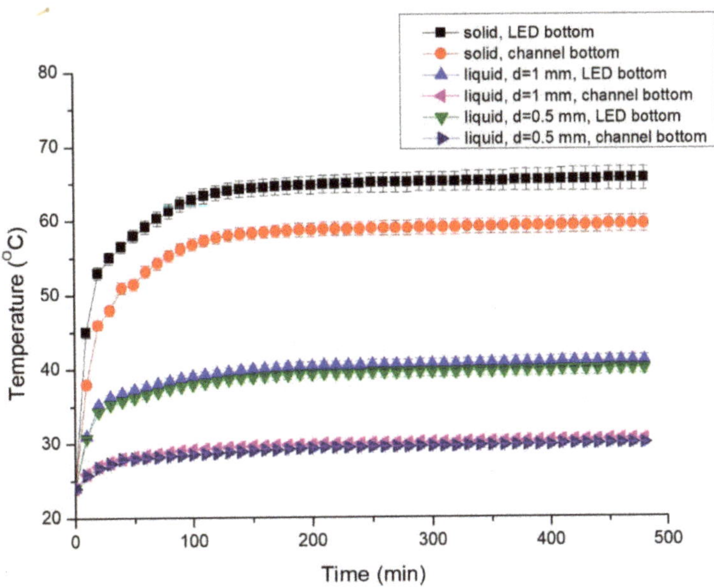

**Figure 7.** Temperatures of the bottom of the LED bottom and channel bottom with various conductors.

The temperature differences of the LED bottom between initial and 480 min using solid conductor and graphene solution with various flow rates are shown in Figure 8. The temperature difference using the solid conductor was 35.1 °C. The temperature differences using the graphene solution were 17.2, 13.4, 11.5 and 10.1 °C at flow rates of 0, 0.05, 0.2 and 1 mL/min, respectively. When the flow rate was increased, the temperature difference was lower and the heat dissipation of the graphene solution was excellent. The heat dissipation of the graphene solution was apparently higher than that of the solid conductor. At static state, the heat dissipation of the graphene solution was still higher than that of the solid conductor.

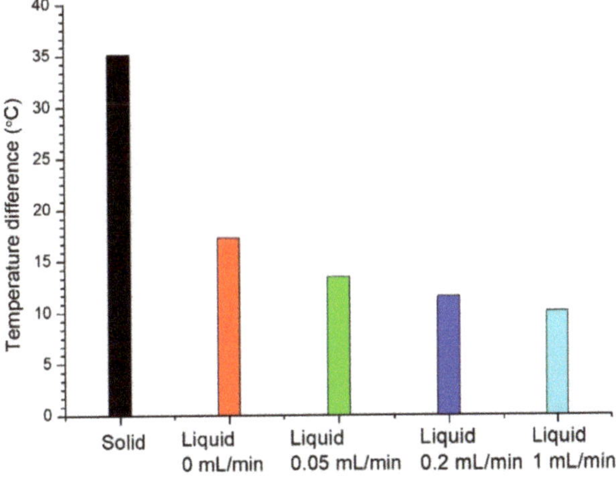

**Figure 8.** Temperature differences of the LED bottom between initial and 480 min using solid conductor and graphene solution with various flow rates.

The temperature decreases of the LED using various cooling methods are shown in Figure 9. The temperature decrease in the LED with fin only was 3 °C, that of the LED with fan only was 8 °C, and that of the LED with fin and fan could be 10 °C [22,23]. The temperature decrease in the LED with heat pipe was larger than 7 °C [24–27], that of the LED with dielectric layer could reach 20 °C [29], and it may be larger than 25 °C using a ferrofluid [31]. In this study, the temperature decrease in the LED reached 30 °C using a graphene solution. The temperature decrease in the LED in this study was equal to or larger than that using other cooling methods.

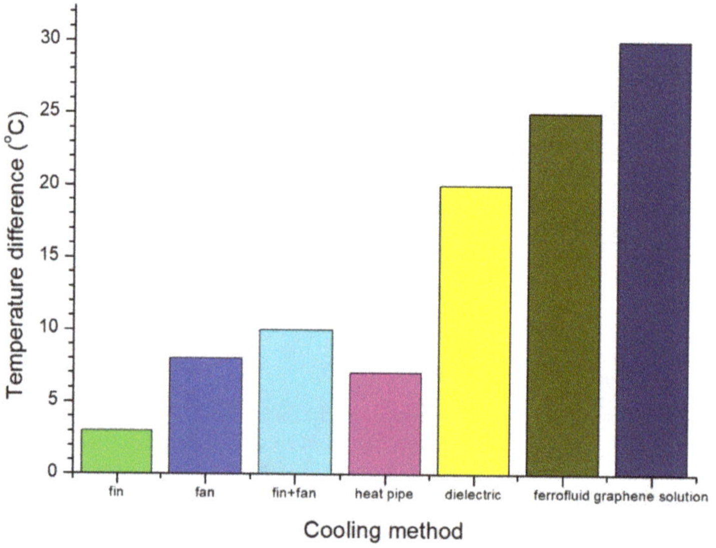


**Figure 9.** Temperature differences of LED using various cooling methods.

The thermal conductivity $k_s$ of the solid conductor was about 15 W/m·°C. The thermal conductivity of graphene was about 5300 W/m·°C. The thermal conductivity of the graphene solution $k_g$ was about 20–100 W/m·°C for various concentration for 200–1000 ppm (from graphene solution company, Golden Innovation Business Co., Ltd., New Taipei, Taiwan.).

The thermal resistance of a solid conductor can be expressed as follows [64,65]:

$$R_{th,s} = 1/(k_s A/\Delta x) \tag{1}$$

where $\Delta x$ is the distance between two locations, and A is the area between two locations. Consider the convection heat transfer coefficient $h$, the thermal resistance of a graphene solution can be expressed as follows:

$$R_{th,g} = 1/(k_g A/\Delta x + hA) \tag{2}$$

The value $h$ of water was 1000–35000 W/m²·°C for various flow rates [64,65].

Based on the comparison between $R_{th,s}$ and $R_{th,g}$, the thermal resistance of the graphene solution with flow rate was smaller than that of the solid conductor, and the heat transfer of the LED was improved.

### 3.3. Optical Power Variation Using Different Conductors

The optical power at various distances between the LED and the power meter is displayed in Figure 10. The optical power of the LED did not decrease noticeably when the distance was lower than 5 mm. However, it decreased markedly when the distance was greater than 10 mm. The optical power at a distance of 3 mm from the LED was selected as the representative value. The optical power of the LEDs measured using solid and liquid conductors is displayed in Figure 11. The maximum optical power of the LED with the solid conductor was 30.5 mW at 10 min, and it decreased gradually. The optical power decreased noticeably after 120 min, and that of the LED with the liquid conductor with $d = 0.5$ mm increased gradually before 30 min and stabilized at 29 mW by 200 min. The optical power of the LED with the liquid conductor with $d - 1$ mm increased gradually before 30 min and stabilized at 28.8 mW by 180 min. The optical power of the LED with the solid conductor was greater than that of the liquid conductor until 90 min. The optical power difference of the LED with liquid conductors with various $d$ values was small. The temperature of the LED with the liquid conductor was much lower than that of the LED with the solid conductor after 30 min (Figure 7). The temperature differences between the LED bottoms with the liquid and solid conductors increased over time during the first 120 min and then stabilized at approximately 24 °C. The optical power of the LED with the liquid conductor was higher than that of the solid conductor after 120 min, and the optical power difference was approximately 3 mW after 150 min. This result was attributable to the higher temperature of the LED with the solid conductor.

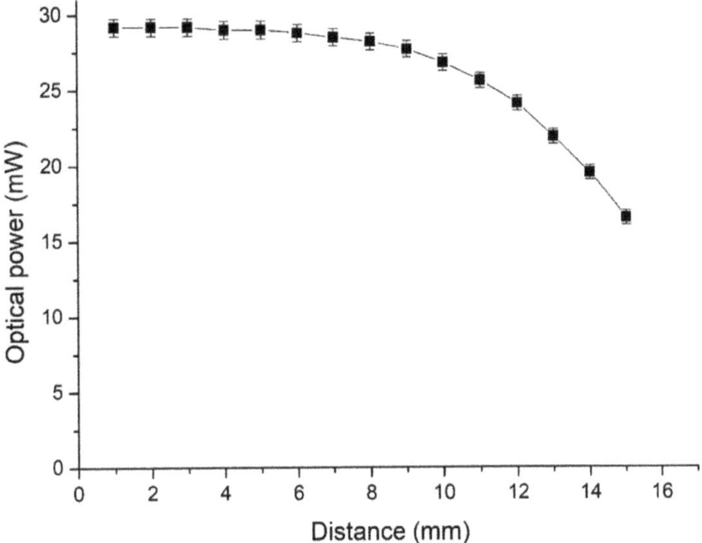

**Figure 10.** Optical power of the LED chip at various distances between the LED and the power meter.

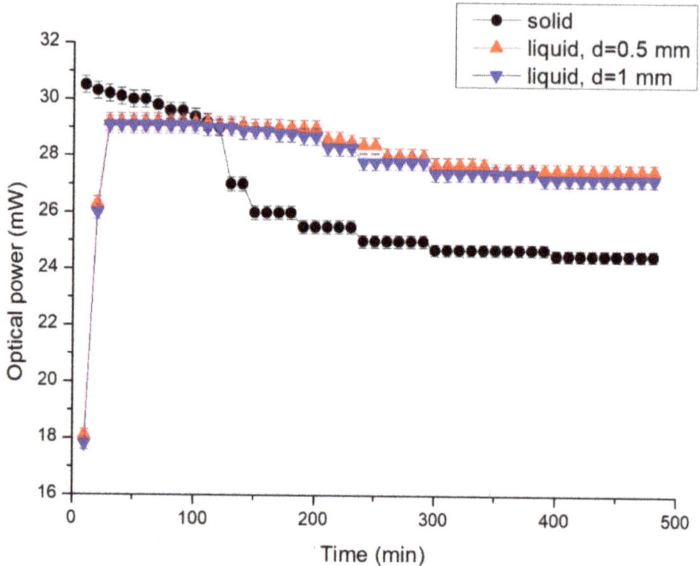

**Figure 11.** Optical power of the LED chip with various conductors and distances between channels.

*3.4. Comparison of Power Transfer*

The liquid pumping power (additional energy) required to pump the graphene solution is discussed in this section. The liquid pumping power, $P_l$ (W) can be expressed as follows [44]:

$$P_l = (\Delta P) \times Q \tag{3}$$

where $\Delta P$ is the pressure drop (kPa), and $Q$ is the flow rate (m$^3$/s).

The maximum flow rate in this study was 1 mL/min (1.67 $\times$ 10$^{-8}$ m$^3$/s) and the pressure drop was approximately 50 kPa. The pumping power was estimated to be $8.3 \times 10^{-4}$ mW. Compared with the optical power of the LED (29 mW), the pumping power was much lower. The overall operational expenditure of the device was unaffected in this study. Moreover, the volume of the graphene solution in the recycling system was 30 mL and the cost of the graphene solution was approximately USD 60. Thus, the cost of the graphene solution was not high, and the graphene solution could be automatically recycled. The cost of the graphene solution did not increase noticeably because it required no replenishment in this study. Such a system is highly convenient and avoids substantial increases in the total device cost. Therefore, a graphene-solution microfluidic channel is a favorable conductor for use in LEDs.

The optical power of the LED using the solid conductor and graphene solution with various flow rates at 480 min are shown in Figure 12. The optical power of the LED using the solid conductor was 24.5 mW. The optical power of the LED using the graphene solution were 25.1, 27.5, 28.2 and 28.8 mW at flow rates of 0, 0.05, 0.2 and 1 mL/min, respectively. The optical power of the LED using the graphene solution was higher than that of the solid conductor. When the flow rate was increased, the optical power of the LED was higher, but the difference was not large.

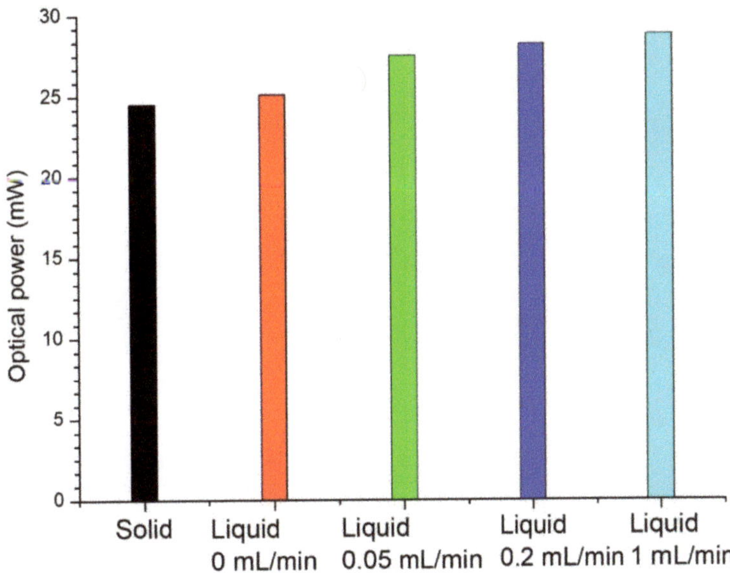

**Figure 12.** Optical power of the LED at 480 min using solid conductor and graphene solution with various flow rates.

The optical/electric power transfer rate of the LED was also investigated. The electric currents of the LED using various conductors were measured at 10, 40, 90, 150, and 480 min of LED operation. The supplied electric power was calculated using the following equation:

$$Pe = I \times V \tag{4}$$

where $Pe$ is the supplied electric power (mW), $I$ is the electric current (mA), and $V$ is the voltage (V). The relative optical/electric power transfer rate was calculated using the following equation:

$$Rt = Po/Pe \tag{5}$$

where $Rt$ is the optical/electric power transfer rate, $Po$ is the measured optical power, and $Pe$ is the supplied electric power. The results are presented in Figure 13. After 10 min of LED operation, the optical/electric power transfer rates were 50.1%, 35.3%, and 34.7% for the solid conductor, the liquid conductor with $d = 0.5$ mm, and the liquid conductor with $d = 1$ mm, respectively. At 60 min, the transfer rates were 45.0%, 42.7%, and 41.9%, respectively. At 120 min, the transfer rates were 40.7%, 46.1%, and 44.8%, respectively. The optical/electric power transfer rate of the liquid conductor was higher than that of the solid conductor after 120 min of LED operation. Therefore, the optical/electric power transfer rate was affected by the temperature of the LED and was improved by the graphene-solution liquid conductor.

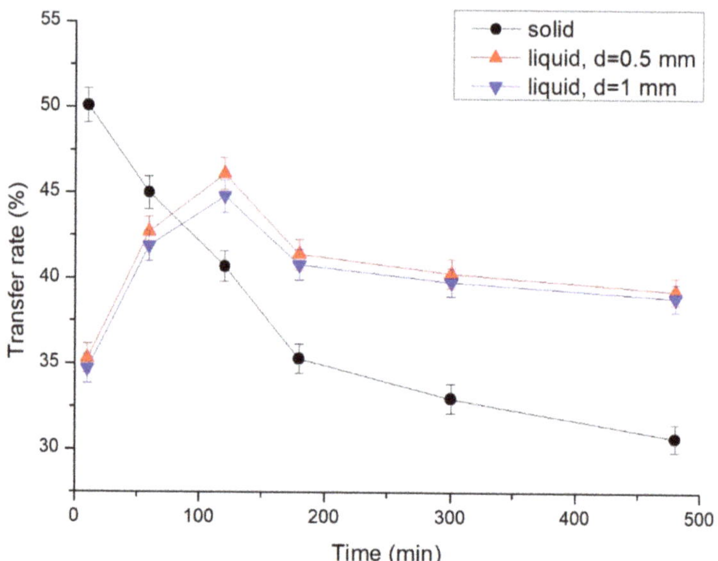

**Figure 13.** Optical/electric power transfer rate of the LED chip with various conductors and distances between channels.

## 4. Conclusions

This study proposes an automated graphene-solution circulating system that can efficiently dissipate heat and conduct electricity. The temperature and optical power of an LED were measured over an extended period. The thermal resistance of graphene solution with flow rate was smaller than that of the solid conductor, and the heat transfer of the LED was improved. The difference in temperature of the LED bottom between the LEDs using liquid and solid conductors reached 25 °C. After 120 min of LED operation, the optical power of the LED with the liquid conductor was higher than that of the solid conductor. When the flow rate was increased, the temperature difference of the LED bottom between initial and 480 min was lower, and the optical power of the LED was higher. This result is attributable to the higher temperature of the LED with the solid conductor. Furthermore, after 120 min of LED operation, the optical/electric power transfer rate of the liquid conductor was higher than that of the solid conductor, and the difference between them increased over time. The graphene-solution automated circulating system is a suitable conductor for LEDs and is particularly useful for operation over long periods.

**Author Contributions:** Conceptualization, Y.-C.C.; methodology, Y.-C.C. and H.-H.C.; validation, H.-H.C. and S.-H.L.; formal analysis, Y.-C.C., H.-H.C. and S.-H.L.; investigation, H.-H.C. and S.-H.L.; resource, Y.-C.C. and S.-H.L.; data curation, Y.-C.C., H.-H.C. and S.-H.L.; writing—original draft preparation, Y.-C.C.; writing—review and editing, Y.-C.C.; supervision, Y.-C.C.; project administration, Y.-C.C.; funding acquisition, Y.-C.C. All authors have read and agreed to the published version of the manuscript.

**Funding:** This research received no external funding.

**Data Availability Statement:** Authors ensure that data shared are in accordance with consent provided by participants on the use of confidential data. The data presented in this study are available on request from the corresponding author. The data are not publicly available due to the consideration of commercialization.

**Conflicts of Interest:** The authors declare no conflict of interest.

## References


1. Colangelo, G.; Favale, E.; Milanese, M.; de Risi, A.; Laforgia, D. Cooling of electronic devices: Nanofluids contribution. *Appl. Therm. Eng.* **2017**, *127*, 421–435. [CrossRef]
2. Bahiraei, M.; Heshmatian, S. Electronics cooling with nanofluids: A critical review. *Energy Convers. Manag.* **2018**, *172*, 438–456. [CrossRef]
3. Kheirabadi, A.C.; Groulx, D. Experimental evaluation of a thermal contact liquid cooling system for server electronics. *Appl. Therm. Eng.* **2018**, *129*, 1010–1025. [CrossRef]
4. Heidari, N.; Rahimi, M.; Azimi, N. Experimental investigation on using ferrofluid and rotating magnetic field (RMF) for cooling enhancement in a photovoltaic cell. *Int. Commun. Heat Mass Transf.* **2018**, *94*, 32–38. [CrossRef]
5. Zheng, Y.; Shi, Y.; Huang, Y. Optimisation with adiabatic interlayers for liquid-dominated cooling system on fast charging battery packs. *Appl. Therm. Eng.* **2019**, *147*, 636–646. [CrossRef]
6. Naphon, P.; Wiriyasart, S. Liquid cooling in the mini-rectangular fin heat sink with and without thermoelectric for CPU. *Int. Commun. Heat Mass Transf.* **2009**, *36*, 166–171. [CrossRef]
7. Ma, H.-K.; Chen, B.-R.; Gao, J.-J.; Lin, C.-Y. Development of an OAPCP-micropump liquid cooling system in a laptop. *Int. Commun. Heat Mass Transf.* **2009**, *36*, 225–232. [CrossRef]
8. Siahkamari, L.; Rahimi, M.; Azimi, N.; Banibayat, M. Experimental investigation on using a novel phase change material (PCM) in micro structure photovoltaic cooling system. *Int. Commun. Heat Mass Transf.* **2019**, *100*, 60–66. [CrossRef]
9. Nasef, H.A.; Nada, S.A.; Hassan, H. Integrative passive and active cooling system using PCM and nanofluid for thermal reg-ulation of concentrated photovoltaic solar cells. *Energy Convers. Manag.* **2019**, *109*, 112065. [CrossRef]
10. Zhou, F.; Joshi, S.N.; Liu, Y.; DeDe, E.M. Near-junction cooling for next-generation power electronics. *Int. Commun. Heat Mass Transf.* **2019**, *108*, 104300. [CrossRef]
11. Lu, W.; Meng, Z.; Sun, Y.; Zhong, Q.; Zhu, H. Improved energy performance of ammonia recycling system using floating con-densing temperature control. *Appl. Therm. Eng.* **2016**, *102*, 1011–1018. [CrossRef]
12. Gong, M.-H.; Yi, Q.; Huang, Y.; Wu, G.-S.; Hao, Y.-H.; Feng, J.; Li, W.-Y. Coke oven gas to methanol process integrated with CO2 recycle for high energy efficiency, economic benefits and low emissions. *Energy Convers. Manag.* **2017**, *133*, 318–331. [CrossRef]
13. Jin, H.; Fan, C.; Guo, L.; Liu, S.; Cao, C.; Wang, R. Experimental study on hydrogen production by lignite gasification in super-critical water fluidized bed reactor using external recycle of liquid residual. *Energy Convers. Manag.* **2017**, *145*, 214–219. [CrossRef]
14. Abou-Ziyan, H.; Ibrahim, M.; Abdel-Hameed, H. Characteristics enhancement of one-section and two-stepwise microchannels for cooling high-concentration multi-junction photovoltaic cells. *Energy Convers. Manag.* **2020**, *206*, 112488. [CrossRef]
15. Ma, K.; Liu, M.; Zhang, J. A method for determining the optimum state of recirculating cooling water system and experimental investigation based on heat dissipation efficiency. *Appl. Therm. Eng.* **2020**, *176*, 115398. [CrossRef]
16. Chen, P.; Harmand, S.; Ouenzerfi, S. Immersion cooling effect of dielectric liquid and self-rewetting fluid on smooth and porous surface. *Appl. Therm. Eng.* **2020**, *180*, 115862. [CrossRef]
17. Yang, H.; Yang, B.; Li, J.; Yang, P. Failure analysis and reliability reinforcement on gold wire in high-power COB-LED under current and thermal shock combined loading. *Appl. Therm. Eng.* **2019**, *150*, 1046–1053. [CrossRef]
18. Yung, K.C.; Liem, H.; Choy, H.S. Heat transfer analysis of a high-brightness LED array on PCB under different displacement configurations. *Int. Commun. Heat Mass Transf.* **2014**, *53*, 79–86. [CrossRef]
19. Lai, Y.; Cordero, N.; Barthel, F.; Tebbe, F.; Kuhn, J.; Apfelbeck, R.; Würtenberger, D. Liquid cooling of bright LEDs for automotive applications. *Appl. Therm. Eng.* **2009**, *29*, 1239–1244. [CrossRef]
20. Ramos-Alvarado, B.; Feng, B.; Peterson, G.P. Comparison and optimization of single-phase liquid cooling system devices for the heat dissipation of high-power LED arrays. *Appl. Therm. Eng.* **2013**, *59*, 648–659. [CrossRef]
21. Kim, D.; Lee, J.; Kim, J.; Choi, C.-H.; Chung, W. Enhancement of heat dissipation of LED module with cupric-oxide composite coating on aluminum-alloy heat sink. *Energy Convers. Manag.* **2015**, *106*, 958–963. [CrossRef]
22. Jang, D.; Yook, S.J.; Lee, K.S. Optimum design of a radial heat sink with a fin-height profile for high-power LED lighting ap-plications. *Appl. Energy* **2014**, *116*, 260–268. [CrossRef]
23. Cheng, H.H.; Huang, D.-S.; Lin, M.-T. Heat dissipation design and analysis of high power LED array using the finite element method. *Microelectron. Reliab.* **2012**, *52*, 905–911. [CrossRef]
24. Xiao, C.; Liao, H.; Wang, Y.; Li, J.; Zhu, W. A novel automated heat-pipe cooling device for high-power LEDs. *Appl. Therm. Eng.* **2017**, *111*, 1320–1329. [CrossRef]
25. Wang, Y.; Cen, J.; Jiang, F.; Cao, W. Heat dissipation of high-power light emitting diode chip on board by a novel flat plate heat pipe. *Appl. Therm. Eng.* **2017**, *123*, 19–28. [CrossRef]
26. Wang, M.; Tao, H.; Sun, Z.; Zhang, C. The development and performance of the high-power LED radiator. *Int. J. Therm. Sci.* **2017**, *113*, 65–72. [CrossRef]
27. Huang, D.-S.; Chen, T.-C.; Tsai, L.-T.; Lin, M.-T. Design of fins with a grooved heat pipe for dissipation of heat from high-powered automotive LED headlights. *Energy Convers. Manag.* **2019**, *180*, 550–558. [CrossRef]
28. Liang, F.; Gao, J.; Xu, L. Investigation on a grinding motorized spindle with miniature-revolving-heat-pipes central cooling structure. *Int. Commun. Heat Mass Transf.* **2020**, *112*, 104502. [CrossRef]
29. Jeong, M.W.; Jeon, S.W.; Lee, S.H.; Kim, Y. Effective heat dissipation and geometric optimization in an LED module with aluminum nitride (AlN) insulation plate. *Appl. Therm. Eng.* **2015**, *76*, 212–219. [CrossRef]

30. Deng, X.; Luo, Z.; Xia, Z.; Gong, W.; Wang, L. Active-passive combined and closed-loop control for the thermal management of high-power LED based on a dual synthetic jet actuator. *Energy Convers. Manag.* **2017**, *132*, 207–212. [CrossRef]
31. Seo, J.-H.; Lee, M.-Y. Illuminance and heat transfer characteristics of high power LED cooling system with heat sink filled with ferrofluid. *Appl. Therm. Eng.* **2018**, *143*, 438–449. [CrossRef]
32. Lin, X.; Mo, S.; Jia, L.; Yang, Z.; Chen, Y.; Cheng, Z. Experimental study and Taguchi analysis on LED cooling by thermoelectric cooler integrated with microchannel heat sink. *Appl. Energy* **2019**, *242*, 232–238. [CrossRef]
33. Tsai, H.-L.; Le, P.T. Self-sufficient energy recycling of light emitter diode/thermoelectric generator module for its active-cooling application. *Energy Convers. Manag.* **2016**, *118*, 170–178. [CrossRef]
34. Sadeghinezhad, E.; Mehrali, M.; Saidur, R.; Mehrali, M.; Latibari, S.T.; Akhiani, A.R.; Metselaar, H.S.C. A comprehensive review on graphene nanofluids: Recent research, development and applications. *Energy Convers. Manag.* **2016**, *111*, 466–487. [CrossRef]
35. Joseph, M.; Sajith, V. Graphene enhanced paraffin nanocomposite based hybrid cooling system for thermal management of electronics. *Appl. Therm. Eng.* **2019**, *163*, 114342. [CrossRef]
36. Wang, Y.; Al-Saaidi, H.A.I.; Kong, M.; Alvarado, J.L. Thermophysical performance of graphene based aqueous nanofluids. *Int. J. Heat Mass Transf.* **2018**, *119*, 408–417. [CrossRef]
37. Taherialekouhi, R.; Rasouli, S.; Khosravi, A. An experimental study on stability and thermal conductivity of water-graphene oxide/aluminum oxide nanoparticles as a cooling hybrid nanofluid. *Int. Commun. Heat Mass Transf.* **2019**, *145*, 118751. [CrossRef]
38. Zhao, X.; Jiaqiang, E.; Wu, G.; Deng, Y.; Han, D.; Zhang, B.; Zhang, Z. A review of studies using graphenes in energy conversion, energy storage and heat transfer development. *Energy Convers. Manag.* **2019**, *184*, 581–599. [CrossRef]
39. Bahiraei, M.; Heshmatian, S. Graphene family nanofluids: A critical review and future research directions. *Energy Convers. Manag.* **2019**, *196*, 1222–1256. [CrossRef]
40. Kazemi, I.; Sefid, M.; Afrand, M. Improving the thermal conductivity of water by adding mono & hybrid nano-additives con-taining graphene and silica: A comparative experimental study. *Int. Commun. Heat Mass Transf.* **2020**, *116*, 104648.
41. Gan, J.S.; Yu, H.; Tan, M.K.; Soh, A.K.; Wu, H.A.; Hung, Y.M. Performance enhancement of graphene-coated micro heat pipes for light-emitting diode cooling. *Int. J. Heat Mass Transf.* **2020**, *154*, 119687. [CrossRef]
42. Naghash, A.; Sattari, S.; Rashidi, A. Experimental assessment of convective heat transfer coefficient enhancement of nanofluids prepared from high surface area nanoporous graphene. *Int. Commun. Heat Mass Transf.* **2016**, *78*, 127–134. [CrossRef]
43. Akhavan-Zanjani, H.; Saffar-Avval, M.; Mansourkiaei, M.; Sharif, F.; Ahadi, M. Experimental investigation of laminar forced convective heat transfer of Graphene–water nanofluid inside a circular tube. *Int. J. Therm. Sci.* **2016**, *100*, 316–323. [CrossRef]
44. Sarafraz, M.; Yang, B.; Pourmehran, O.; Arjomandi, M.; Ghomashchi, R. Fluid and heat transfer characteristics of aqueous graphene nanoplatelet (GNP) nanofluid in a microchannel. *Int. Commun. Heat Mass Transf.* **2019**, *107*, 24–33. [CrossRef]
45. Keklikcioglu, O.; Dagdevir, T.; Ozceyhan, V. Heat transfer and pressure drop investigation of graphene nanoplatelet-water and titanium dioxide-water nanofluids in a horizontal tube. *Appl. Therm. Eng.* **2019**, *162*, 114256. [CrossRef]
46. Alous, S.; Kayfeci, M.; Uysal, A. Experimental investigations of using MWCNTs and graphene nanoplatelets water-based nanofluids as coolants in PVT systems. *Appl. Therm. Eng.* **2019**, *162*, 114265. [CrossRef]
47. Bahiraei, M.; Salmi, H.K.; Safaei, M.R. Effect of employing a new biological nanofluid containing functionalized graphene nanoplatelets on thermal and hydraulic characteristics of a spiral heat exchanger. *Energy Convers. Manag.* **2019**, *180*, 72–82. [CrossRef]
48. Purbia, D.; Khandelwal, A.; Kumar, A.; Sharma, A.K. Graphene-water nanofluid in heat exchanger: Mathematical modelling, simulation and economic evaluation. *Int. Commun. Heat Mass Transf.* **2019**, *108*, 104327. [CrossRef]
49. Askari, S.; Rashidi, A.; Koolivand, H. Experimental investigation on the thermal performance of ultra-stable kerosene-based MWCNTs and Graphene nanofluids. *Int. Commun. Heat Mass Transf.* **2019**, *108*, 104334. [CrossRef]
50. Naddaf, A.; Heris, S.Z. Experimental study on thermal conductivity and electrical conductivity of diesel oil-based nanofluids of graphene nanoplatelets and carbon nanotubes. *Int. Commun. Heat Mass Transf.* **2018**, *95*, 116–122. [CrossRef]
51. Alizadeh, J.; Moraveji, M.K. An experimental evaluation on thermophysical properties of functionalized graphene nano-platelets ionanofluids. *Int. Commun. Heat Mass Transf.* **2018**, *98*, 31–40. [CrossRef]
52. Girish, T. Some suggestions for photovoltaic power generation using artificial light illumination. *Sol. Energy Mater. Sol. Cells* **2006**, *90*, 2569–2571. [CrossRef]
53. Foti, M.; Tringali, C.; Battaglia, A.; Sparta, N.; Lombardo, S.; Gerardi, C. Efficient flexible thin film silicon module on plastics for indoor energy harvesting. *Sol. Energy Mater. Sol. Cells* **2014**, *130*, 490–494. [CrossRef]
54. Aoki, Y. Photovoltaic performance of Organic Photovoltaics for indoor energy harvester. *Org. Electron.* **2017**, *48*, 194–197. [CrossRef]
55. Apostolou, G.; Reinders, A.; Verwaal, M. Comparison of the indoor performance of 12 commercial PV products by a simple model. *Energy Sci. Eng.* **2016**, *4*, 69–85. [CrossRef]
56. Minnaert, B.; Veelaert, P. Efficiency simulations of thin films chalcogenide photovoltaic cells for different indoor lighting con-ditions. *Thin Solid Films* **2011**, *519*, 7537–7540. [CrossRef]
57. De Rossi, F.; Pontecorvo, T.; Brown, T.M. Characterization of photovoltaic devices for indoor light harvesting and customization of flexible dye solar cells to deliver superior efficiency under artificial lighting. *Appl. Energy* **2015**, *156*, 413–422. [CrossRef]
58. Reynaud, C.A.; Clerc, R.; Lechêne, P.B.; Hébert, M.; Cazier, A.; Arias, A.C. Evaluation of indoor photovoltaic power production under directional and diffuse lighting conditions. *Sol. Energy Mater. Sol. Cells* **2019**, *200*, 110010. [CrossRef]

59. Rühle, S. Tabulated values of the Shockley–Queisser limit for single junction solar cells. *Sol. Energy* **2016**, *130*, 139–147. [CrossRef]
60. Jarosz, G.; Marczyński, R.; Signerski, R. Effect of band gap on power conversion efficiency of single-junction semiconductor photovoltaic cells under white light phosphor-based LED illumination. *Mater. Sci. Semicond. Process.* **2020**, *107*, 104812. [CrossRef]
61. Yuruker, S.U.; Tamdogan, E.; Arik, M. An Experimental and Computational Study on Efficiency of White LED Packages With a Thermocaloric Approach. *IEEE Trans. Components, Packag. Manuf. Technol.* **2017**, *7*, 201–207. [CrossRef]
62. Cohen, A.B.; Kraus, A.D.; Davidson, S.F. Thermal frontiers in the design and packaging of microelectronic equipment. *J. Mech. Eng.* **1983**, *105*, 53–59.
63. Chung, Y.-C.; Chung, H.-H.; Lee, Y.-H.; Yang, L.-Q. Heat dissipation and electrical conduction of an LED by using a microfluidic channel with a graphene solution. *Appl. Therm. Eng.* **2020**, *175*, 115383. [CrossRef]
64. Holman, J.P. (Ed.) *Heat Transfer*, 8th ed.; The McGraw-Hill Companies, Inc.: New York, NY, USA, 2000.
65. *Principles of Heat and Mass Transfer*, 7th ed.; Incropera, F.P.; DeWitt, D.P.; Bergman, T.L.; Lavine, A.S. (Eds.) John Wiley & Sons, Inc.: New York, NY, USA, 2016.

 *nanomaterials*

*Article*

# Screening in Graphene: Response to External Static Electric Field and an Image-Potential Problem

Vyacheslav M. Silkin [1,2,3,*], Eugene Kogan [4] and Godfrey Gumbs [5]

1 Donostia International Physics Center (DIPC), Paseo de Manuel Lardizabal 4, E-20018 San Sebastián, Basque Country, Spain
2 Departamento de Polímeros y Materiales Avanzados: Física, Química y Tecnología, Facultad de Ciencias Químicas, Universidad del País Vasco (UPV-EHU), Apdo. 1072, E-20080 San Sebastián, Basque Country, Spain
3 IKERBASQUE, Basque Foundation for Science, E-48011 Bilbao, Basque Country, Spain
4 Department of Physics, Jack and Pearl Resnick Institute, Bar-Ilan University, Ramat-Gan 52900, Israel; Eugene.Kogan@biu.ac.il
5 Department of Physics and Astronomy, Hunter College of the City University of New York, 695 Park Avenue, New York, NY 10065, USA; ggumbs@hunter.cuny.edu
* Correspondence: vyacheslav.silkin@ehu.es

**Abstract:** We present a detailed first-principles investigation of the response of a free-standing graphene sheet to an external perpendicular static electric field $E$. The charge density distribution in the vicinity of the graphene monolayer that is caused by $E$ was determined using the pseudopotential density-functional theory approach. Different geometries were considered. The centroid of this extra density induced by an external electric field was determined as $z_{im} = 1.048$ Å at vanishing $E$, and its dependence on $E$ has been obtained. The thus determined $z_{im}$ was employed to construct the hybrid one-electron potential which generates a new set of energies for the image-potential states.

**Keywords:** graphene; electric field; valence charge density; image potential; image-plane position; image-potential states

**Citation:** Silkin, V. M.; Kogan, E.; Gumbs, G. Screening in Graphene: Response to External Static Electric Field and an Image-Potential Problem. *Nanomaterials* **2021**, *11*, 1561. https://doi.org/10.3390/nano11061561

Academic Editor: Werner Blau

Received: 4 May 2021
Accepted: 5 June 2021
Published: 13 June 2021

**Publisher's Note:** MDPI stays neutral with regard to jurisdictional claims in published maps and institutional affiliations.



## 1. Introduction

The numerous properties of graphene have been intensively investigated after its experimental realization. Thousands of papers on this material were published. However, there still remains a simple unanswered question regarding the way in which the induced charge density is distributed in the vicinity of a graphene monolayer when an external electric field is applied to the graphene sheet. This topic was addressed, to some degree, by considering the problem of screening of the electric field induced by point charges in graphite [1–5]. Specifically, the in-plane distribution of the induced charge has been actively discussed [5–10]. As for the charge distribution in the direction perpendicular to the plane of carbon atoms, it was considered as being localized on it [5].

The perpendicular charge distribution was studied by considering two- and multi-layer graphene films [11–13], though to the best of our knowledge, not for monolayer graphene. Moreover, regarding the question around the location of its center of mass with respect to the carbon atoms position, we are unaware of such work for a graphene film of any thickness. As a matter of fact, this question is important since, for instance, the position of the centroid of the induced density determines the so-called image-plane position $z_{im}$, (here we define the $z$ axis as pointing in the direction perpendicular to the carbon atoms basal plane) that is a "real position" of a solid surface for many phenomena occurring there. It determines a "physical" position of a metal surface when an external perturbation is applied. This problem was widely studied in the case of metal surfaces. In general, this "real" surface position is different from the spatial localization of the top atomic layer or a geometrical crystal edge, staying towards the vacuum side [14–18].

It is usually assumed for a quasi two-dimensional (2D) system that the excess charge is confined within an infinitesimally thin 2D layer [5,19]. Certainly, this assumption is reasonable if the relevant distance largely exceeds the atomic scale. However, it is critical to take into consideration what occurs on the atomic scale. For instance, if one intends to construct a capacitor by adopting graphene sheets, it would be helpful to determine its "physical size" which defines its electrical properties and may be different from the geometrical distance between two graphene layers. Addionally, determination of the spatial localization of the charge induced by an external electric field can be important in understanding the phenomena occurring in field-effect transistors based on 2D materials [20–22].

Knowledge of the position of the center of mass of the induced charge density is important in many fields of surface science. Thus, it determines the reference plane for the image-potential felt by an external charge placed in front of a surface. If this charge is an excited electron with energy below the vacuum level, it can be trapped by this image potential in a state belonging to an infinite Rydberg -like series [23,24]. The members of this series are referred to as image-potential states (IPSs).

In the previous work devoted to the IPSs in graphene, it was assumed [25,26] that $z_{im}$ is located at the carbon atom plane, which seems reasonable owing to the mirror symmetry of the system. Consequently, all the quantum states should be symmetric or anti-symmetric with respect to the $z = 0$ plane. As a result, a double Rydberg -like series of IPSs was predicted [25] to exist in a free-standing graphene monolayer since two surfaces are separated by a single atomic layer of matter only.

Up to now, IPSs for a free-standing graphene were not studied experimentally. On the other hand, numerous measurements were performed on the graphene supported on various metallic or semiconducting substrates. Usually, the interface distance between the graphene sheet and the surface atomic layer is such that the conventional single Rydberg series of a whole system is observed. Thus, in the graphene/metal systems where the graphene atomic layer is placed closer to the substrate, only a single series of IPSs was observed [27–36]. Nevertheless, there are cases where the distance separating the graphene and the top surface atomic layer is sufficiently large so as to realize the two lowest members of the graphene double-IPS series. In scanning-tunneling microscopy (STM) measurements, evidence for the Stark-shifted first two members (symmetric and antisymmetric ones) of this series was reported in the Gr/SiC(0001) system [37,38]. These states were also clearly observed in two-photon photoemission spectroscopy experiments [39]. However, in the same system, the splitting of the IPS series was not confirmed in the Ref. [40]. In the very recent experimental paper, the arguments in favor of the splitting were presented [41].

For a description of the IPSs in the graphene/substrate systems, a number of potentials have been developed. Indeed, an accurate description of IPSs is a challenge since the conventional density-functional theory (DFT) calculations do not accurately account for a correct long-range interaction in front of solid surfaces. One of the approaches consists of constructing the nonlocal van der Waals functional [42]. Although it does not yield the correct image potential behavior at long distances away from the 2D sheet, it improves the IPS description. Another input employing a conventional DFT scheme based on the local-density approximation (LDA) consists of the construction of a hybrid potential with the same computational cost. Some others use totally model potentials [26,35,43]. Since the binding energies of IPSs are sensitive to the long-range behavior of an effective potential, a key point is the image-plane position $z_{im}$ with respect to the carbon atom plane. Upon construction of the model potential in the Ref. [43], the fitting procedure gave $z_{im} = 0.99$ Å. This is significantly different from $z_{im} = 0$ assumed in other publications [25,26].

Our goal in this work is to determine the $z_{im}$ value for free-standing monolayer graphene from the direct DFT calculations of redistribution of its valence charge density upon application of an external electric field. Subsequently, the thus obtained $z_{im}$ is employed for the construction of a new hybrid "LDA+image−tail" potential. With this potential a new set of binding energies for IPSs is obtained and compared with the previous ones.

The rest of this paper is organized as follows. In Section 2, a brief description of our calculation method and some computational details are given. In Section 3, we present our calculated results. A summary and concluding remarks are presented in Section 4.

## 2. Calculational Methods and Details

The band structure of a graphene monolayer in the absence and presence of an external electric field of varying intensity was obtained within the LDA by solving the Kohn-Sham equations employing a home-made band structure computer code [44]. We used norm-conserving Troullier-Martin pseudopotentials to describe the electron-ion interaction for the carbon ions [45]. At the iteration stage, the exchange-correlation potential was taken in the form given in the Refs. [46,47]. For the expansion of the wave functions, a plane-wave basis set with an energy cutoff of 50 Rydberg was employed. In a self-consistent procedure, the summation over wave vectors in the irreducible part of the first Brillouin zone (BZ) was performed over a $48 \times 48 \times 1$ **k** mesh.

The self-consistent procedure was realized by considering a repeated-slab geometry with the lateral lattice constant of 2.424 Å. The external electric field applied in the direction perpendicular to the graphene plane has no translation symmetry. In order to implement it in the repeated-slab geometry, we added to the Hamiltonian a term corresponding to the extra charge $-\sigma(z)$ constant in the $x$-$y$ plane as shown in Figure 1a. Its $z$-dependence is defined by a Gaussian with a decay length of 1 a.u. This extra charge was placed at a distance of 10 Å from the graphene plane. In order to ensure the neutrality of the system, the charge $+\sigma$ was removed from the graphene system. The $z$ variation of the extra potential added to the system is schematically shown in Figure 1a. One can see that in the gap between the graphene and the extra charge position, this potential varies linearly from $V_g$ to $V_e$ with the $E = 2\pi\sigma$ slope. The problem with such a geometry is that there is a discontinuity in the potential between the left and right sides. In order to employ the repeated-slab geometry, we double the unit cell by mirror reflection of the picture of Figure 1a and establishing the distance between the graphene sheets in 20 Å. The resulting lattice constant in the perpendicular direction is 40 Å. We performed calculations considering the electric fields applied to the graphene sheet ranging from $-0.4$ to $0.5$ V/Å with a step of 0.1 V/Å and keeping the in-plane $(1 \times 1)$ geometry for carbon ion positions.

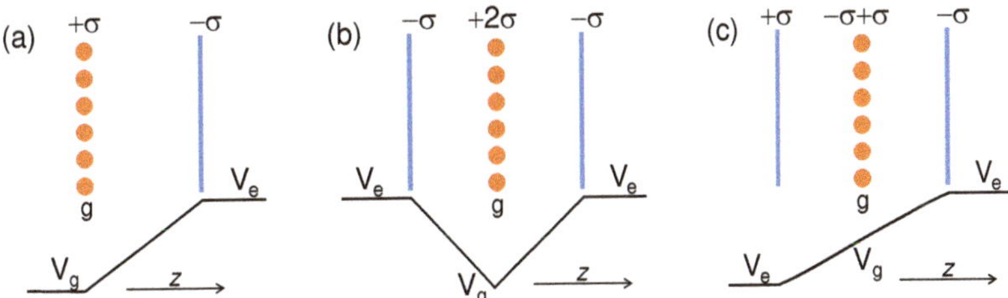


**Figure 1.** Schematic illustration of three geometries considered in this work of a graphene sheet (solid circles) interacting with an external electric charge uniformly distributed in the $x$-$y$ plane with density $-\sigma$ at a chosen distance in the $z$ direction. In the geometry (**a**) this extra charge is placed on the right. In the case (**b**) the charges of the same signs are located on the left and right sides. Panel (**c**) illustrates the geometry when the charges of the opposite sings placed on each side.

In other sets of calculations, we considered a geometry when the external electric field is applied from both sides of the graphene sheet as shown in Figure 1b. In this case, the unit cell contains only one graphene sheet and a lattice parameter of 20 Å is chosen. This geometry allows us to investigate the scale on which the charge density distribution established in graphene can be considered additively. On the other hand, this geometry is not suitable for the determination of the image-plane position since the resulting system is symmetric by construction. The third geometry considered in this study is schematically presented in Figure 1c. In this case, the two planes charged with $+\sigma$ and $-\sigma$ are located on each side of the graphene sheet that, in turn, is kept neutral. Since the total induced charge of the graphene is zero, this geometry cannot be used for the determination of the $z_{im}$ position. Nevertheless, the polarization induced in the carbon atom plane by the external field can be represented by two charged planes. In such a way, each surface can be considered as the covers of the different capacitors and charged oppositely.

## 3. Calculation Results

The electronic structure of graphene around the Fermi level at zero external electric field is presented in Figure 2 by thick black lines. The carbon-derived bonding and antibonding $\pi$ bands are marked as $\pi$ and $\pi^*$, respectively. The two lowest energy bands above the Fermi level characterized by strong expansion into the vacuum are marked as $1^+$ and $1^-$. At energies above the vacuum level, one can notice the quantization of the bands representing a free-electron continuum due to the finite size of the vacuum interval. In the same figure, we show how the energy position of all these bands changes when the external electric field of $-0.4$ V/Å (blue curves) or $0.4$ V/Å (red curves) is applied. One can notice that the $\pi$ and $\pi^*$ bands experience a shift of almost the same magnitude from the bare dispersion upon changing the sign of the electric field. On the contrary, the position of the upper energy bands with a strong expansion of its wave functions into the vacuum side changes differently for opposite signs.

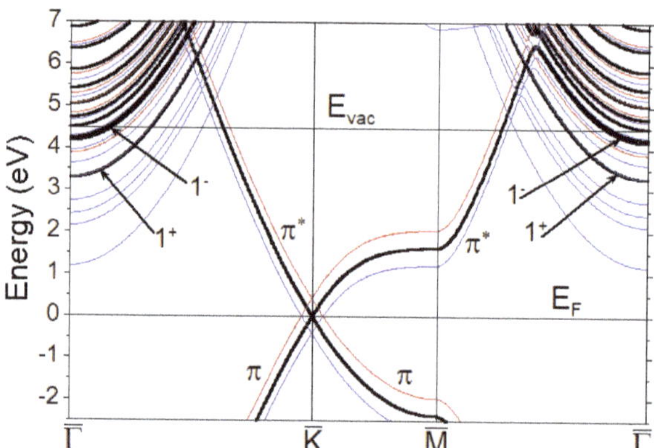


**Figure 2.** Electronic band structure of graphene when $E = 0$ (thick black lines), $0.4$ V/Å (thin red lines), and $-0.4$ V/Å (thin blue lines) obtained with application of the geometry of Figure 1a. The Fermi level, $E_F$, is placed at zero energy. The position of the vacuum level, $E_{vac}$, is shown for the zero field. The $\pi$ and $\pi^*$ bands are marked by corresponding symbols. The two unoccupied lowest energy states around the $\bar{\Gamma}$ point with strong localization in the vacuum are marked as $1^+$ and $1^-$ according to the Ref. [25].

### 3.1. Electric Field Effects

We have examined the way in which the induced charge density profile $n_{ind}(\mathbf{r}, E)$ varies with the strength and direction of the applied electric field $E$. Figure 3 reports $n_{ind}(z, E)$ obtained by averaging $n_{ind}(\mathbf{r}, E)$ in the $x$-$y$ plane for the values of $E$ ranging from $-0.4$ to $0.5$ V/Å. In order to perform a comparison, $n_{ind}(z, E)$ is normalized by the amplitude of $E$. One can see that its shape deviates qualitatively from the total valence density depicted by the green solid line. This can be understood, since the total density is dominated by the $\sigma$ bands that have a maximum at $z = 0$. On the contrary, the induced density is generated mainly by $\pi$ bands. Additionally, one can observe that the shape of the induced density only slightly depends on the sign and the magnitude of $E$. In general, we observe that at larger $E$, the shapes of $n_{ind}(z, E)$ are almost the same. However, upon reduction of the $E$ amplitude, the variations in $n_{ind}(z, E)$ gradually increase (hardly noticeable in Figure 3). This has consequences in the calculated centroid of the induced charge density versus $E$, defined as

$$z_{\text{im}}(E) = \frac{\int z\, n_{\text{ind}}(z, E)\, dz}{\int n_{\text{ind}}(z, E)\, dz} \tag{1}$$

and presented in Figure 4. Linear interpolation gives a value of 1.048 Å for $z_{\text{im}}(E = 0)$. Curiously, by constructing a model potential to describe IPSs measured experimentally in graphene/substrate systems, a very close value of 0.99 Å was established for $z_{\text{im}}$ in graphene monolayer [43]. A similar value was chosen for the crystal border in graphene in the Ref. [48]. We expect that the value of $z_{\text{im}}$ obtained here should not be affected significantly by the presence of the substrate once the valence electronic structure of graphene is not modified strongly by the substrate. In Figure 4, one can notice that upon approaching the $E = 0$ limit, $z_{\text{im}}(E)$ starts to deviate from the linear behavior. Moreover, this deviation is different for negative and positive $E$. In the former case, $z_{\text{im}}$ shifts downward, whereas in the latter case it is shifted upward. This can be explained by the fact that with reduction in the magnitude of $E$, the size of the Fermi surface shrinks and the possible calculation oscillations increase. For comparison, in the insert of Figure 4, we present the way in which $z_{\text{im}}(E)$ varies with $E$ in a free-standing Al(111) monolayer. Since, for Al, the Fermi surface is large because there are three valence electrons, the deviation from the linear behavior is small.

When we apply an external electric field to the graphene sheet from both sides according to the scheme depicted in Figure 1b, the induced charge density has a symmetric shape owing to the mirror symmetry. Its shape can be reproduced very well by superimposing that of Figure 3 onto a reflection of itself, thereby demonstrating that the response is additive. It means that once one knows how the electronic system of a graphene sheet responds to an external electric field applied from one side, the response to a more complex external perturbing field can be readily evaluated.

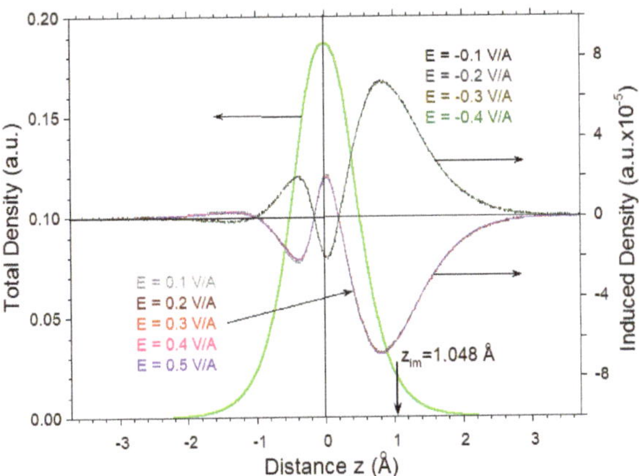


**Figure 3.** The valence charge density of graphene averaged in the *x-y* plane (green thick line) and induced charge densities generated by an applied electric field $E$ for the color-coded values shown in the insets. The induced density for $E = \theta \times 0.1$ V/Å is normalized by the value of $|\theta|$. The origin of the $z$ direction is taken as the carbon atom position. The image plane position $z_{\text{im}}$ at 1.048 Å is marked by vertical arrow (the positive value is due to the application of the electric field from the right side).

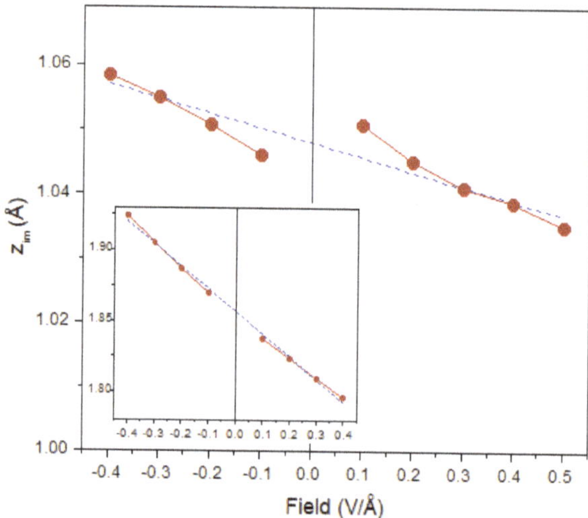

**Figure 4.** Dependence of the image plane position $z_{im}(E)$ in graphene versus the electric field $E$ amplitude (red circles and solid line). The linear interpolation is shown by a blue dashed line. The insert shows the way in which $z_{im}(E)$ depends on $E$ in the case of an Al(111) monolayer.

Clearly, in the symmetric case (Figure 1b), the calculated centroid of the induced charge density is placed at $z = 0$. However, knowing that each side responds independently to external electric fields applied from both the respective sides, the resulting charge distribution can be described in the electrostatic limit by two planes charged with the same signs and located at $z = -z_{im}$ and $z = z_{im}$. We believe that this picture should hold for a case when the valence electronic system is perturbed in a photoemission experiment, for example. In this case, an excited electron is promoted above the Fermi level. If its kinetic energy is lower than the work function, it can be trapped in the discrete IPSs whose number is two times larger than in the conventional Rydberg series of the hydrogen atom [25].

In the case of the geometry described in Figure 1c, the charge redistribution in the neutral graphene caused by placing it inside a capacitor can be represented at the electrostatic level by two charged planes with the opposite signs located at $z = -z_{im}$ and $z = +z_{im}$. Moreover, we found that the shape of the calculated induced density is also reproduced very well by employing the charge density distributions obtained for the positive and negative $E$s reported in Figure 1a. Since the calculated induced densities and the fitting results are very similar we do not include such a figure.

### 3.2. Image-Potential States

In our numerical calculations devoted to IPSs, the perpendicular lattice constant was increased up to 80 Å which allowed us to obtain convergent energies for the six lowest-energy members of the series. As it was mentioned previously, IPSs cannot be properly described with the use of conventional DFT calculations since the long-ranged image-potential on the vacuum side is not reproduced correctly. Additionally, the tight-binding methods are not inherently desired for its description [49,50]. Indeed, such states are a result of screening by the valence electron system of an external point charge placed in front of a system. This many-body information is not contained in the one-particle DFT Hamiltonian. In order to overcome this problem, maintaining the computational cost at the DFT level, we constructed a hybrid "LDA+image−tail" potential $V(\mathbf{r})$ which replaces the LDA local exchange-correlation potential term $V_{xc}(\mathbf{r})$ in the DFT Hamiltonian. At $|z|$ smaller than a certain $z_0$ value this potential coincides with $V_{xc}(\mathbf{r})$. For $|z| > z_0$, it has the following form:

$$V(\mathbf{r}) = -\frac{1 - A(x,y) \cdot e^{-\lambda(x,y)\cdot|z-\mathrm{sgn}(z)\cdot z_{\mathrm{im}}|}}{4|z - \mathrm{sgn}(z) \cdot z_{\mathrm{im}}|}. \tag{2}$$

The parameters $A(x,y)$ and $\lambda(x,y,)$ are defined from the smoothness conditions for $V(\mathbf{r})$ and its derivative at the matching planes $|z| = z_0$. In this work, these parameters depend on the $x$ and $y$ coordinates, since $V_{\mathrm{xc}}(x,y,z)$ still has a small corrugation at the matching plane. The only parameter left is $z_0$ which is unknown. In the following, we present results for three values of $z_0$ to show the sensitivity of the image-potential state energies to it.

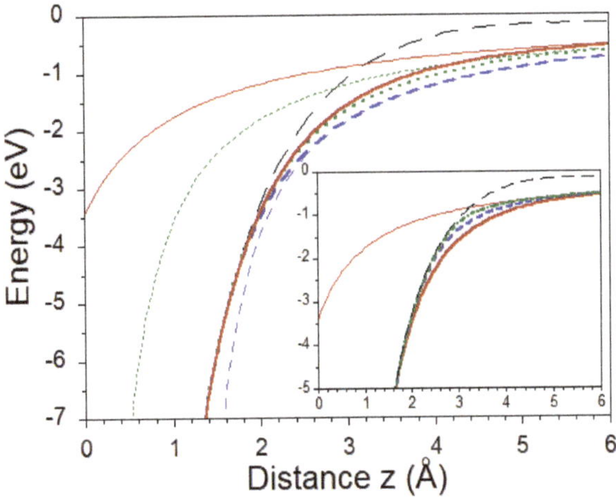

**Figure 5.** The LDA potential averaged in the $x$-$y$ plane as a function of the $z$ distance is shown by the thin dashed line. Hybrid "LDA+image−tail" potentials for $z_{\mathrm{im}} = 0$, 1.048, −1.048 Å with the matching plane at $z_0$=1.6 a.u. are presented as thick dotted, dashed, and solid lines, respectively. The corresponding bare image potentials are shown by thin dotted, dashed, and solid lines, respectively. Insert: Hybrid "LDA+image−tail" potentials constructed for $z_{\mathrm{im}} = -1.048$ Å with the matching planes $z_0 = 1.6$, 2.1, and 2.6 Å are represented by thick solid, dashed, and dashed-dotted lines, respectively.

In Figure 5, the thick dashed line shows the hybrid "LDA+image−tail" potential averaged in the $x$-$y$ plane constructed with $z_{\mathrm{im}} = 1.048$ Å and $z_0 = 1.6$ Å. One can see how at distances $z$ larger than $z_0$ it evolves from the averaged LDA potential (thin long-dashed line) to the image-potential defined as $-1/4(z - z_{\mathrm{im}})$ (thin dashed line). Notice that the potentials we construct here and employ for the band structure calculations are symmetric according to the $z = 0$ plane. Here, we show its behavior for positive $z$ only. For comparison, in Figure 5 by thick dotted line, we show the hybrid potential constructed for $z_{\mathrm{im}} = 0$ and $z_0 = 1.6$ Å of the Ref. [25]. One can see that the hybrid potential constructed with $z_{\mathrm{im}} = 1.048$ Å is noticeably lower for $z$ larger than $z_0$. This results in larger binding energies of IPSs. This is confirmed by the values obtained at the center of the BZ as reported in Table 1. One can see that the binding energy of the lowest-energy symmetric $1^+$ state increases from the 1.47 eV of the Ref. [25] to 1.58 eV here. Almost the same change is experienced by the antisymmetric $1^-$ state. For the states with larger numbers, this shift is notably smaller. Certainly, as $n$ is increased this difference is gradually reduced. With $z_{\mathrm{im}} = 1.048$ Å by employing $z_0$ larger than 1.6 Å we encountered a problem with the construction of the hybrid potential. Beyond this value for $z_0$, the two matching conditions for the hybrid potential cannot be fulfilled since the image potential with $z_{\mathrm{im}} = 1.048$ Å is located too far away on the right-hand side of the LDA potential, as seen in Figure 5. Notice that the downward shift of the IPSs is observed over a whole BZ.

Nevertheless, this does not significantly affect the interaction of IPSs with the scattering resonances [51] around the $\overline{K}$ point.

**Table 1.** Binding energies (in eV) of the image-potential states in graphene obtained with the hybrid potentials constructed with $z_0 = 1.6$ Å and $z_{im} = 1.048$ and $z_{im} = 0$ Å. In the case of $z_{im}$ placed at $-1.048$ Å the energies are obtained for three values of the matching plane position $z_0$. Last line presents the values of the states obtained in the LDA calculation [25].

| $z_{im}$(Å) | $z_o$(Å) | $1^+$ | $1^-$ | $2^+$ | $2^-$ | $3^+$ |
|---|---|---|---|---|---|---|
| 1.048 | 1.6 | 1.58 | 0.84 | 0.29 | 0.21 | 0.12 |
| 0 | 1.6 | 1.47 | 0.72 | 0.25 | 0.19 | 0.11 |
| −1.048 | 1.6 | 1.43 | 0.64 | 0.21 | 0.16 | 0.10 |
| | 2.1 | 1.30 | 0.52 | 0.19 | 0.15 | 0.11 |
| | 2.6 | 1.27 | 0.49 | 0.20 | 0.15 | 0.10 |
| LDA | | 1.17 | 0.25 | - | - | - |

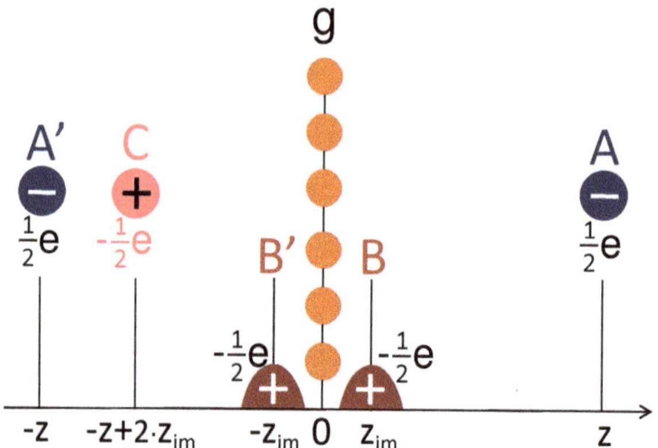


**Figure 6.** Schematic illustration of the charges created in the vicinity of a graphene monolayer. The carbon atom plane is located at $z = 0$. The charge distribution in an image-potential state with the centers of gravity on the positive and negative sides according to the graphene plane located at $z$ and $-z$ are represented by two point charges as shown by blue circles A and A'. The positive charge density generated in graphene in response to this external perturbation is represented by red areas B and B' centered at $z_{im}$ and $-z_{im}$, respectively. As a result, the negative point charge A interacts with its own image charge C, negative charge A', and positive charge B'.

The approach described above for construction of a hybrid "LDA+image−tail" potential is an adoption of the conventional image-potential picture employed for solid surfaces [52]. It can also be safely applied for sufficiently thick films as well. However, in a film consisting of just a single atomic layer, the situation might be different. In such a system, in a photoemission experiment, an excited electron can occupy a quantum state with the charge density symmetrical with respect to the atom plane, contrary to what occurs for solids where only a single surface is involved. The presence of two independent surfaces was indeed taken into account in our model presented above. Nevertheless, let us consider the situation from another point of view by applying a simple image-potential picture in a different way. In this case, we will replace an excited electron with some spatial charge density distribution by two point charges having $\frac{1}{2}$e located at distances $z$ and $-z$ as denoted by A and A', respectively, in Figure 6. In the graphene sheet, these two point charges create the screening charges B and B' whose centers of gravity are located at $z_{im}$

and $-z_{im}$, respectively. The spacial arrangement of these screening charges should be such to ensure efficient screening and avoid any charge current. Assuming the distance $z$ is large, let us account for the interaction between the charge A with real charges A$'$, B, and B$'$. The interaction of A with the charge B can be replaced by interaction with its image point charge C with positive sign located at $-z + 2z_{im}$, like it occurs at a metal surface. The interaction with a point charge A$'$ is obviously a Coulomb-like one. However, the interaction with a charge B$'$ is not obvious. The space distribution of this part of the total screening charge is such as to screen the point charge A$'$. Therefore, since $z$ is large, for a point charge A, it can be considered as a point charge located at $-z_{im}$. Counting all these three contributions at the first order in $1/z$, the resulting potential takes the form $V(z) = -1/4|z + z_{im}|$, that is, it looks like a charge A interacting with a point charge located at $z = -z_{im}$. A factor of four in the denominator is due to a fractionally charged electron with half its charge representing the A and B$'$ charges. Clearly, this latter model might be reasonable for IPSs with high numbers $n$. However, it may not be good for $n = 1^+$ since the maximum of its wave function is localized [25] around 2 Å, that is, being very close to the $z_{im} = 1.048$ Å position.

Based on this picture, we constructed a hybrid "LDA+image−tail" potential for which $z_{im}$ is placed at $-1.048$ Å. This potential with the matching plane at $z_0 = 1.6$ Å is shown as the thick solid line in Figure 5. Our calculated energies for the five lowest image-potential states are reported in Table 1. Comparing them with those obtained for $z_{im} = 0$ and $z_{im} = 1.048$ Å we observe a significant reduction of the binding energies, especially for $n = 1$. Varying $z_0$, we do not encounter problems with construction of the hybrid potential, contrary to the situation with $z_{im} = 1.048$ Å. For completeness, in Table 1, we report the image-potential energies obtained with $z_0 = 2.1$ and 2.6 Å as well. The respective hybrid potentials are reported in the insert of Figure 5. One can see that the effect of the variation in the potential caused by changing $z_0$ on the states $n = 1^+$ and $n = 1^-$ is substantial. Thus, for the lowest-energy image-potential state the binding energy may vary from 1.58 to 1.27 eV depending on the values of $z_{im}$ and $z_0$. Indeed, one can see how by increasing $z_0$, the value for the state $1^+$ is approaching the energy of 1.17 eV for the surface state [53,54] obtained in the LDA calculation [25]. However, for the state $1^-$, it is not the case.

We believe, the measurements of energies of free-standing graphene will provide important information on its screening properties. It may contribute to establishing a detailed picture of what is going on there due to an external perturbation. So far, all the experiments on IPSs were performed on supported graphene. Sensitivity of the image-potential states to the environment where graphene was kept was significant. Our findings point out that they can also provide important information about free-standing graphene screening properties.

## 4. Conclusions

In this theoretical study we have reported the detailed charge density distribution produced in free-standing graphene by external static electric fields with three geometries. The image-plane position was established. Surprisingly, it is rather large, located at 1.048 Å outside the carbon atom plane. Using this information, we constructed a new potential felt by an electron excited to the image-potential states. We checked several kinds of such a potential, demonstrating sensitivity of the energies of the lowest image-potential states to the details of this potential. It would be of interest to obtain the experimental information, such as from the photoemission spectroscopy, on the image-potential state energies for free-standing graphene. We believe that the experimentally determined image-potential energies will be extremely helpful for development of a more detailed picture for the graphene potential and how it reacts to the external perturbation on the atomic scale.

Our data on $z_{im}$ gives support to the value used for construction of an effective potential in the graphene/substrate systems [43]. Such potentials can be developed for the study of IPSs and interface states in a large class of molecular layers with the $\pi$-$\pi$ interaction similar to graphene [55–59]. Moreover, the information on the image-plane position can be useful for the construction of effective potentials in the systems with more complex geometries like fullerens and nanotubes, where the nearly-free states and the super-atomic orbitals, a subject of intense ongoing research, are inherently linked to IPSs in a flat graphene layer [60–67]. We believe that such a study as ours will not only be restricted to the carbon atoms case, since image-potential states can be realized in many other quasi-2D systems of current interest, like phosphorene, silicene and germanene [68], borophene [69], MXenes [70–72], and molecular overlayers on graphene [73].

**Author Contributions:** Conceptualization, V.M.S., E.K. and G.G.; methodology, V.M.S.; software and calculations, V.M.S.; writing—original draft preparation, V.M.S.; writing—review and editing, V.M.S., E.K., and G.G. All authors have read and agreed to the published version of the manuscript.

**Funding:** V.M.S. acknowledges support from the Project of the Basque Government for consolidated groups of the Basque University, through the Department of Universities (Q-NANOFOT IT1164-19) and from the Spanish Ministry of Science and Innovation (Grant No. PID2019–105488GB–I00). G.G. would like to acknowledge the support from the Air Force Research Laboratory (AFRL) through Grant No. FA9453-21-1-0046.



## Abbreviations

The following abbreviations are used in this manuscript:

| | |
|---|---|
| 2D | Two-dimensional |
| IPS | Image-potential state |
| STM | Scanning-tunneling microscopy |
| DFT | Density-functional theory |
| LDA | Local-density approximation |
| BZ | Brillouin zone |

## References

1. Visscher, P.B.; Falicov, L.M. Dielectric screening in a layered electron gas. *Phys. Rev. B* **1971**, *3*, 2541–2547. [CrossRef]
2. Pietronero, L.; Strässler, S.; Zeller, H.R.; Rice, M.J. Charge distribution in *c* direction in lamellar graphite acceptor intercalation compounds. *Phys. Rev. Lett.* **1978**, *41*, 763–767. [CrossRef]
3. Pietronero, L.; Strässler, S.; Zeller, H.R. Nonlinear screening in layered semimetals. *Solid State Commun.* **1979**, *30*, 399–401. [CrossRef]
4. Safran, S.A.; Hamann, D.R. Electrostatic interactions and staging in graphite intercalation compounds. *Phys. Rev. B* **1980**, *22*, 606–612. [CrossRef]
5. DiVincenzo, D.P.; Mele, E.J. Self-consistent effective-mass theory for intralayer screening in graphite intercalation compounds. *Phys. Rev. B* **1984**, *29*, 1685–1694. [CrossRef]
6. DiCenzo, S.D.; Wertheim, G.K.; Basu, S.; Fischer, J.E. Charge distribution in potassium graphite. *Phys. Rev. B* **1981**, *24*, 2270–2273. [CrossRef]
7. DiCenzo, S.D.; Basu, S.; Wertheim, G.K.; Buchanan, D.N.E.; Fischer, J.E. In-plane charge distribution in potassium-intercalated graphite. *Phys. Rev. B* **1982**, *25*, 620–626. [CrossRef]
8. Grunes, L.A.; Ritsko, J.J. Valence and core excitation spectra in K, Rb, and Cs alkali-metal stage-1 intercalated graphite. *Phys. Rev. B* **1983**, *28*, 3439–3446. [CrossRef]
9. Peres, N.M.R.; Guinea, F.; Neto, A.H.C. Electronic properties of disordered two-dimensional carbon. *Phys. Rev. B* **2006**, *73*, 125411. [CrossRef]
10. Polini, M.; Tomadin, A.; Asgari, R.; MacDonald, A.H. Density functional theory of graphene sheets. *Phys. Rev. B* **2008**, *78*, 115426. [CrossRef]
11. Yu, E.K.; Stewart, D.A.; Tiwari, S. *Ab initio* study of polarizability and induced charge densities in multilayer graphene films. *Phys. Rev. B* **2008**, *77*, 195406. [CrossRef]
12. Wang, R.-N.; Dong, G.-Y.; Wang, S.-F.; Fu, G.-S.; Wang, J.-L. Intra- and inter-layer charge redistribution in biased bilayer graphene. *AIP Adv.* **2016**, *6*, 035213. [CrossRef]
13. Gao, Y.L.; Okada, S. Carrier distribution control in bilayer graphene under a perpendicular electric field by interlayer stacking arrangements. *Appl. Phys. Express* **2021**, *14*, 035001. [CrossRef]
14. Lang, N.D.; Kohn, W. Theory of metal surfaces: Induced surface charge and image potential. *Phys. Rev. B* **1973**, *7*, 3541–3550. [CrossRef]

15. Serena, P.A.; Soler, J.M.; Garcia, N. Self-consistent image potential in a metal surface. *Phys. Rev. B* **1986**, *34*, 6767–6769. [CrossRef] [PubMed]
16. Inglesfield, J.E. The screening of an electric field at an Al(001) surface. *Surf. Sci.* **1987**, *188*, L701–L707. [CrossRef]
17. Eguiluz, A.G.; Hanke, W. Evaluation of the exchange-correlation potential at a metal surface from many-body perturbation theory. *Phys. Rev. B* **1989**, *39*, 10433–10436. [CrossRef] [PubMed]
18. Kiejna, A. Image plane position at a charged surface of stabilized jellium. *Surf. Sci.* **1993**, *287–288*, 618–621. [CrossRef]
19. McCann, E. Asymmetry gap in the electronic band structure of bilayer graphene. *Phys. Rev. B* **2006**, *74*, 161403(R). [CrossRef]
20. Novoselov, K.S.; Geim, A.K.; Morozov, S.V.; Jiang, D.; Zhang, Y.; Dubonos, S.V.; Grigorieva, I.V.; Firsov, A.A. Electric field effect in atomicall thin carbon films. *Science* **2004**, *306*, 666–669. [CrossRef] [PubMed]
21. Friori, G.; Bonaccorso, F.; Iannaccone, G.; Palacios, T.; Neumaier, D.; Seabaugh, A.; Banerjee, S.K.; Colombo, L. Electronics based on two-dimensional materials. *Nat. Nanotechnol.* **2014**, *9*, 768–779. [CrossRef] [PubMed]
22. Grillo, A.; Di Bartolomeo, A.; Urban, F.; Passacantando, M.; Caridad, J.M.; Sun, J.; Camilli, L. Observation od 2D conduction in ultrathin germanium arsenide field-effect transistors. *ACS Appl. Mater. Interfaces* **2020**, *12*, 12998–13004. [CrossRef]
23. Echenique, P.M.; Pendry, J.B. Existence and detection of Rydbeg states at surfaces. *J. Phys. C Solid State Phys.* **1978**, *11*, 2065–2075. [CrossRef]
24. Echenique, P.M.; Pendry, J.B. Theory of image states at metal surfaces. *Prog. Surf. Sci.* **1989**, *32*, 111–159. [CrossRef]
25. Silkin, V.M.; Zhao, J.; Guinea, F.; Chulkov, E.V.; Echenique, P.M.; Petek, H. Image potential states in graphene. *Phys. Rev. B* **2009**, *80*, 121408(R). [CrossRef]
26. de Andres, P.L.; Echenique, P.M.; Niesner, D.; Fauster, T.; Rivacova, A. One-dimensional potential for image-potential states on graphene. *New J. Phys.* **2014**, *16*, 023012. [CrossRef]
27. Borca, B.; Barja, S.; Garnica, M.; Sánchez-Portal, D.; Silkin, V.M.; Chulkov, E.V.; Hermanns, C.F.; Hinarejos, J.J.; Vxaxzquez de Parga, A.L.; Arnau, A.; et al. Potential energy landscape for hot electrons in periodically nanostructured graphene. *Phys. Rev. Lett.* **2010**, *105*, 036804. [CrossRef]
28. Zhang, H.G.; Hu, H.; Pan, Y.; Mao, J.H.; Gao, M.; Guo, H.M.; Du, S.X.; Greber, T.; Gao, H.-J. Graphene based quantum dots. *J. Phys. Condens. Matter* **2010**, *22*, 302001. [CrossRef]
29. Niesner, D.; Fauster, T.; Dadap, J.I.; Zaki, N.; Knox, K.R.; Yeh, P.-C.; Bhandari, R.; Osgood, R.M.; Petrović, M.; Kralj, M. Trapping surface electrons on graphene layers and islands. *Phys. Rev. B* **2012**, *85*, 081402(R). [CrossRef]
30. Armbrust, N.; Güdde, J.; Jakob, P.; Höfer, U. Time-resolved two-photon photoemission of unoccupied electronic states of periodically rippled graphene on Ru(0001). *Phys. Rev. Lett.* **2012**, *108*, 056801. [CrossRef]
31. Nobis, D.; Potenz, M.; Niesner, D.; Fauster, T. Image-potential states of graphene on noble-metal surfaces. *Phys. Rev. B* **2013**, *88*, 195435. [CrossRef]
32. Niesner, D.; Fauster, T. Image-potential states and work function of graphene. *J. Phys. Condens. Matter* **2014**, *26*, 393001. [CrossRef]
33. Craes, F.; Runte, S.; Klinkhammer, J.; Kralj, M.; Michely, T.; Busse, C. Mapping image potential states on graphene quantum dots. *Phys. Rev. Lett.* **2013**, *111*, 056804. [CrossRef] [PubMed]
34. Achilli, S.; Tognolini, S.; Fava, E.; Ponzoni, S.; Drera, G.; Cepek, C.; Patera, L.L.; Africh, C.; del Castillo, E.; Trioni, M.I.; et al. Surface states characterization in the strongly interacting graphene/Ni(111) system. *New J. Phys.* **2018**, *20*, 103039. [CrossRef]
35. Lin, Y.; Li, Y.-Z.; Sadowski, J.T.; Jin, W.-C.; Dadap, J.I.; Hybertsen, M.S.; Osgood, R.M., Jr. Excitation and characterization of image potential state electrons on quasi-free-standing graphene. *Phys. Rev. B* **2018**, *97*, 165413. [CrossRef]
36. Tognolini, S.; Achilli, S.; Ponzoni, S.; Longetti, L.; Mariani, C.; Trioni, M.I.; Pagliara, S. On- and off-resonance measurement of the Image State lifetime at the graphene/Ir(111) interface. *Surf. Sci.* **2019**, *679*, 11–16. [CrossRef]
37. Bose, S.; Silkin, V.M.; Ohmann, R.; Brihuega, I.; Vitali, L.; Michaelis, C.H.; Mallet, P.; Veuillen, J.Y.; Schneider, M.A.; Chulkov, E.V.; et al. Image potential states as a quantum probe of graphene interfaces. *New J. Phys.* **2010**, *12*, 023028. [CrossRef]
38. Sandin, A.; Pronschinske, A.; Rowe, J.E.; Dougherty, D.B. Incomplete screening by epitaxial graphene on the Si face of 6H-SiC(0001). *Appl. Phys. Lett.* **2010**, *97*, 113104. [CrossRef]
39. Takahashi, K.; Imamura, M.; Yamamoto, I.; Azuma, J.; Kamada, M. Image potential states in monolayer, bilayer, and trilayer epitaxial graphene studied with time- and angle-resolved two-photon photoemission spectroscopy. *Phys. Rev. B* **2014**, *89*, 155303. [CrossRef]
40. Gugel, D.; Niesner, D.; Eikhoff, C.; Wagner, S.; Weinelt, M.; Fauster, T. Two-photon photoemission from image-potential states of epitaxial graphene. *2D Mater.* **2015**, *2*, 045001. [CrossRef]
41. Ambrosio, G.; Achilli, S.; Pagliara, S. Resonance intensity of the $n = 1$ image potential state of graphene on SiC via two-photon photoemission. *Surf. Sci.* **2021**, *703*, 121722. [CrossRef]
42. Hamada, I.; Hamamoto, Y.; Morikawa, Y. Image potential states from the van der Waals density functional. *J. Phys. Chem.* **2017**, *147*, 044708. [CrossRef] [PubMed]
43. Armbrust, N.; Güdde, J.; Höfer, U. Formation of image-potential states at the graphene/metal interface. *New J. Phys.* **2015**, *17*, 103043. [CrossRef]
44. Silkin, V.M.; Chulkov, E.V.; Sklyadneva, I.Y.; Panin, V.E. Self-consistent pseudopotential calcualtion of the aluminum energy spectrum. *Soviet Phys. J.* **1984**, *27*, 762–767. [CrossRef]
45. Troullier, N.; Martins, J.L. Efficient pseudopotentials for plane-wave calculations. *Phys. Rev. B* **1991**, *43*, 1993–2006. [CrossRef]
46. Ceperley, D.M.; Alder, B.J. Ground state of the electron gas by a stochastic method. *Phys. Rev. Lett.* **1980**, *45*, 566–569. [CrossRef]

47. Perdew, J.P.; Zunger, A. Self-interaction correction to density-functional approximations for many-electron systems. *Phys. Rev. B* **1981**, *23*, 5048–5079. [CrossRef]
48. Krasovskii, E.E. Ab initio theory of photoemission from graphene. *Nanomaterials* **2021**, *11*, 1212. [CrossRef]
49. Kogan, E.; Nazarov, V.U.; Silkin, V.M.; Kaveh, M. Energy bands in graphene: Comparison between the tight-binding model and ab initio calculations. *Phys. Rev. B* **2014**, *89*, 165430. [CrossRef]
50. Kogan, E.; Silkin, V.M. Electronic structure of graphene: (Nearly) free electron bands versus tight-binding bands. *Phys. Status Solidi B* **2017**, *254*, 1700035. [CrossRef]
51. Nazarov, V.U.; Krasovskii, E.E.; Silkin, V.M. Scattering resonances in two-dimensional crystals with application to graphene. *Phys. Rev. B* **2013**, *87*, 041405. [CrossRef]
52. Chulkov, E.V.; Silkin, V.M.; Echenique, P.M. Image potential states on metal surfaces: binding energies and wave functions. *Surf. Sci.* **1999**, *437*, 330–352. [CrossRef]
53. Posternak, M.; Balderischi, A.; Freeman, A.J.; Wimmer, E.; Weinelt, M. Prediction of electronic interlayer states in graphite and reinterpretation of alkali bands in graphite intercalation compounds. *Phys. Rev. Lett.* **1983**, *50*, 761–764. [CrossRef]
54. Posternak, M.; Balderischi, A.; Freeman, A.J.; Wimmer, E. Prediction of electronic surface states in layered materials: Graphite. *Phys. Rev. Lett.* **1984**, *52*, 863–866. [CrossRef]
55. Tsirkin, S.S.; Zaitsev, N.L.; Nechaev, I.A.; Tonner, R.; Höfer, U.; Chulkov, E.V. Inelastic decay of electrons in Shockley-type metal-organic interface states. *Phys. Rev. B* **2015**, *92*, 235434. [CrossRef]
56. Armbrust, N.; Schiller, F.; Güdde, J.; Höfer, U. Model potential for the description of metal/organic interface states. *Sci. Rep.* **2017**, *7*, 46561. [CrossRef]
57. Eschmann, L.; Sabitova, A.; Temirov, R.; Tautz, F.S.; Kruger, P.; Rohlfing, M. Electric and thermoelectric transport in graphene and helical metal in finite magnetic fields. *Phys. Rev. B* **2019**, *100*, 125155. [CrossRef]
58. Marks, M.; Armbrust, N.; Güdde, J.; Höfer, U. Impact of interface-state formation on the charge-carrier dynamics at organic-metal interfaces. *New J. Phys.* **2020**, *22*, 093042. [CrossRef]
59. Stalberg, K.; Shibuta, M.; Höfer, U. Temperature effects on the formation and the relaxation dynamics of metal-organic interface states. *Phys. Rev. B* **2020**, *102*, 121401. [CrossRef]
60. Feng, M.; Zhao, J.; Petek, H. Atomlike, hollow-core-bound molecular orbitals of $C_{60}$. *Science* **2008**, *320*, 359–362. [CrossRef]
61. Zhao, J.; Feng, M.; Yang, J.; Petek, H. The superatom states of fullerenes and their hybridization into the nearly free electron bands of fullerites. *ACS Nano* **2009**, *3*, 853–864. [CrossRef] [PubMed]
62. Dutton, G.J.; Dougherty, D.B.; Jin, W.; Reutt-Robey, J.E.; Robey, S.W. Superatom orbitals of $C_{60}$ on Ag(111): Two-photon photoemission and scanning tunneling spectroscopy. *Phys. Rev. B* **2011**, *84*, 195435. [CrossRef]
63. Zhao, J.; Zheng, Q.J.; Petek, H.; Yang, J.L. Nonnuclear nearly free electron conduction channels induced by doping charge in nanotube-molecular sheet composites. *J. Phys. Chem. A* **2014**, *118*, 7255–7260. [CrossRef]
64. Gumbs, G.; Balassis, A.; Iurov, A.; Fekete, P. Strongly localized image states of spherical graphitic particles. *Sci. World J.* **2014**, *2014*, 726303. [CrossRef]
65. Knorzer, J.; Fey, C.; Sadeghpour, H.R.; Schmelcher, P. Control of multiple excited image states around segmented carbon nanotubes. *J. Chem. Phys.* **2015**, *143*, 204309. [CrossRef]
66. Johansson, J.O.; Bohl, E.; Campbell, E.E.B. Super-atom molecular orbital excited states of fullerenes. *Philos. Trans. R. Soc. A* **2016**, *374*, 20150322. [CrossRef]
67. Shibuta, M.; Yamamoto, K.; Guo, H.L.; Zhao, J.; Nakajima, A. Highly dispersive nearly free electron bands at a 2D-assembled $C_{60}$ monolayer. *J. Phys. Chem. C* **2020**, *124*, 734–741. [CrossRef]
68. Borca, B.; Castenmiller, C.; Tsvetanova, M.; Sotthewes, K.; Rudenko, A.N.; Zandvliet, H.J.W. Image potential states of germanene. *2D Mater.* **2020**, *7*, 035021. [CrossRef]
69. Kong, L.J.; Liu, L.R.; Chen, L.; Zhong, Q.; Cheng, P.; Li, H.; Zhang, Z.H.; Wu, K.H. One-dimensional nearly free electron states in borophene. *Nanoscale* **2019**, *11*, 15605–15611. [CrossRef]
70. Khazaei, M.; Ranjbar, A.; Ghorbani-Asl, M.; Arai, M.; Sasaki, T.; Liang, Y.Y.; Yunoki, S. Nearly free electron states in MXenes. *Phys. Rev. B* **2016**, *93*, 205125. [CrossRef]
71. Jiang, X.; Kuklin, A.V.; Baev, A.; Ge, Y.Q.; Ågren, H.; Zhang, H.; Prasad, P.N. Two-dimensional MXenes: From morphological to optical, electric, and magnetic properties and applications. *Phys. Rep.* **2020**, *848*, 1–58. [CrossRef]
72. Wang, M.Y.; Khazaei, M.; Kawazoe, Y.; Liang, Y.Y. First-principles study of a topological phase transition induced by image potential states in MXenes. *Phys. Rev. B* **2021**, *103*, 035433. [CrossRef]
73. Wella, S.A.; Sawada, H.; Kawaguchi, N.; Muttaqien, F.; Inagaki, K.; Hamada, I.; Morikawa, Y.; Hamamoto, Y. Hybrid image potential states in molecular overlayers on graphene. *Phys. Rev. Mater.* **2017**, *1*, 061001. [CrossRef]

 *nanomaterials*

*Article*

# Oblique and Asymmetric Klein Tunneling across Smooth NP Junctions or NPN Junctions in 8-*Pmmn* Borophene

Zhan Kong [1], Jian Li [1], Yi Zhang [1], Shu-Hui Zhang [2,*] and Jia-Ji Zhu [1,*]

1   School of Science and Laboratory of Quantum Information Technology, Chongqing University of Posts and Telecommunications, Chongqing 400065, China; kongz2021@163.com (Z.K.); jianli@cqupt.edu.cn (J.L.); zhangyia@cqupt.edu.cn (Y.Z.)
2   College of Mathematics and Physics, Beijing University of Chemical Technology, Beijing 100029, China
*   Correspondence: shuhuizhang@mail.buct.edu.cn (S.-H.Z.); zhujj@cqupt.edu.cn (J.-J.Z.)

**Abstract:** The tunneling of electrons and holes in quantum structures plays a crucial role in studying the transport properties of materials and the related devices. 8-*Pmmn* borophene is a new two-dimensional Dirac material that hosts tilted Dirac cone and chiral, anisotropic massless Dirac fermions. We adopt the transfer matrix method to investigate the Klein tunneling of massless fermions across the smooth NP junctions and NPN junctions of 8-*Pmmn* borophene. Like the sharp NP junctions of 8-*Pmmn* borophene, the tilted Dirac cones induce the oblique Klein tunneling. The angle of perfect transmission to the normal incidence is 20.4°, a constant determined by the Hamiltonian of 8-*Pmmn* borophene. For the NPN junction, there are branches of the Klein tunneling in the phase diagram. We find that the asymmetric Klein tunneling is induced by the chirality and anisotropy of the carriers. Furthermore, we show the oscillation of electrical resistance related to the Klein tunneling in the NPN junctions. One may analyze the pattern of electrical resistance and verify the existence of asymmetric Klein tunneling experimentally.

**Keywords:** Klein tunneling; borophene; Dirac fermions

**Citation:** Kong, Z.; Li, J.; Zhang, Y.; Zhang, S.-H.; Zhu, J.-J. Oblique and Asymmetric Klein Tunneling across Smooth NP Junctions or NPN Junctions in 8-*Pmmn* Borophene. *Nanomaterials* **2021**, *11*, 1462. https://doi.org/10.3390/nano11061462

Academic Editors: Filippo Giannazzo and Eugene Kogan

Received: 10 April 2021
Accepted: 28 May 2021
Published: 31 May 2021

**Publisher's Note:** MDPI stays neutral with regard to jurisdictional claims in published maps and institutional affiliations.



## 1. Introduction

Two-dimensional (2D) materials have been the superstars for their novel properties in condensed matter physics since its first isolation of graphene in 2004 [1]. Right now, the booming 2D materials family includes not just graphene and the derivatives of graphene but also transition metal dichalcogenides (TMDs) [2–4], black phosphorus [5–8], indium selenide [9–11], stanene [12,13], and many other layered materials [14,15]. Among these 2D materials, the so-called Dirac materials host massless Dirac fermions, always in the spotlight. Carriers in 2D Dirac materials usually have chirality or pseudospin from two atomic sublattices. Together with chirality, the linear Dirac dispersion gives rise to remarkable transport properties, including the absence of backscattering [1,16,17]. Due to the suppression of backscattering, massless Dirac fermions could tunnel a single square barrier with 100% transmission probability. This surprising result has been known as Klein tunneling [16,18–21]. Klein tunneling is the basic electrical conduction mechanism through the interface between *p*-doped and *n*-doped regions. Klein tunneling's elucidation plays a key role in designing and inventing electronic devices based on 2D Dirac materials.

Recently, several 2D boron structures have been predicted and experimentally fabricated [22–25]. The 8-*Pmmn* borophene belongs to the space group *Pmmn*, which means an orthorhombic lattice has an *mmm* symmetric point group (three-mirror symmetry planes perpendicular to each other) combine with a glide plane at one of the mirror symmetry planes [22,26]. This kind of structure is the most stable symmetric phase of borophene and may be kinetically stable at ambient conditions. It revealed the tilted Dirac cone and anisotropic massless Dirac fermions by first-principles calculations [27,28].These unique Dirac fermions attracts people to explore the various physical properties such as strain-induced pseudomagnetic field [29], anisotropic density–density response [30–33], optical conductivity [34,35], modified Weiss oscillation [36,37], borophane and its tight-binding model [37], nonlinear optical polarization rotation [38], oblique Klein tunneling [39–41], few-layer borophene [42,43], intense light response [44,45], RKKY interaction [46,47], anomalous caustics [48], electron–phonon coupling [49], valley–contrast behaviors [50,51], Andreev reflection [52], and so on. The oblique Klein tunneling, the deviation of the perfect transmission direction to the normal direction of the interface, is induced by the anisotropic massless Dirac fermions or the tilted Dirac cone [39,53]. However, the on-site disorder or smoothing of the NP junction interface or the square potential may destroy the ideal Klein tunneling, which means the sharp interface strongly depends on high-quality fabrication state-of-the-art technology [54]. Therefore, the detailed discussion of the smooth NP junction and the tunable trapezoid potential would be helpful for the promising electronic devices based on 2D Dirac materials.

In this paper, we study the transmission properties of anisotropic and tilted massless Dirac fermions across smooth NP junctions and NPN junctions in 8-*Pmmn* borophene. Similar to the sharp NP junction, the oblique Klein tunneling retains due to the tilted Dirac cone. This conclusion does not depend on the NP junctions' doping levels as the normal Klein tunneling but depends on the junction direction. We show the angle of oblique Klein tunneling is 20.4°, a constant determined by the Hamiltonian parameters of 8-*Pmmn* borophene. For the NPN junction, there are branches of the Klein tunneling in the phase diagram. We find that the asymmetric Klein tunneling is induced by the chirality and anisotropy of the carriers [55]. The indirect consequence of the asymmetric Klein tunneling lies in the oscillation of the electrical resistance. The analysis of the pattern of the oscillation of electrical resistance would help verify the existence of asymmetric Klein tunneling experimentally.

The rest of the paper is organized as follows. In Section 2, we introduce the Hamiltonian and the energy spectrum for the 8-*Pmmn* borophene, the NP and NPN junction's potential, and present the transfer matrix method for the detailed derivation of transmissions across the junctions. In Section 3, we demonstrate perfect transmission numerically, showing that the oblique Klein tunneling in NP junctions and the asymmetric Klein tunneling in NPN junctions. Then, we calculate the electrical resistance from the Landauer formula for the NPN junction. Finally, we give a brief conclusion in Section 4.

## 2. Theoretical Formalism

### 2.1. Model

The crystal structure of 8-*Pmmn* borophene has two sublattices, as illustrated in Figure 1a by different colors. It is made of buckled triangular layers where each unit cell has eight atoms under the symmetry of space group *Pmmn* (No. 59 in [56]), the so-called 8-*Pmmn* structure. The tilted Dirac cone emerges from the hexagonal lattice formed by the inner atoms (yellow in Figure 1a) [28]. This hexagonal structure is topologically equivalent to uniaxially strained graphene, and the Hamiltonian of 8-*Pmmn* borophene around one Dirac point is given by [29,30,36,37].

$$\hat{H}_0 = v_x \sigma_x \hat{p}_x + v_y \sigma_y \hat{p}_y + v_t \mathbf{I}_{2\times2}\hat{p}_y \tag{1}$$

where $\hat{p}_{x,y}$ are the momentum operators, $\sigma_{x,y}$ are $2 \times 2$ Pauli matrices, and $\mathbf{I}_{2\times2}$ is a $2 \times 2$ unit matrix. The anisotropic velocities are $v_x = 0.86\,v_F, v_y = 0.69\,v_F, v_t = 0.32\,v_F, v_F = 10^6$ m/s [29].The energy dispersion and the corresponding wave functions of $\hat{H}_0$ are

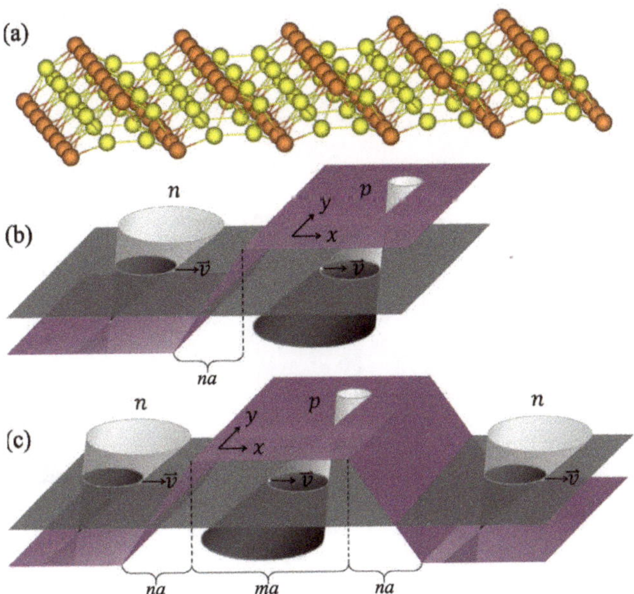

**Figure 1.** (**a**) Crystal structure of 8-*Pmmn* borophene. The unit cell of 8-*Pmmn* borophene contains two types of nonequivalent boron atoms, the ridge atoms (orange) and the inner atoms (yellow). (**b**) The schematic diagram of the smooth NP junction in 8-*Pmmn* borophene. Note that the true tilted Dirac cone is along $y$ direction but $x$ direction. (**c**)The schematic diagram of the smooth NPN junction in 8-*Pmmn* borophene. Here, we choose $n = 6.25$ and $m = 12.5$ for the numerical calculations.

$$E_{\lambda,\mathbf{k}} = v_t p_y + \lambda v_x \sqrt{p_x^2 + \gamma_1^2 p_y^2}, \quad \gamma_1 = \frac{v_y}{v_x} \tag{2}$$

$$\psi_{\lambda,\mathbf{k}}(\mathbf{r}) = \frac{1}{\sqrt{2}} \begin{bmatrix} 1 \\ \lambda \dfrac{k_x + i\gamma_1 k_y}{\sqrt{k_x^2 + \gamma_1^2 k_y^2}} \end{bmatrix} e^{i\mathbf{k}\cdot\mathbf{r}} \tag{3}$$

Here, $\lambda = \pm 1$, denoting the conduction $(+1)$ and valence $(-1)$ band, respectively. For 8-*Pmmn* borophene, the shape of Fermi surface for the fixing energy is elliptical with eccentricity $e$ determined by $v_x$, $v_y$ and $v_t$, which differs from the circular shape with radius $E_F/\hbar v_F$ of graphene. We can rewrite Equation (2) in following way [39,57]:

$$\frac{p_x^2}{a_{\lambda,E}} + \frac{\left(p_y + c_{\lambda,E}\right)^2}{b_{\lambda,E}} = 1 \tag{4}$$

$$a_{\lambda,E} = \frac{v_y^2 E_{\lambda,\mathbf{k}}^2}{v_x^2 \left(v_y^2 - v_t^2\right)}, \quad b_{\lambda,E} = \frac{v_y^2 E_{\lambda,\mathbf{k}}^2}{\left(v_y^2 - v_t^2\right)^2}, \quad c_{\lambda,E} = \frac{v_t E_{\lambda,\mathbf{k}}}{\left(v_y^2 - v_t^2\right)} \tag{5}$$

The eccentricity of the Fermi surface can be determined by $e = \sqrt{v_x^2 - v_y^2 + v_t^2}/v_x$. As a direct consequence, the eccentricity is not depend on the energy and the center of ellipse is at

$$\hbar k_x \;=\; 0, \quad \hbar k_y = -\frac{v_t E_{\lambda,\mathbf{k}}}{\left(v_y^2 - v_t^2\right)} \tag{6}$$

Notice that the center of ellipse is not at the origin and it moves with increasing the Fermi levels. In a NP junction setup, the translation symmetry preserves along the $y$

axis, so the $k_y$ is always a good quantum number. When the momentum $p_y$ is given, the $p_x$ in different regions of 8-*Pmmn* borophene NP junction is

$$p_x = \pm \frac{1}{v_x} \sqrt{\left(E_{\lambda,\mathbf{k}} - v_t p_y\right)^2 - \left(v_y p_y\right)^2} \tag{7}$$

Like the graphene NP junction, one can implement a bipolar NP junction or tunable NPN-type potential barriers in 8-*Pmmn* borophene by top/back gate voltages, and the potential function of the NP junction (as depicted in Figure 1b) has the form:

$$U_{NP}(x) = \begin{cases} V_0 & , & x > na/2 \\ 2V_0 x/na & , & na/2 \leq x \leq na/2 \\ -V_0 & , & x < -na/2 \end{cases} \tag{8}$$

where $a = \hbar v_F / 0.04$ eV is a unit length and $n > 0 \wedge n \in \mathbb{R}$. The NPN junction depicted in Figure 1c has the form

$$U_{NPN}(x) = \begin{cases} -V_0 & , & 3na/2 + ma & < & x \\ -2V_0(x - ma - na)/na & , & na/2 + ma & \leq & x & \leq & 3na/2 + ma \\ V_0 & , & na/2 & < & x & < & na/2 + ma \\ 2V_0 x/na & , & -na/2 & \leq & x & \leq & na/2 \\ -V_0 & , & & & x & < & -na/2 \end{cases} \tag{9}$$

where $m > 0 \wedge m \in \mathbb{R}$. Next, we will utilize the transfer matrix method to solve the ballistic transport problem in smooth NP/NPN junctions of 8-*Pmmn* borophene.

## 2.2. Transfer Matrix Method

The transfer matrix method is a powerful tool in the analysis of quantum transport of the massless fermions in 2D Dirac materials [18,58,59]. The central idea lies in that the wave function in one position can be related to those in other positions through a transfer matrix [60].

We adopt a transfer matrix method to study quantum transport in the smooth NP or NPN junction in 8-*Pmmn* borophene. There are two different matrices in transfer matrix method: one is the transmission matrix and the other is the propagating matrix. Transmission matrix connects the electrons across an interface and the propagating matrix connects the electrons propagating over a distance in the homogeneous regions. As we can see below, the propagating matrix can be derived by the transmission matrix. We define the transmission matrix $T$ as follows:

$$T \begin{pmatrix} A_{R_{m+1}} \\ A_{L_{m+1}} \end{pmatrix} = \begin{pmatrix} A_{R_m} \\ A_{L_m} \end{pmatrix} \tag{10}$$

where $A_{R_m}$ ($A_{L_m}$) represents the right (left) traveling wave amplitude in $m$ region. The transmission matrix connects the wave function's amplitude of two different regions. The condition of connecting amplitude coefficients between adjacent regions is the continuity of the wave functions at the interface. We can treat the smooth potential as the sum of infinite slices of junctions and figure out the wave function from the Schrödinger equation. Since the energy dispersion of 8-*Pmmn* borophene is linear, we only need the continuity condition of the wave functions at the interface. Then, the transmission matrices $T$ can be constructed from matrices $M$ of each slice,

$$M(k_{m+1}, x_m) \begin{pmatrix} A_{R_{m+1}} \\ A_{L_{m+1}} \end{pmatrix} = M(k_m, x_m) \begin{pmatrix} A_{R_m} \\ A_{L_m} \end{pmatrix}$$

$$M(k_m, x_m)^{-1} M(k_{m+1}, x_m) \begin{pmatrix} A_{R_{m+1}} \\ A_{L_{m+1}} \end{pmatrix} = \begin{pmatrix} A_{R_m} \\ A_{L_m} \end{pmatrix}$$

$$M(k_m, x_m)^{-1} M(k_{m+1}, x_m) = T$$

Suppose that an $n$-doped region $m$ is next to a $p$-doped region $m+1$ and carriers go through from $n$-doped region to $p$-doped region like in Figure 2a, the wave functions at interface can be connected in the way of

$$\frac{A_{R_{m+1}}}{\sqrt{2}}\begin{pmatrix} 1 \\ e^{i\theta_{m+1}} \end{pmatrix}e^{ik_{z,m+1}x_m+ik_yy} + \frac{A_{L_{m+1}}}{\sqrt{2}}\begin{pmatrix} 1 \\ -e^{-i\theta_{m+1}} \end{pmatrix}e^{-ik_{z,m+1}x_m+ik_yy}$$

$$= \frac{A_{R_m}}{\sqrt{2}}\begin{pmatrix} 1 \\ e^{-i\theta_{m+1}} \end{pmatrix}e^{-ik_{z,m}x_m+ik_yy} + \frac{A_{L_m}}{\sqrt{2}}\begin{pmatrix} 1 \\ -e^{i\theta_{m+1}} \end{pmatrix}e^{ik_{z,m}x_m+ik_yy}. \tag{11}$$

Here, we define $k_{x,m}(x)$ and $\theta_m$ as

$$k_{x,m}(x) = \frac{1}{\hbar v_x}\sqrt{\left(-U_m(x)+\hbar v_t k_y\right)^2 - \left(\hbar v_y k_y\right)^2} \tag{12}$$

$$e^{i\theta_m} = \frac{k_{x,m}+i\gamma_1 k_y}{\sqrt{k_{x,m}^2+\gamma_1^2 k_y^2}} \tag{13}$$

where $U_m(x)$ is the doping level in $m$ region and $k_x$ may take positive or negative imaginary values when $\left(-U_m(x)+\hbar v_t k_y\right)^2 - \left(\hbar v_y k_y\right)^2 < 0$. The phase $e^{i\theta_m}$ in Equation (11) is defined as the wave function phase difference between the two sublattices. The sign of the $k_x$ defines the propagating direction of the carriers. Without loss of generality, we can take only positive imaginary value for the transmission matrix, which means the positive propagating direction of electrons is defined on right-going state. Here, the potential profile $U_m(x)$ in adjacent regions within NP junction is linear but not rectangular; we treat the potential as a series of step potential to solve the tunneling problems by the transmission matrices. For convenience, we choose $a = \hbar v_F/0.04$ eV to be the length unit and 0.01 eV to be the energy unit, where 0.04 eV is the maximum of the doping level.

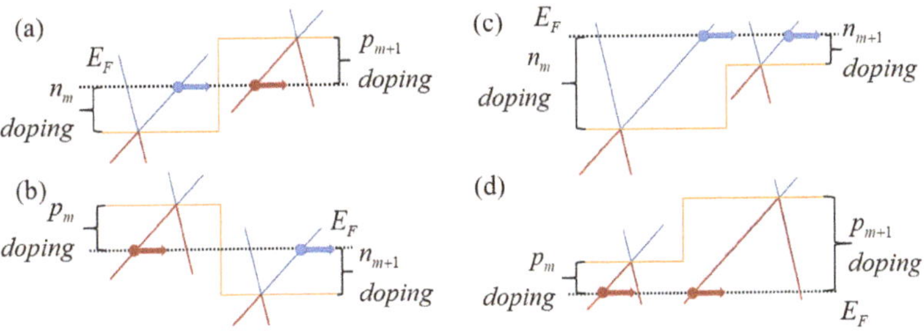

**Figure 2.** Potential profile of (**a**) NP junction, (**b**) PN junction, (**c**) NN junction, and (**d**) PP junction in each slice of the junctions.

Then, we rewrite the Equation (11) to construct the transmission matrices

$$\begin{pmatrix} e^{-ik_{x,m+1}x_m} & e^{ik_{x,m+1}x_m} \\ e^{-i\theta_{m+1}}e^{-ik_{x,m+1}x_m} & -e^{i\theta_{m+1}}e^{ik_{x,m+1}x_m} \end{pmatrix}\begin{pmatrix} A_{R_{m+1}} \\ A_{L_{m+1}} \end{pmatrix}$$

$$= \begin{pmatrix} e^{ik_{x,m}x_m} & e^{-ik_{x,m}x_m} \\ e^{i\theta_m}e^{ik_{x,m}x_m} & -e^{-i\theta_m}e^{-ik_{x,m}x_m} \end{pmatrix}\begin{pmatrix} A_{R_m} \\ A_{L_m} \end{pmatrix}.$$

Therefore, the transmission matrix between $m$ and $m+1$ region is

$$T_{m,m+1}^{n\to p} = \begin{pmatrix} e^{ik_{x,m}x_m} & e^{-ik_{x,m}x_m} \\ e^{i\theta_m}e^{ik_{x,m}x_m} & -e^{-i\theta_m}e^{-ik_{x,m}x_m} \end{pmatrix}^{-1} \begin{pmatrix} e^{-ik_{x,m+1}x_m} & e^{ik_{x,m+1}x_m} \\ e^{-i\theta_{m+1}}e^{-ik_{x,m+1}x_m} & -e^{i\theta_{m+1}}e^{ik_{x,m+1}x_m} \end{pmatrix} \tag{14}$$

while the transmission matrices of the carriers going through from $p$-doped region $m$ to $n$-doped region $m+1$ and between two $n$-doped or $p$-doped region (shown in Figure 2) are

$$
T^{p\to n}_{m,m+1} = \begin{pmatrix} e^{-ik_{x,m}x_m} & e^{ik_{x,m}x_m} \\ e^{-i\theta_m}e^{-ik_{x,m}x_m} & -e^{i\theta_m}e^{ik_{x,m}x_m} \end{pmatrix}^{-1} \begin{pmatrix} e^{ik_{x,m+1}x_m} & e^{-ik_{x,m+1}x_m} \\ e^{i\theta_{m+1}}e^{ik_{x,m+1}x_m} & -e^{-i\theta_{m+1}}e^{-ik_{x,m+1}x_m} \end{pmatrix} \tag{15}
$$

$$
T^{n\to n}_{m,m+1} = \begin{pmatrix} e^{ik_{x,m}x_m} & e^{-ik_{x,m}x_m} \\ e^{i\theta_m}e^{ik_{x,m}x_m} & -e^{-i\theta_m}e^{-ik_{x,m}x_m} \end{pmatrix}^{-1} \begin{pmatrix} e^{ik_{x,m+1}x_m} & e^{-ik_{x,m+1}x_m} \\ e^{i\theta_{m+1}}e^{ik_{x,m+1}x_m} & -e^{-i\theta_{m+1}}e^{-ik_{x,m+1}x_m} \end{pmatrix} \tag{16}
$$

$$
T^{p\to p}_{m,m+1} = \begin{pmatrix} e^{-ik_{x,m}x_m} & e^{ik_{x,m}x_m} \\ e^{-i\theta_m}e^{-ik_{x,m}x_m} & -e^{i\theta_m}e^{ik_{x,m}x_m} \end{pmatrix}^{-1} \begin{pmatrix} e^{-ik_{x,m+1}x_m} & e^{ik_{x,m+1}x_m} \\ e^{-i\theta_{m+1}}e^{-ik_{x,m+1}x_m} & -e^{i\theta_{m+1}}e^{ik_{x,m+1}x_m} \end{pmatrix} \tag{17}
$$

For the case of NPN junction, a trapezoidal potential profile as in Figure 1c, we can also treat the trapezoidal potential into infinite slices of connected step potentials. The transmission matrices define at the interface between each step potentials. Multiplying all the transmission matrices would give the propagation matrices,

$$
T_{all} = T^{n\to n}_{0,1}T^{n\to n}_{1,2}\dots T^{n\to n}_{k-1,k}T^{n\to p}_{k,k+1}T^{p\to p}_{k+1,k+2}\dots \times
$$
$$
T^{p\to p}_{k'-1,k'}T^{p\to p}_{k',k'+1}\dots T^{p\to p}_{k''-1,k}T^{p\to n}_{k'',k''+1}T^{n\to n}_{k''+1,k''+2}\dots T^{n\to n}_{m-2,m-1}T^{n\to n}_{m-1,m} \tag{18}
$$

Then, we reach the formula

$$
T_{all}\begin{pmatrix} A_{R_m} \\ A_{L_m} \end{pmatrix} = \begin{pmatrix} A_{R_0} \\ A_{L_0} \end{pmatrix} \tag{19}
$$

When incident electrons go from the leftmost side of the NPN junction to the rightmost side, there are no reflection states in the rightmost side, i.e., $A_{L_m} = 0$. We can connect the amplitude of incident states to the amplitude of reflection states

$$
\begin{pmatrix} T_{11} & T_{12} \\ T_{21} & T_{22} \end{pmatrix}\begin{pmatrix} A_{R_m} \\ 0 \end{pmatrix} = \begin{pmatrix} A_{R_0} \\ A_{L_0} \end{pmatrix}
$$
$$
\frac{A_{R_m}}{A_{R_0}} = \frac{1}{T_{11}}
$$

Finally, the transmission probability is $T = |t|^2 = |A_{R_m}/A_{R_0}|^2 = |1/T_{11}|^2$.

There is a trick in constructing the propagation matrices from the transmission matrices. As shown in Figure 3, the incident states at the left-hand side of the junction have a different Fermi surface from the transmitted states at the right-hand side in the NP junction. Suppose the NP junction is sharp. The good quantum number $k_y$ should be restricted between the top dotted green line and the middle dotted green line, since the incident states and the transmitted states are propagating only in this scenario. While supposing the NP junction is smooth, the Fermi surface in the region of varying potential would shrink to the Dirac point, and the $E_{\lambda,k}$, $a_{\lambda,E}$, $b_{\lambda,E}$, and $c_{\lambda,E}$ from Equation (5) reduce to zero as well. Therefore, $k_x$ vanishes to diverge the transmission matrices when the carriers approaching the NP junction center. However, we could play a trick by properly segmenting the region of varying potential and jumping the diverging point. The trick lies in the fact that the carriers would not experience any singularity when going through an infinitesimal interval around the diverging point. For instance, the transmission matrix at the Dirac point cannot be well defined with incident states $k_y = 0$, whereas the carriers are well-defined decay states at the Dirac point. We can ignore the decay states of the carriers going through infinitesimal intervals around the Dirac point, and it would eliminate any possible ambiguity.

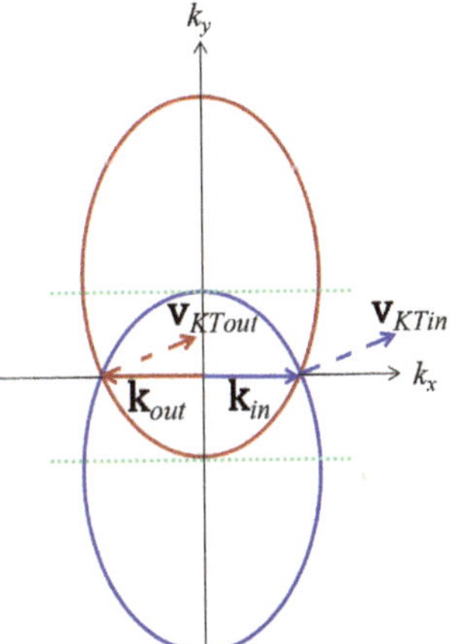


**Figure 3.** Fermi surface at different doped regions $\pm\varepsilon_{doping}$. The blue (red) ellipse represents the electron (hole) Fermi surface in $n$-doped ($p$-doped) region. The solid vectors $\mathbf{k}_{in}$ (blue) and $\mathbf{k}_{out}$ (red) are the wave vector of incident carriers and transmitted carriers, respectively. The dashed vectors $\mathbf{v}_{KTin}$ (blue) and $\mathbf{v}_{KTout}$ (red) are the group velocity of incident carriers and transmitted carriers, respectively. The green dotted lines indicate the values of the good quantum number $k_y$ posed restrictions for the NP junction and the NPN junction.

## 3. Results and Discussions

In this section, we present the numerical results for the transmission probability and electrical conduction of the massless Dirac fermions across the borophene NP junction and NPN junction.

### 3.1. The Oblique Klein Tunneling in Smooth NP Junctions

Various smooth NP junctions with fixing $n/p$ doping level but different slopes are depicted in Figure 4a. We set the length of the varying region in different NP junctions as 6.25 $a$, 12.5 $a$, 25 $a$, and 50 $a$, respectively, where $a = \hbar v_F/0.04$ eV, and plot the angular transmission probability for different NP junctions. As shown in Figure 4b, the shaper the NP junction is, the wider the angular transmission probability spans. This phenomenon is caused by the decay states in the varying region and is similar to the graphene smooth NP junction. In the varying region, $(-U_m(x) + \hbar v_t k_y)^2 - (\hbar v_y k_y)^2 < 0$, so that the propagating states degenerate to the decaying states when the carriers gradually approach the junction's center. Therefore, the transmission probability increases with increasing the slope of potential in the varying region. If we take $k_y = 0$, i.e., the normal incident case, we can see the perfect transmission, the Klein tunneling.

Figure 5 shows that the angular transmission amplitude of the $k$ vector is different from the one of group velocity. The actual incident angle across the junction is based on the group velocity of carriers. The actual angular transmission probability for group velocity shown in Figure 5 indicates a rotation of the Klein tunneling, the oblique Klein tunneling. It means that the perfect transmission does not occur in the normal incident but with a nonzero angle $\theta_K$.

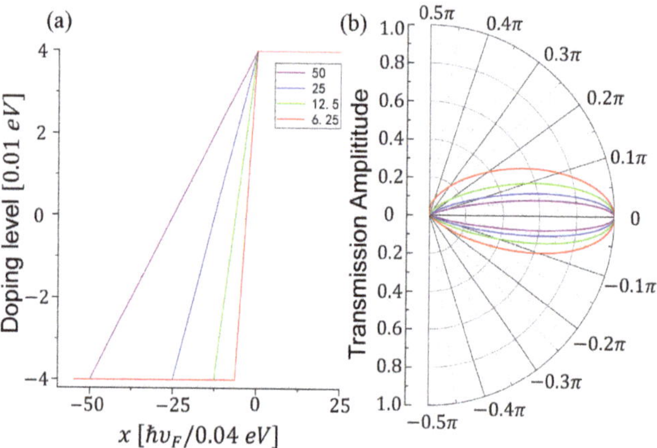

**Figure 4.** (a) Potential profile of smooth NP junctions and (b) the angular behavior of the transmission probability for different NP junctions corresponding to different colors at (a).

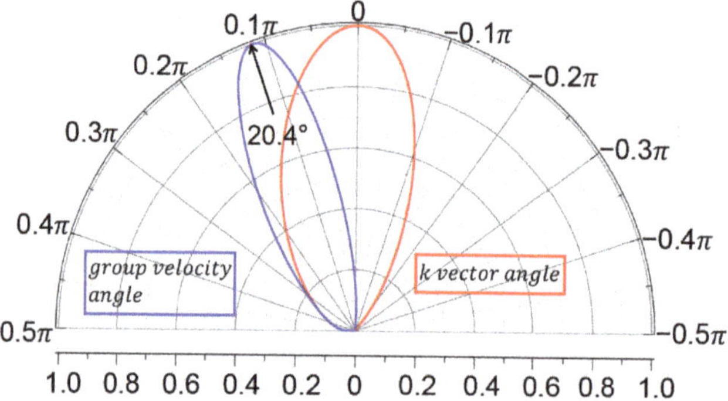


**Figure 5.** Angular transmission amplitude for *k* vector (red) and for group velocity (blue). The doping level is 0.04 eV and the length of varying region is 6.25 *a*.

The value of $\theta_K$ can be determined from the elliptical Fermi surface of 8-*Pmmn* borophene. The angle for the group velocity is $\theta_v = \arctan\left[v_y(\varepsilon, k_y)/v_x(\varepsilon, k_y)\right]$, where $v_y(\varepsilon, k_y)$ and $v_x(\varepsilon, k_y)$ can be obtained by

$$v_x(\varepsilon, k_y) \;=\; \frac{\partial E_{\lambda,\mathbf{k}}}{\hbar \partial k_x} = \frac{\lambda k_x v_x}{\sqrt{k_x^2 + \gamma_1^2 k_y^2}} \tag{20}$$

$$v_y(\varepsilon, k_y) \;=\; \frac{\partial E_{\lambda,\mathbf{k}}}{\hbar \partial k_y} = v_t + \frac{\lambda \gamma_1^2 k_y v_x}{\sqrt{k_x^2 + \gamma_1^2 k_y^2}} \tag{21}$$

Combined with above equations and let $k_y = 0$, we can find the angle of Klein tunneling for group velocity,

$$\theta_K = \arctan\left(\frac{v_t}{v_x}\right) \approx 20.4° \tag{22}$$

This oblique Klein tunneling can also be found in sharp NP junctions of 8-*Pmmn* borophene [39,53].

### 3.2. The Asymmetric Klein Tunneling in the Smooth NPN Junctions

The NPN junction, as shown in the Figure 1c, can be seen as a trapezoid potential barrier. We set the length of the varying regions as 6.25 *a* and the length of the flat potential barrier as 12.5 *a*.

In Figure 6, we plot the transmission probability depending on different doping levels and $k_y$. Note that *n*- and *p*-regions have the same absolute value of doping level. We can see the Klein tunneling in several branches. The number of branches increases by lifting the doping level, which could also be observed in the graphene NPN junctions [16,18] We can see the Klein tunneling is asymmetric. The asymmetric Klein tunneling results from the carriers' chirality and anisotropy [61]. It is not surprising to see it here because the carriers of 8-*Pmmn* have both chirality and anisotropy.

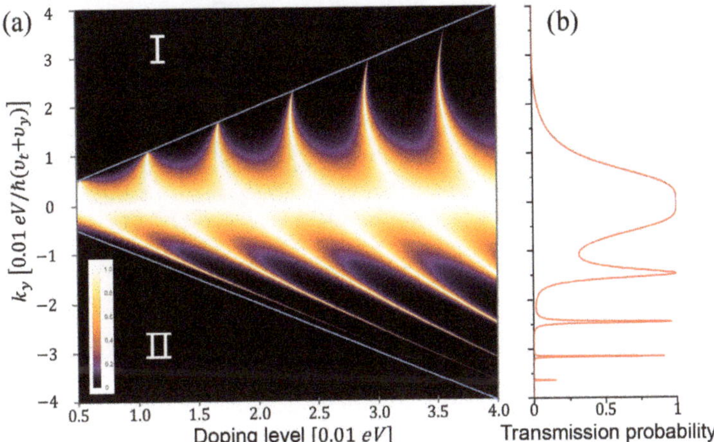

**Figure 6.** (**a**) Transmission probability versus the doping level and the $k_y$ in NPN junction. Blue lines denote the forbidden zone, where transmission probability vanishes, and there are only the decaying states in the *p*-doped region. (**b**) The transmission probability depending on $k_y$ when the doping level is $4 \times 0.01$ eV.

The blue lines in Figure 6a denote the forbidden zones, where the transmission probability vanishes. The equation of the boundary of the forbidden zone is $k_y = \pm \varepsilon_{doping}/\hbar(v_t + v_y)$. There are two types of the forbidden zone: (I) the no-incident zone and (II) the vanishing transmitted zone. In the no-incident zone $k_y \geq \varepsilon_{doping}/\hbar(v_t + v_y)$, there is no incident states since the parameters $k_y$ and doping level is beyond the Dirac cone; in the vanishing transmitted zone $k_y \leq -\varepsilon_{doping}/\hbar(v_t + v_y)$, the transmitted carriers severely decay in the region of barrier.

Next, we fix the bottom edge and the height of the trapezoid potential (NPN junction) and plot the transmission probability versus the potential's top edge. When the top edge's length varies from 0 to the bottom edge's length, the NPN junction experiences a change from triangle potential to trapezoid potential and finally to a square potential. We can see from Figure 7, the number of branches increases with increasing the top edge's length. It is somehow counterintuitive that the square potential favors Klein tunneling more than the triangle potential. The reason is that the carriers would have more chances to degenerate to decaying states when incident into a slope of potential, in fact, a smooth NP junction.

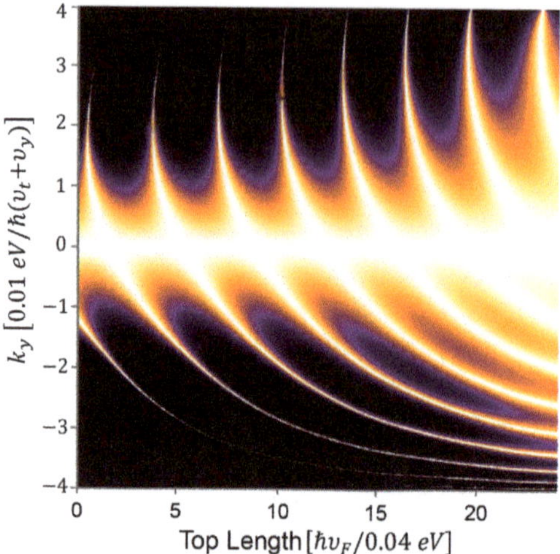

**Figure 7.** Transmission probability depends on top edge of the trapezoid potentials. The top edge varies from 0 to 24 *a* and the bottom edge is fixed as 25 *a*. The height of the trapezoid potentials or the absolute value of n/p doping level is fixed as 0.04 eV.

### 3.3. The Electrical Resistance of the Smooth NPN Junctions

One can create the NPN junction by implementing a design with two electrostatic gates, a global back gate and a local top gate. A back voltage applied to the back gate could tune the carrier density in the borophene sheet, whereas a top voltage applied to the top gate could tune the density only in the narrow strip below the gate. These two gates can be controlled independently [62].

To clarify the effect of the Klein tunneling on the transport property, here, we discuss the electrical conduction of the NPN junction in 8-*Pmmn* borophene. In the ballistic regime, we apply the Landauer–Buttiker formula $G = 2e^2 MT/h$ to calculate the electrical conductance [63]. In our setup, the Landauer formula can be written as [64]

$$G_{fet} = \frac{4e^2}{h}\sum_{ch.} T_{ch} \approx \frac{4e^2}{h}\int_{k_{y\,min}}^{k_{y\,max}} \frac{dk_y}{2\pi/W} T(k_y) \tag{23}$$

where $k_{y\,max} = \varepsilon_{doping}/\hbar(v_t + v_y)$ and $k_{y\,min} = \varepsilon_{doping}/\hbar(v_t - v_y)$.

We choose the width of the junction $W = 10$ μm and calculate the electrical resistance by the Landauer formula. To reveal the link of the resistance with the Klein tunneling, we plot the transmission probability versus the doping level in Figure 8a and the resistance depending on doping levels in Figure 8b. We can see the resistance oscillation when increasing the doping level from 0 to 0.08 eV. The oscillation pattern indicates the effect of the Klein tunneling. When the doping level varies from 0 to −0.08 eV, the NPN junction becomes a NNN junction so that the curves of resistance are flat in the negative doping regime.

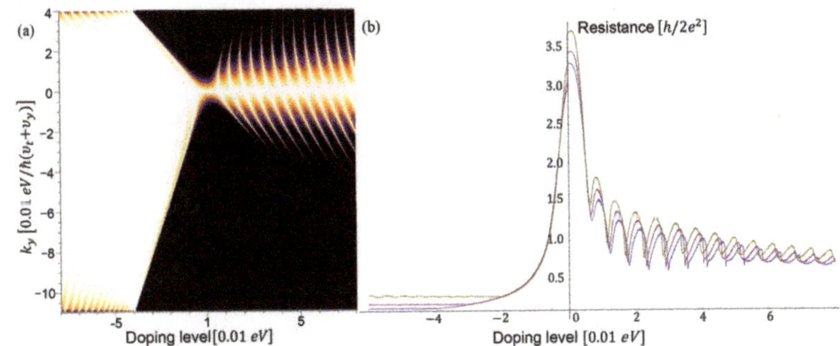

**Figure 8.** (a) Transmission probability depends on $k_y$ and the height of the trapezoid potentials (doping levels of the NPN junctions). The top edge's length is 12.5 $a$ and the bottom edge's length is 25 $a$. The doping level of $n$-doped region (outside the NPN junction) is set $-0.04$ eV. (b) The electrical resistance of the NPN junction depending on the doping level.

## 4. Conclusions

This work investigates the transport properties of massless fermions in the smooth 8-*Pmmn* borophene NP and NPN junctions by the transfer matrix method. Compare with the sharp junction, the smooth NP junction also shows that the oblique Klein tunneling induced by the tilted Dirac cones. We can calculate from the parameters of the Hamiltonian that the angle of oblique Klein tunneling is 20.4°. We also show the branches of the NPN tunneling in the phase diagram, which indicates the asymmetric Klein tunneling. The physical origin of the asymmetric Klein tunneling lies in the chirality and anisotropy of the carriers, and we can verify the asymmetric Klein tunneling experimentally by analyzing the pattern of the electrical resistance oscillation. For the oblique Klein tunneling, we have discussed the experimental feasibility in detail in our previous study [39]. The present numerical demonstration in smooth junctions proves the effectiveness of our previous discussion and favors the observation in future experiments.

**Author Contributions:** Conceptualization, J.-J.Z. and S.-H.Z.; investigation, Z.K., S.-H.Z., and J.-J.Z.; writing—original draft preparation, Z.K.; writing—review and editing, all authors; visualization, Z.K.; supervision, J.-J.Z. and S.-H.Z.; funding acquisition, J.-J.Z., J.L., and Y.Z. All authors have read and agreed to the published version of the manuscript.

**Funding:** This work was supported by the Scientific Research Program from Science and Technology Bureau of Chongqing City (Grant No. cstc2020jcyj-msxm0925, cstc2020jcyj-msxmX0810), the Science and Technology Research Program of Chongqing Municipal Education Commission (Grant No. KJQN202000639), and the key technology innovations project to industries of Chongqing (cstc2016zdcy-ztzx0067).

**Data Availability Statement:** The study did not report any data.

**Conflicts of Interest:** The authors declare no conflict of interest.

## References

1. Neto, A.H.C.; Guinea, F.; Peres, N.M.R.; Novoselov, K.S.; Geim, A.K. The electronic properties of graphene. *Rev. Mod. Phys.* **2009**, *81*, 109–162. [CrossRef]
2. Xiao, D.; Liu, G.-U.; Feng, W.-A.; Xu, X.-I.; Yao, W. Coupled Spin and Valley Physics in Monolayers of MoS2 and Other Group-VI Dichalcogenides. *Phys. Rev. Lett.* **2012**, *108*, 196802. [CrossRef]
3. Wang, Q.-I.; Kalantar-Zadeh, K.; Kis, A.; Coleman, J.N.; Strano, M.S. Electronics and optoelectronics of two-dimensional transition metal dichalcogenides. *Nat. Nanotechnol.* **2012**, *7*, 699–712. [CrossRef]
4. Zhang, K.-E.; Zhang, T.-I.; Cheng, G.-U.; Li, T.-I.; Wang, S.; Wei, W.; Zhou, X.-I.; Yu, W.-E.; Sun, Y.; Wang, P.; et al. Interlayer Transition and Infrared Photodetection in Atomically Thin Type-II MoTe2/MoS2 van der Waals Heterostructures. *ACS Nano* **2016**, *10*, 3852–3858. [CrossRef]

5. Du, Y.-L.; Ouyang, C.-Y.; Shi, S.-Q.; Lei, M.-S. Electronic Ab initio studies on atomic and electronic structures of black phosphorus. *J. Appl. Phys.* **2010**, *107*, 093718. [CrossRef]

6. Li, L.-K.; Kim, J.; Jin, C.-H.; Ye, G.-J.; Qiu, D.Y.; Felipe, H.; Shi, Z.-W.; Chen, L.; Zhang, Z.-C.; Yang, F.-Y.; et al. Direct observation of the layer-dependent electronic structure in phosphorene. *Nat. Nanotechnol.* **2017**, *12*, 21–25. [CrossRef] [PubMed]

7. Li, L.-K.; Yang, F.-Y.; Ye, G.-J.; Zhang, Z.-C.; Zhu, Z.-W.; Lou, W.-K.; Zhou, X.-Y.; Li, L.; Watanabe, K.; Taniguchi, T.; et al. Quantum Hall effect in black phosphorus two-dimensional electron system. *Nat. Nanotechnol.* **2016**, *11*, 593–597.

8. Zhou, X.-Y.; Lou, W.-K.; Zhai, F.; Chang, K. Anomalous magneto-optical response of black phosphorus thin films. *Phys. Rev. B* **2015**, *92*, 165405. [CrossRef]

9. Tamalampudi, S.R.; Lu, Y.-Y.; Sankar, R.K.U.R.; Liao, C.-D.; Cheng, K.M.B.C.; Chou, F.-C.; Chen, Y.-T. High Performance and Bendable Few-Layered InSe Photodetectors with Broad Spectral Response. *Nano Lett.* **2014**, *14*, 2800–2806. [CrossRef] [PubMed]

10. Brotons-Gisbert, M.; Andres-Penares, D.; Suh, J.; Hidalgo, F.; Abargues, R.; Rodríguez-Cantó, P.J.; Segura, A.; Cros, A.; Tobias, G.; Canadell, E.; et al. Nanotexturing To Enhance Photoluminescent Response of Atomically Thin Indium Selenide with Highly Tunable Band Gap. *Nano Lett.* **2016**, *16*, 3221–3229. [CrossRef]

11. Bandurin, D.A.; Tyurnina, A.V.; Yu, G.L.; Mishchenko, A.; Zólyomi, V.; Morozov, S.V.; Kumar, R.K.; Gorbachev, R.V.; Kudrynskyi, Z.R.; Pezzini, S.; et al. High electron mobility, quantum Hall effect and anomalous optical response in atomically thin InSe. *Nat. Nanotechnol.* **2017**, *12*, 223–227. [CrossRef]

12. Xu, Y.; Yan, B.-H.; Zhang, H.-J.; Wang, J.; Xu, G.; Tang, P.-Z.; Duan, W.-H.; Zhang, S.-C. Large-Gap Quantum Spin Hall Insulators in Tin Films. *Phys. Rev. Lett.* **2013**, *111*, 136804. [CrossRef] [PubMed]

13. Zhu, F.-F.; Chen, W.-J.; Xu, Y.; Gao, C.-L.; Guan, D.-D.; Liu, C.-H.; Qian, D.; Zhang, S.-C.; Jia, J.-F. Epitaxial growth of two-dimensional stanene. *Nat. Mater.* **2015**, *14*, 1020–1025. [CrossRef] [PubMed]

14. Novoselov, K.S.; Mishchenko, A.; Carvalho, A.; Neto, A.H.C. 2D materials and van der Waals heterostructures. *Science* **2016**, *353*, 9493. [CrossRef] [PubMed]

15. Schaibley, J.R.; Yu, H.-Y.; Clark, G.; Rivera, P.; Ross, J.S.; Seyler, K.L.; Yao, W.; Xu, X.-D. Valleytronics in 2D materials. *Nat. Rev. Mater.* **2016**, *1*, 1–15. [CrossRef]

16. Beenakker, C.W.J. Colloquium: Andreev reflection and Klein tunneling in graphene. *Rev. Mod. Phys.* **2008**, *80*, 1337–1354. [CrossRef]

17. Wu, Z.-H.; Peeters, F.M.; Chang, K. Electron tunneling through double magnetic barriers on the surface of a topological insulator. *Phys. Rev. B* **2010**, *82*, 115211. [CrossRef]

18. Bai, C.-X.; Zhang, X.-X. Klein paradox and resonant tunneling in a graphene superlattice. *Phys. Rev. B* **2007**, *76*, 075430. [CrossRef]

19. Pereira, J.M., Jr.; Peeters, F.M.; Chaves, A.; Farias, G.A. Klein tunneling in single and multiple barriers in graphene. *Semicond. Sci. Technol.* **2010**, *25*, 033002. [CrossRef]

20. Oh, H.; Coh, S.; Son, Y.-W.; Cohen, M.L. Inhibiting Klein Tunneling in a Graphene p-n Junction without an External Magnetic Field. *Phys. Rev. Lett.* **2016**, *117*, 016804. [CrossRef]

21. Cheianov, V.V.; Fal'ko, V.I. Selective transmission of Dirac electrons and ballistic magnetoresistance of n-p junctions in graphene. *Phys. Rev. B* **2006**, *74*, 041403. [CrossRef]

22. Zhou, X.-F.; Dong, X.; Oganov, A.R.; Zhu, Q.; Tian, Y.-J.; Wang, H.-T. Semimetallic Two-Dimensional Boron Allotrope with Massless Dirac Fermions. *Phys. Rev. Lett.* **2014**, *112*, 085502. [CrossRef]

23. Mannix, A.J.; Zhou, X.-F.; Kiraly, B.; Wood, J.D.; Alducin, D.; Myers, B.D.; Liu, X.-L.; Fisher, B.L.; Santiago, U.; Guest, J.R.; et al. Synthesis of borophenes: Anisotropic, two-dimensional boron polymorphs. *Science* **2015**, *350*, 1513–1516. [CrossRef]

24. Jiao, Y.-L.; Ma, F.-X.; Bell, J.; Bilic, A.; Du, A.-J. Two-Dimensional Boron Hydride Sheets: High Stability, Massless Dirac Fermions, and Excellent Mechanical Properties. *Angew. Chem.* **2016**, *128*, 10448–10451. [CrossRef]

25. Kunstmann, J.; Quandt, A.r. Broad boron sheets and boron nanotubes: An ab initio study of structural, electronic, and mechanical properties. *Phys. Rev. B* **2006**, *74*, 035413. [CrossRef]

26. Dressellhaus, M.S.; Dresselhaus, G.; Jorio, A. *Group Theory: Application to the Physics of Condensed Matter*; Springer: Berlin/Heidelberg, Germany, 2008; pp. 190–191.

27. Feng, B.-J.; Sugino, O.; Liu, R.-Y.; Zhang, J.; Yukawa, R.; Kawamura, M.; Iimori, T.; Kim, H.-W.; Hasegawa, Y.; Li, H.; et al. Dirac Fermions in Borophene. *Phys. Rev. Lett.* **2017**, *118*, 096401. [CrossRef]

28. Lopez-Bezanilla, A.; Littlewood, P.B. Electronic properties of 8-*Pmmn* borophene. *Phys. Rev. B* **2016**, *93*, 241405. [CrossRef]

29. Zabolotskiy, A.D.; Lozovik, Y.E. Strain-induced pseudomagnetic field in the Dirac semimetal borophene. *Phys. Rev. B* **2016**, *94*, 165403. [CrossRef]

30. Sadhukhan, K.; Agarwal, A. Anisotropic plasmons, Friedel oscillations, and screening in 8-*Pmmn* borophene. *Phys. Rev. B* **2017**, *96*, 035410. [CrossRef]

31. Jalali-Mola, Z.; Jafari, S.A. Tilt-induced kink in the plasmon dispersion of two-dimensional Dirac electrons. *Phys. Rev. B* **2018**, *98*, 195415. [CrossRef]

32. Jalali-Mola, Z.; Jafari, S.A. Kinked plasmon dispersion in borophene-borophene and borophene-graphene double layers. *Phys. Rev. B* **2018**, *98*, 235430. [CrossRef]

33. Lian, C.; Hu, S.-Q.; Zhang, J.; Cheng, C.; Yuan, Z.; Gao, S.-W.; Meng, S. Integrated Plasmonics: Broadband Dirac Plasmons in Borophene. *Phys. Rev. Lett.* **2020**, *125*, 116802. [CrossRef] [PubMed]

34. Verma, S.; Mawrie, A.; Ghosh, T.K. Effect of electron-hole asymmetry on optical conductivity in 8-*Pmmn* borophene. *Phys. Rev. B* **2017**, *96*, 155418. [CrossRef]

35. Mojarro, M.A.; Carrillo-Bastos, R.; Maytorena, J.A. Optical properties of massive anisotropic tilted Dirac systems. *Phys. Rev. B* **2021**, *103*, 165415. [CrossRef]

36. Islam, S.K.F.; Jayannavar, A.M. Signature of tilted Dirac cones in Weiss oscillations of 8-*Pmmn* borophene. *Phys. Rev. B* **2017**, *96*, 235405. [CrossRef]

37. Nakhaee, M.; Ketabi, S.A.; Peeters, F.M. Tight-binding model for borophene and borophane. *Phys. Rev. B* **2018**, *97*, 125424. [CrossRef]

38. Singh, A.; Ghosh, S.; Agarwal, A. Nonlinear and anisotropic polarization rotation in two-dimensional Dirac materials. *Phys. Rev. B* **2018**, *97*, 205420. [CrossRef]

39. Zhang, S.-H.; Yang, W. Oblique Klein tunneling in 8-*Pmmn* borophene $p - n$ junctions. *Phys. Rev. B* **2018**, *97*, 235440. [CrossRef]

40. Zhou, X.-F. Valley-dependent electron retroreflection and anomalous Klein tunneling in an 8-*pmm* borophene-based $n - p - n$ junction. *Phys. Rev. B* **2019**, *100*, 195139. [CrossRef]

41. Zhou, X.-F. Valley splitting and anomalous Klein tunneling in borophane-based $n - p$ and $n - p - n$ junctions. *Phys. Lett. A* **2020**, *384*, 126612. [CrossRef]

42. Zhong, H.-X.; Huang, K.-X.; Yu, G.-D.; Yuan, S.-J. Electronic and mechanical properties of few-layer borophene. *Phys. Rev. B* **2018**, *98*, 054104. [CrossRef]

43. Nakhaee, M.; Ketabi, S.A.; Peeters, F.M. Dirac nodal line in bilayer borophene: Tight-binding model and low-energy effective Hamiltonian. *Phys. Rev. B* **2018**, *98*, 115413. [CrossRef]

44. Champo, A.E.; Naumis, G.G. Metal-insulator transition in 8-*Pmmn* borophene under normal incidence of electromagnetic radiation. *Phys. Rev. B* **2019**, *99*, 035415. [CrossRef]

45. Ibarra-Sierra, V.G.; Sandoval-Santana, J.C.; Kunold, A.; Naumis, G.G. Dynamical band gap tuning in anisotropic tilted Dirac semimetals by intense elliptically polarized normal illumination and its application to 8-*Pmmn* borophene. *Phys. Rev. B* **2019**, *100*, 125302. [CrossRef]

46. Paul, G.C.; Islam, S.K.F.; Saha, A. Fingerprints of tilted Dirac cones on the RKKY exchange interaction in 8-*Pmmn* borophene. *Phys. Rev. B* **2019**, *99*, 155418. [CrossRef]

47. Zhang, S.-H.; Shao, D.-I.; Yang, W. Velocity-determined anisotropic behaviors of RKKY interaction in 8-*Pmmn* borophene. *J. Magn. Magn. Mater.* **2019**, *491*, 165631. [CrossRef]

48. Zhang, S.-H.; Yang, W. Anomalous caustics and Veselago focusing in 8-*Pmmn* borophene p–n junctions with arbitrary junction directions. *New J. Phys.* **2019**, *21*, 103052. [CrossRef]

49. Gao, M.; Yan, X.-U.; Wang, J.; Lu, Z.-H.; Xiang, T. Electron-phonon coupling in a honeycomb borophene grown on Al (111) surface. *Phys. Rev. B* **2019**, *100*, 024503. [CrossRef]

50. Kapri, P.; Dey, B.; Ghosh, T.K. Valley caloritronics in a photodriven heterojunction of Dirac materials. *Phys. Rev. B* **2020**, *102*, 045417. [CrossRef]

51. Zheng, J.-I.; Lu, J.-U.; Zhai, F. Anisotropic and gate-tunable valley filtering based on 8-*Pmmn* borophene. *Phys. Rev. B* **2020**, *32*, 025205.

52. Zhou, X.-I. Anomalous Andreev reflection in an 8-*pmmn* borophene-based superconducting junction. *Phys. Rev. B* **2020**, *102*, 045132. [CrossRef]

53. Nguyen, V.H.; Charlier, J.C. Klein tunneling and electron optics in Dirac-Weyl fermion systems with tilted energy dispersion Superlattices. *Phys. Rev. B* **2018**, *97*, 235113. [CrossRef]

54. Zhang, S.-H.; Zhu, J.-I.; Yang, W.; Lin, H.-A.; Chang, K. Hidden quantum mirage by negative refraction in semiconductor P-N junctions. *Phys. Rev. B* **2016**, *94*, 085408. [CrossRef]

55. Zhang, S.-H.; Yang, W.; Peeters, F.M. Veselago focusing of anisotropic massless Dirac fermions. *Phys. Rev. B* **2018**, *97*, 205437. [CrossRef]

56. Hahn, T. *International Tables for Crystallography, Volume A*; Springer: Berlin/Heidelberg, Germany, 2005; p. 59.

57. Napitu, B.D. Photoinduced Hall effect and transport properties of irradiated 8-*Pmmn* borophene monolayer. *J. Appl. Phys.* **2020**, *128*, 039901. [CrossRef]

58. Li, H.-A.; Wang, L.; Lan, Z.-H.; Zheng, Y.-I. Generalized transfer matrix theory of electronic transport through a graphene waveguide. *Phys. Rev. B* **2009**, *79*, 155429. [CrossRef]

59. Zhang, L.B.; Chang, K.; Xie, X.-I.; Buhmann, H.; Molenkamp, L.W. Quantum tunneling through planar p–n junctions in HgTe quantum wells. *New J. Phys.* **2010**, *12*, 083058. [CrossRef]

60. Zhan, T.-I.; Shi, X.; Dai, Y.Y.; Liu, X.H.; Zi, J. Transfer matrix method for optics in graphene layers. *J. Phys. Condens. Matter.* **2013**, *25*, 215301. [CrossRef]

61. Li, Z.-H.; Cao, T.; Wu, M.; Louie, S.G. Generation of Anisotropic Massless Dirac Fermions and Asymmetric Klein Tunneling in Few-Layer Black Phosphorus Superlattices. *Nano Lett.* **2017**, *17*, 2280–2286. [CrossRef]

62. Huard, B.; Sulpizio, J.A.; Stander, N.; Todd, K.; Yang, B.; Goldhaber-Gordon, D. Transport Measurements Across a Tunable Potential Barrier in Graphene. *Phys. Rev. Lett.* **2007**, *98*, 236803. [CrossRef]

63. Datta, S. *Electronic Transport in Mesoscopic Systems*; Cambridge University Press: Cambridge, UK, 1999; pp. 48–110.

64. Allain, P.E.; Fuchs, J.N. Klein tunneling in graphene: Optics with massless electrons. *Eur. Phys. J. B* **2011**, *83*, 301–317. [CrossRef]

*nanomaterials*



*Article*

# Feature-Rich Geometric and Electronic Properties of Carbon Nanoscrolls

Shih-Yang Lin [1], Sheng-Lin Chang [2,*], Cheng-Ru Chiang [3], Wei-Bang Li [3], Hsin-Yi Liu [3] and Ming-Fa Lin [3,*]

1   Department of Physics, National Chung Cheng University, Chiayi 621, Taiwan; sylin.1985@gmail.com
2   Department of Electrophysics, National Chiao Tung University, Hsinchu 300, Taiwan
3   Department of Physics, National Cheng Kung University, Tainan 701, Taiwan;
    davidgo86@livemail.tw (C.-R.C.); weibang1108@gmail.com (W.-B.L.); buttid41@gmail.com (H.-Y.L.)
*   Correspondence: lccs38@hotmail.com (S.-L.C.); mflin@mail.ncku.edu.tw (M.-F.L.)

**Abstract:** How to form carbon nanoscrolls with non-uniform curvatures is worthy of a detailed investigation. The first-principles method is suitable for studying the combined effects due to the finite-size confinement, the edge-dependent interactions, the interlayer atomic interactions, the mechanical strains, and the magnetic configurations. The complex mechanisms can induce unusual essential properties, e.g., the optimal structures, magnetism, band gaps and energy dispersions. To reach a stable spiral profile, the requirements on the critical nanoribbon width and overlapping length will be thoroughly explored by evaluating the width-dependent scrolling energies. A comparison of formation energy between armchair and zigzag nanoscrolls is useful in understanding the experimental characterizations. The spin-up and spin-down distributions near the zigzag edges are examined for their magnetic environments. This accounts for the conservation or destruction of spin degeneracy. The various curved surfaces on a relaxed nanoscroll will create complicated multi-orbital hybridizations so that the low-lying energy dispersions and energy gaps are expected to be very sensitive to ribbon width, especially for those of armchair systems. Finally, the planar, curved, folded, and scrolled graphene nanoribbons are compared with one another to illustrate the geometry-induced diversity.

**Keywords:** graphene; nanoscroll; first-principle

**Citation:** Lin, S.-Y.; Chang, S.-L.; Chiang, C.-R.; Li, W.-B.; Liu, H.-Y.; Lin, M.-F. Feature-Rich Geometric and Electronic Properties of Carbon Nanoscrolls. *Nanomaterials* **2021**, *11*, 1372. https://doi.org/10.3390/nano11061372

Academic Editor: Eugene Kogan

Received: 8 April 2021
Accepted: 18 May 2021
Published: 22 May 2021

**Publisher's Note:** MDPI stays neutral with regard to jurisdictional claims in published maps and institutional affiliations.

## 1. Introduction

Condensed-matter systems purely made up of carbon atoms comprise diamond [1,2], few-layer graphenes [3–5], carbon nanotubes [6–8], graphene nanoribbons (GNR) [9–12], nanoscrolls [13–16] and C60-related fullerenes [17–19]. These systems exhibit very rich physical, chemical, and material properties, mainly owing to their special structural symmetries and varying dimensionality. Recently, one-dimensional carbon nanoscrolls (1D CNSs) have attracted much attention for their special geometric structure and electronic properties [13,14,17,18,20–30]. Each CNS can be regarded as a spirally-wrapped 2D graphene sheet with a 1D scroll structure. Unlike a carbon nanotube, which is a closed cylinder, a CNS is open at two edges. Clearly, CNSs possess flexible interlayer spaces to intercalate or to be susceptible to doping, indicating the high application potentials in hydrogen storage [24,25,31,32], lithium batteries [26,29,33,34], aluminum batteries [29], and mechanical devices [26,27]. However, regarding nanoscroll structures, the question remains whether they are perfectly spiral or not. The previous studies [35,36] on carbon nanotubes show that the non-cylindrical structures are more prone to exist in the large diameter cases due to the layer–layer interactions. Such effects are expected to play an important role in plastic CNSs. In this paper, we investigate the geometric and electronic properties of non-ideal CNSs, and these predicted results are innovative and interesting.

CNSs have been successfully produced by the different physical and chemical methods [13,14,17,18,21,22], including arc-discharge [17], high-energy ball milling of graphite [21]. and the chemical route. However, theoretical researches are rare and only focus on the ideal CNSs. Their electronic properties are predicted to be similar to those of the flat graphene nanoribbons, depending on edge structures and ribbon widths. In addition to the edge and quantum-confinement effects [37], the non-ideal CNSs are significantly affected by the curvature and stacking effects in terms of the structure stability and the electronic properties. In the past, many studies about curved ribbons [38–41] and few-layer graphene [42–45] have shown that the geometric structure is the key factor for the change of physical properties. The curved surface in CNSs leads to non-parallel $2p_z$ orbitals between the adjacent carbon atoms in the direction of bending, which results in hybridizations of carbon four orbitals [40,46,47]. However, the orbital hybridizations between two locally parallel surfaces are not found. Different stackings will have an impact on the layer–layer interactions and the free charge carriers [42,48,49]. These hybrid features in CNSs enable the possession of versatile and enhanced properties that are more adaptable in future electronic applications.

In this paper, the geometric and electronic properties of non-ideal carbon nanoscrolls are investigated by the first-principles calculations. This work is the first systematic study on two different kinds of CNSs with various widths and internal lengths. A new theoretical framework of charge distribution and the multi-orbital hybridization is implemented to explain the results. The dependences of formation energy and energy gaps on the internal length and the width are first obtained. A thorough discussion on electronic properties has not been published before. The essential properties, including the optimal geometric, formation energy, charge density, spin configuration, band structure, energy gap, and density of states, are determined by the completion and cooperation between the curvature and stacking effects. They possess basic properties similar to that of the flat nanoribbon, such as the zigzag systems being magnetic materials, three types of energy gaps classified in the armchair system, and the decreasing energy gaps resulting from the increased ribbon width. However, the hybrid structure accounts for the distinct properties. For instance, zigzag systems possess special electronic properties associated with the spin arrangements, the rule governing the size order of the energy gap is changed in the armchair system, and disregarding their system types, they all have smaller energy gaps compared with the flat nanoribbon. The predicted results could be verified by experimental measurements. These enriched electronic properties let the CNS have potential suitability not only in energy storage and machine components but also in electronic and spintronic devices.

## 2. Materials and Methods

The geometric and electronic structures of CNSs were studied by the Vienna ab initio simulation package [50,51] in the density-functional theory (DFT). The DFT-D2 method [52] was taken into account in order to describe the weak van der Waals interactions. The projector augmented wave method was utilized to characterize the electron–ion interactions. The exchange-correlation energy of the electron–electron interactions was evaluated within the local-density approximation. The wave functions were expanded by plane waves with the maximum kinetic energy limited to 500 eV. The k-point sampling is outlined by the Monkhorst-Pack scheme [53]. The $12 \times 1 \times 1$ and $300 \times 1 \times 1$ k-grids in the first Brillouin zone are, respectively, the settings used for the geometry optimization and band-structure calculations. The Hellmann–Feynman net force on each atom is smaller than 0.03 eV/Å. The axis of all nanoscrolls was set to be in the x-direction. In order to avoid interactions between the scrolled graphene superlattices of the adjacent unit cells, various vacuum spacings in the z-direction and y-direction are tested and a value of 15 Å is best for accurate and efficient calculations.

## 3. Results and Discussion

### 3.1. Geometric Properties and Formation Energy

A CNS could be regarded as a rolled-up graphene sheet in the vector direction $R = m\mathbf{a_1} + n\mathbf{a_2}$, or the $(m, n)$ notation, where $\mathbf{a_1}$ and $\mathbf{a_2}$ are the basic vectors of a graphene sheet. Two open edges are saturated by hydrogen atoms (green color balls). Two typical longitudinal structures, armchair $(m, m)$ and zigzag $(m, 0)$ CNSs were chosen for a model study, since they exhibit the unique geometric and electronic properties. Moreover, the optimal structures of CNSs are dependent on the initial conditions, including ribbon widths and internal lengths. The initial structures are kept at an arch shape, as shown in Figures 1a and 2a.

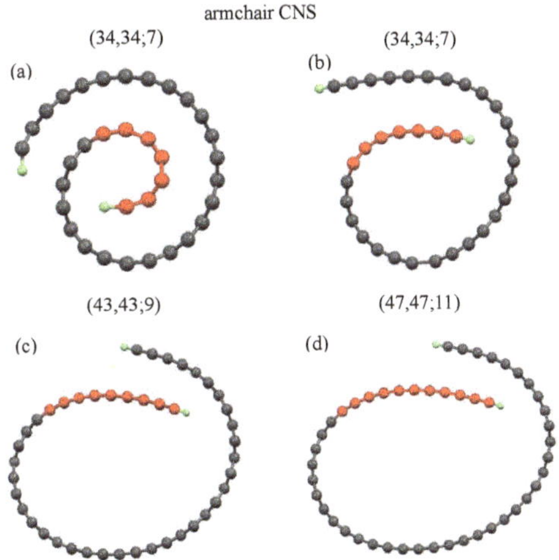


**Figure 1.** For armchair carbon nanoscrolls: the ideal structure of (**a**) $(34, 34; 7)$ and the optimal structures of (**b**) $(34, 34; 7)$, (**c**) $(43, 43; 9)$; (**d**) $(47, 47; 11)$.

The ribbon width ($N_y$) is characterized by the number of dimer or zigzag lines along the transverse, and the internal length ($N_{in}$) only counts the dimer or zigzag lines in the internal lengths (red balls). Armchair and zigzag CNSs, with their geometric characteristics, are defined by ($N_y,N_y;N_{in}$) and ($N_y,0;N_{in}$), respectively.

Before the self-consistent constraint is imposed, the initial arc structure is set to be an Archimedean spiral, as shown in Figures 1a and 2a, and the carbon atoms on the curved surface are set to be the hexagonal structure. However, the relaxed optimal structure becomes less regular, as displayed in Figures 1b and 2b. The internal length ($N_{in}$) describes the ideal geometric structures before optimization, i.e., they only stand for initial conditions. The various initial conditions can result in different optimized geometric structures. We investigated three kinds of internal lengths for both armchair and zigzag CNSs with various scroll widths. The $N_{in}$-dependent formation energy with various scroll widths was obtained. The results show that wider CNSs need to have a larger internal length as an initial condition to form the scroll shape, as shown by Figure 3a,b (discussed later). In other words, the increased initial $N_{in}$ leads to different formation energy and critical formation width. However, the internal length in the optimized geometric structures can either increase or decrease. Interestingly, different initial $N_{in}$ results in the same internal length in the optimized structure, as shown in Figure 1c,d. The $(43, 43; 9)$ and $(47, 47; 11)$ CNSs, respectively, have their $N_{in}$ set to be 9 and 11, but their optimized structures exhibit the same internal length, i.e., the similar overlapping area. It should be noted that the red balls in Figure 1c,d are referred to their initial conditions.

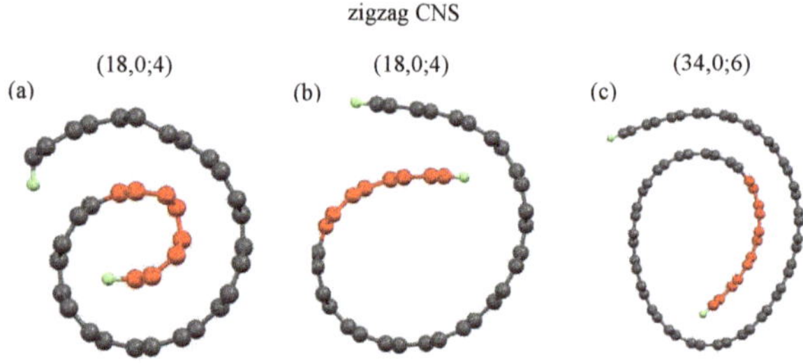



**Figure 2.** For zigzag carbon nanoscrolls: the ideal structure of (**a**) $(18, 0; 4)$ and the optimal structures of (**b**) $(18, 0; 4)$, (**c**) $(34, 0; 6)$.

The scroll geometry is sustained by the layer–layer interactions and simultaneously counterbalanced by strain forces. The reduced overlapping region caused by the insufficient width will hinder the formation of the scroll. The critical formation width of CNSs strongly depends on the internal length. Disregarding the periodic edge shape, all the interlayer distances are between 3.22 and 3.35 Å, which is close to the typical separation of graphene layers. A deeper understanding shows that all the interlayer configurations in CNSs are similar to those of the AB-layered carbon systems, owing to the higher cohesive energy presented in the AB stacking [6,7]. Perceivably, a CNS can be qualified as a stable structure, being determined by the sufficiently large width and overlapping length.

The formation of CNS is mainly dominated by two critical structure parameters: the internal length and the scroll width, as shown in Figure 3.

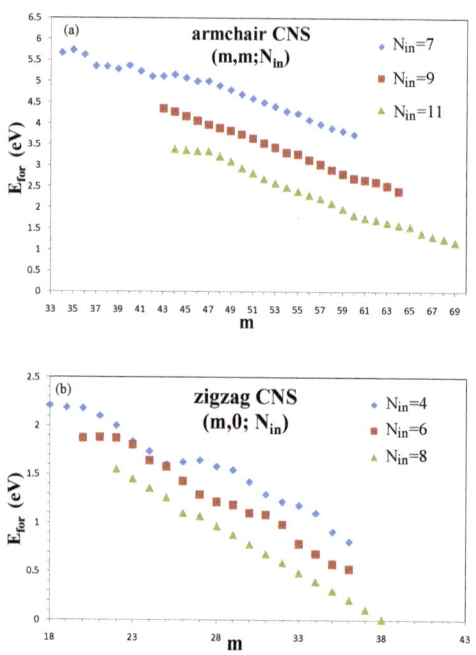



**Figure 3.** Formation energy of the scroll widths for (**a**) armchair CNS and (**b**) zigzag CNS with different internal lengths.

To hold the structure as a scroll, the required formation energy, defined as the energy difference between the total energy of a CNS and that of a flat GNR, is formulated as $E_{for}=E_{int} + E_{cur}$. $E_{int}$ is the energy originating from the interlayer atomic interactions in the overlapping region. This term belongs to the binding force with a negative value. On the other hand, $E_{cur}$ is the restoration force caused by the mechanic strain with a positive value. Given that $E_{for} > 0$, it is obvious that the strain energy is larger than the interlayer interactions. In searching for the minimum-width armchair systems, the critical width, which represents the smallest width to form a CNS with fixed $N_{in}$, is found to be 34 associated with the $(34, 34; 7)$ CNS, and its corresponding internal length is $N_{in} = 7$. As the width becomes larger, the corresponding increase of the overlapping region is responsible for pushing $E_{int}$ negatively, and $E_{cur}$ plunges due to its inverse proportionality to the square of the enlarged effective diameter [40]. Therefore, $E_{for}$ decreases as the width increases, as shown in Figure 3a. Within the width range of $m = 34 \sim 36$, the interlayer distances are relatively large near the end of the overlapping region, leading to the weaker interlayer interactions and thus a smaller and smoother variation in the formation energy. As for $m = 37 \sim 40$, the stacking configurations in the overlapping region are close to a more stable AB stacking near the open end; therefore, the formation energy decreases more dramatically. As the width extends to the ranges $m = 41 \sim 46$ and $m = 47 \sim 52$, they all begin with a slow change but then evolve into a fast decrease in terms of the energy variation, i.e., the slope of the curve is gradually decreased in these two intervals. When the internal length grows, wider critical widths are obtained. $N_{in} = 9$ and 11 correspond to the critical widths $m = 43$ and 47. To counter the decreased overlapping region, the wider critical width can reduce the mechanical strain and thus compensate for the loss of the interlayer interactions in forming the nanoscroll structure. In short, there are two factors taken into consideration in determining the formation energies of armchair nanoscrolls, the internal length and the scroll width. With the same internal length, the formation energy decreases for wider CNSs due to the reduction of mechanical strain. As for the nanoscrolls with the same width, a long internal length is energetically favored owing to a stronger interlayer interaction. These findings support the fact that a larger $N_{in}$ results in a smaller $E_{cur}$, as discussed previously.

The zigzag CNSs are similar to the armchair ones in the width dependence of formation energy. More specifically, when the minimum internal length is $N_{in} = 4$, the smallest zigzag CNS is $(18, 0; 4)$. The scroll-width-dependent formation energy is shown in Figure 4b. Similar to what the armchair system has presented, we find a fluctuation in the dependence on scroll width, signifying that both systems share a common process during the geometric variation. $N_{in} = 6, 8$ and 10 correspond to the critical widths $m = 20, 22$ and 23, respectively. With respect to the minimum-width system, the internal length is smaller in the armchair type than in the zigzag type, meaning that the former can overcome the larger restoration force caused by the mechanic strain. Therefore, the armchair CNSs are formed more easily and become more stable than those of the zigzag type.

### 3.2. Electronic Properties

The electronic properties of CNSs are deeply affected by edge shapes, widths, curvatures, and spin arrangements. An armchair $(38, 38; 7)$ CNS, as shown in Figure 4a, exhibits a lot of 1D parabolic energy dispersions, in which the occupied valence bands are not symmetric to the unoccupied conduction bands about the Fermi level ($E_F = 0$).

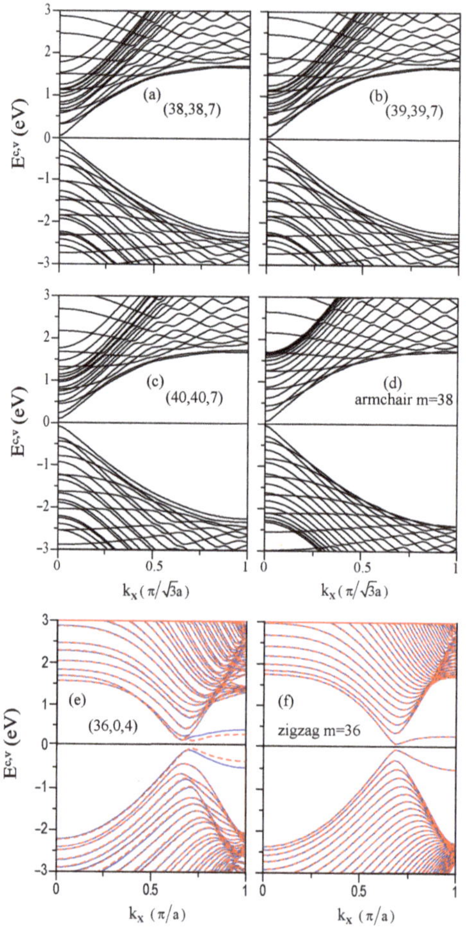

**Figure 4.** Band structures of the armchair system for (**a**) (38,38;7), (**b**) (39,39;7), (**c**) (40,40;7); (**d**) flat GNR with m = 38, and zigzag system for (**e**) (36,0;4); (**f**) flat GNR with m = 36.

The energy bands are all doubly degenerate for both spin states: spin-up and spin-down. Each energy dispersion has the local minimum or maximum at $k_x = 0$ and 1 and also at other wave vectors; that is to say, there are extra band-edge states except those at $k_x = 0$ and 1. In the vicinity of $E_F$, the highest occupied state (HOS) and the lowest unoccupied state (LUS) occur at the same wave vector ($k_x = 0.1$), which, thus, leads to a direct energy gap of $E_g = 0.181$ eV, as shown in Table 1. Contrarily, the armchair (39,39;7) CNS (Figure 4b) possesses an indirect energy gap. Associated with this gap are the highest occupied and the lowest unoccupied states that appear, respectively, at $k_x = 0.01$ and 0.13, and they are separated by a gap size of $E_g = 0.112$ eV. These two energy bands are relatively smooth near $k_x=0$ without obvious dispersions. Such 3N-width characteristic is similar to that of the flat graphene nanoribbon [37]. Both the armchair $(40, 40; 7)$ CNS with a $(3N + 1)$ width (Figure 4c) and the armchair $(38, 38; 7)$ CNS with a $(3N + 2)$ width have direct energy gaps. The important differences between them are that the former has a smaller gap of $E_g = 0.323$ eV and strongly non-monotonous energy dispersions. In addition, energy spacing of the $k_x = 0$ state between two energy bands nearest to $E_F = 0$ is higher than the energy gap. Apparently, there are certain important differences among the 3N-, $(3N + 1)$- and $(3N + 2)$-width systems. On the other hand, the CNSs are in sharp contrast to the flat GNRs (Figure 4d). The latter possesses a pair of monotonous parabolic energy dispersions nearest to $E_F = 0$, in which their $k_x = 0$ states determine a direct energy gap. Moreover, their energy gaps are a bit larger than those of the former, e.g., energy gaps of the $m = 38$ systems.

**Table 1.** The energy gaps, wave vectors of highest occupied state (HOS) and lowest unoccupied state (LUS), and the 3N type of various systems.

| Systems | Type | $K_x$ of HOS/LUS | Energy Gaps |
|---|---|---|---|
| Armchair CNS (38,38;7) | 3N+2 | 0.10/0.10 | direct; 0.181 eV |
| Armchair CNS (39,39;7) | 3N | 0.01/0.13 | indirect; 0.112 eV |
| Armchair CNS (40,40;7) | 3N+1 | 0.10/0.10 | direct; 0.323 eV |
| Zigzag CNS (36,0;4) | N/A | 0.67/0.67 | direct; 0.180/0.230 eV |

Electronic structures of zigzag CNSs are enriched by the anti-ferromagnetic spin configuration at two edges, as shown in Figure 4e for the $(36,0;4)$ CNS. Most of the energy is doubly degenerate, while there exists the spin-up and spin-down splitting bands near $E_F = 0$. This could be clearly understood from the spin-up and spin-down charge distributions in Figure 5 (red and blue colors) since for the distinct spin states, the H-passivated carbon atoms at each end of the open structure tend to interact differently with the surrounding atoms.

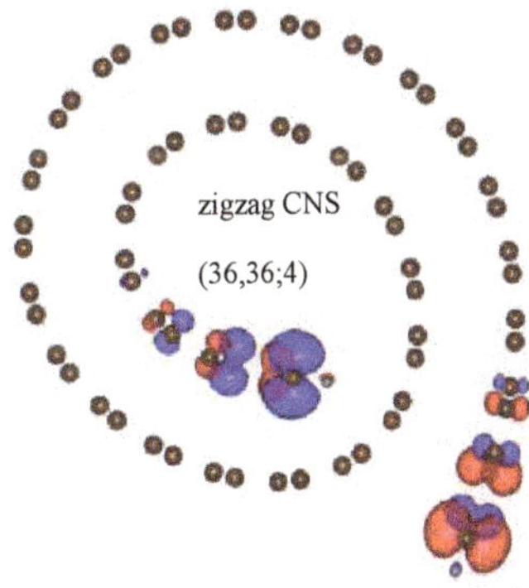


**Figure 5.** The charge distribution of spin-up and spin-down states, indicated by red and blue regions, respectively.

The four splitting bands have weak energy dispersions, in which they are mainly contributed to by the local edge atoms. Such bands will determine two kinds of spin-dependent energy gaps. The energy gaps belonging to the direct type appear approximately at $k_x = 2/3$. Noticeably, the spin-up gap, 0.18 eV, is smaller than the spin-down gap, 0.23 eV. In comparison, for a flat GNR, four flat bands are partially degenerate and form two bands (Figure 4f), since the same end-structure environment in flat GNRs results in no difference for the edge effects from two ends. Again, the energy gap at $k_x = 2/3$ is a direct one, with its size being at 0.14 eV.

### 3.3. Charge Distributions

The charge distribution on CNSs, which is very useful in understanding the hybridizations of orbitals (or the orbital bondings) and the low-lying energy bands, is significantly affected by the curved surface [47,54]. The variation of charge distribution created by subtracting the carrier density of an isolated carbon (a hollow circle) from that of a CNS is clearly illustrated in Figure 6a,b.

As for the planar region of the armchair $(38, 38; 7)$ CNS, resembling a flat GNR as enclosed by the rectangle in Figure 6a, the $2s$, $2p_x$, and $2p_y$ orbitals of one carbon atom interact with those of the nearby carbon atoms to form the $\sigma$ bonds ((I) in orange shades). The charge densities are concentrated at the bond locations in the middle of the two binding atoms and significantly lowered for the remaining parts of the carbon atoms to create the depletion zones, as indicated by the blue shades. Moreover, the $2p_z$ orbitals ((II) in orange shades) perpendicular to the plane can interact with their nearest neighbors to form the $\pi$ bonds. Induced by the curvature effect, there are two main causes that can contribute to the orbital hybridization. One is that the non-parallel $2p_z$ orbitals can lead to the $\sigma$ bonds in addition to the $\pi$ bonds. Another is the hybridization of four orbitals that is also responsible for introducing the complex $\pi$ and $\sigma$ bonds. Associated with these bond formations are the serious hybridizations that take place on the internal side of the curved surface, as shown in (III) and (IV). The direct impact from these hybridizations is reflected in the significant variation of the low-lying band structure, including the non-monotonic energy dispersions associated with the strongly hybridized atomic orbitals and the energy gap due to the $k_x \neq 0$ state (Figure 4a–c). The aforementioned changes to the band structure are in good agreement with the previous studies on the carbon nanotube and curved GNR. A further deduction from the curvature effect is that the dumbbell shape of $2p_z$ orbitals makes the charge distribution thicker on the nanoscroll surface towards the outside but thinner on the surface towards the inside. These larger $2p_z$ orbitals on the outer surface provide an ideal environment to bond with H, Li and other atoms, showing the possibilities in energy storage and electronic nano-devices.

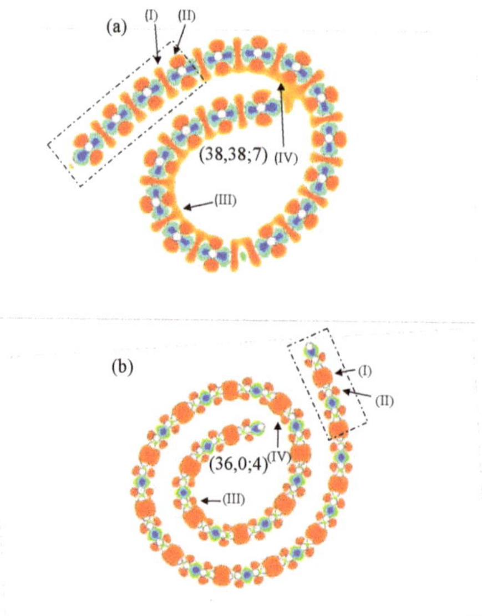


**Figure 6.** The charge distribution for (**a**) armchair (38,38;7) CNS; (**b**) zigzag (36,0;4) CNS. (II) are the 2 pz orbitals and (I) are the other three orbitals. (III) and (IV) are the serious orbital hybridizations.

The zigzag CNS (Figure 6b) and the armchair CNS partially share a similar charge distribution. In the planar region, they both have their $\pi$ bonds created by the $2p_z$ orbitals (II) and their $\sigma$ bonds formed from the $2s$, $2p_x$, and $2p_y$ orbitals (I). Unlike the armchair configuration, the zigzag orbital hybridization in the curved regions (III and IV) appears to be weaker due to the longer distance between carbon atoms. As we move to the outer portion of the nanoscroll, the decreased curvature can reduce the hybridization to a trivial level. That is to say, the very weak hybridization is presented at the outer section. Speaking of the different features of energy bands in the zigzag CNS, the energy bands near $E_F$ are mainly contributed to by the carbon atoms at the two open edges. Given that spin-up and spin-down states are ignored, the distributions of the charge density at these two edges almost remain unchanged. Therefore, the energy bands in the zigzag case are not much different from those presented in the flat GNR.

### 3.4. Density of States

The main features of DOSs in carbon nanoscrolls are mainly determined by the complex cooperation relation among the edge structure, total width, and internal length, as clearly shown in Figure 7a–d.

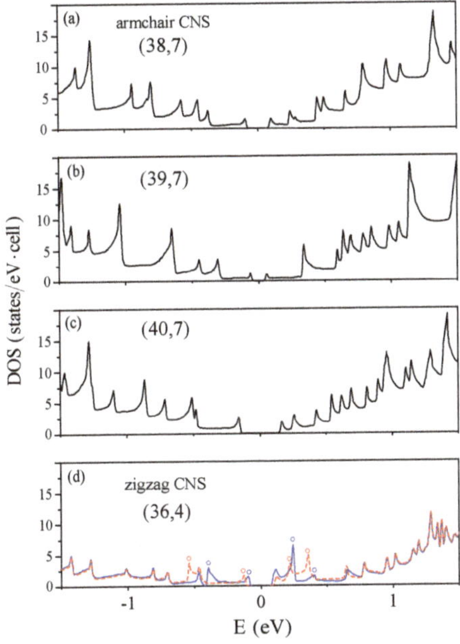


**Figure 7.** Density of states for the (**a**) (38,7), (**b**) (39,7) and (**c**) (40,7) armchair nanoscrolls, and (**d**) for the (36,4) zigzag one.

The van Hove singularities only come from the parabolic energy dispersions (band structures in Figure 4a–c), leading to the square-root pronounced peaks. The valence and conduction peaks closest to the Fermi level form an energy gap corresponding to a semiconducting nanoscroll system. The asymmetric peak structures about $E = 0$ are very apparent; furthermore, a simple relation in energy spacing of two neighboring prominent peaks is absent. That is to say, it is very difficult to identify a specific one-to-one correspondence in peak and geometric structures. Both HRTEM and STS need to be utilized to examine the theoretical predictions on the geometric and electronic properties. There are no spin-split peaks in armchair nanoscrolls (Figure 7a–c), while they are present in zigzag systems (blue and red circles in Figure 7d). The energy splittings, which are due to the partial flat bands at the zone boundary (Figure 7b), are relatively obvious. The SP-STS examinations on them could provide very useful information on the ferromagnetic configurations of zigzag nanoscrolls, being in sharp contrast with degenerate behavior from the anti-ferromagnetic ones of pristine zigzag graphene nanoribbons.

### 3.5. Comparisons among the Planar, Curved/Zipped, Folded and Scrolled Systems

The flexible carbon honeycomb lattice can be presented in various forms under a very strong $\sigma$ bonding. Such structures create the diverse essential properties and thus induce important differences among the planar, curved, folded, and scrolled graphene nanoribbons. For armchair nanoribbons, only parts of the curved systems exhibit the 1D metallic property, mainly owing to the edge–edge interactions [40,41]. Similar behavior is revealed in the even-zAA stacking of the folded zigzag systems [55]. The valence and conduction bands, which determine the metallic or semiconducting properties, are very sensitive to the geometric structure. All the planar and folded armchair systems have the parabolic bands with direct energy gaps at $k_x = 0$ [55]. However, the curved and scrolled ones might possess non-monotonous energy dispersions with direct or indirect energy gaps (Figure 8b).

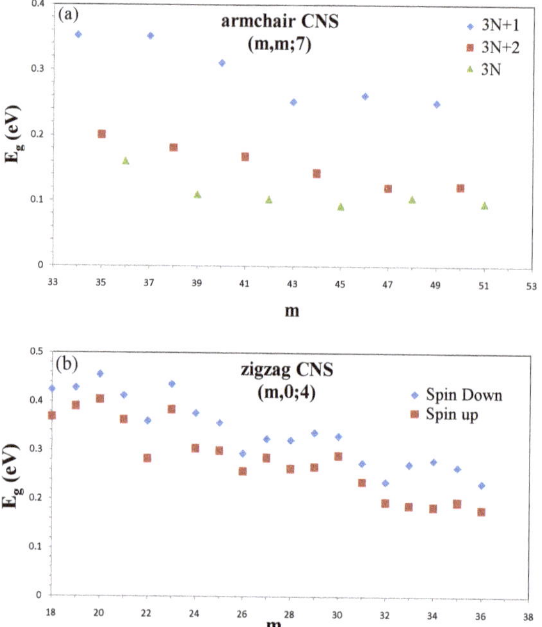


**Figure 8.** The width-dependent energy gaps for (**a**) armchair carbon nanoscrolls with $N_{in} = 7$ and (**b**) zigzag systems with $N_{in} = 4$.

Those of zigzag systems belong to the partially flat edge-localized bands at $k_x > 2/3$. An obvious spin splitting appears near the Fermi level when the magnetic environments are different for spin-up and spin-down states near the open edges, e.g., for folded odd-zAB and scrolled zigzag nanoribbons. Specifically, only the folded even-zAB stacking presents a pair of linearly intersecting energy bands at $k_x \sim 2/3$, as observed in armchair carbon nanotubes.

The width dependences of energy gaps are greatly enriched by the geometric structures. There are three categories in the planar and scrolled armchair nanoribbons (Figure 8a), but six categories in the folded systems. In addition to $N_A = 3I, 3I + 1$, and $3I + 2$, the last ones also depend on the odd/even number of dimer lines, where $N_A$ is the ribbon width and I is an integer. For $N_A = 3I + 2$, the planar systems have the smallest energy gaps because of the finite-size confinement. However, the opposite is true for the scrolled systems under the combined effects. In comparisons among the various pristine systems, the highest energy gaps are revealed in the even-aAA′ folded armchair nanoribbons of $N_A = 6I + 4$. As for zigzag nanoribbons, only the scrolled and odd-zAB folded systems present the spin–split energy gaps. The width-dependent declining behavior is obvious except for the folded even-zAB stacking systems with the strong edge–edge interactions. Furthermore, the wave-like fluctuation comes to exist in the scrolled systems.

## 4. Conclusions

After the self-consistent field is solved, the stable structure is determined by the equilibrium between the ribbon width and the scroll surface. If the internal length or the layer–layer overlapping area is too small, the nanoscroll structure will not be sustainable. In consequence, the typical cross-section of the ideal scroll becomes more oval. For armchair CNSs, the minimum (critical) internal length is $N_{in} = 7$. Given an internal length smaller than the critical length, the optimal structure will be restored back to the shape where the internal length is critical. With respect to each internal length, there exists a stable structure with the minimum ribbon width. Assuming that the ribbon width is less than the minimum value, the scroll structure will collapse to the flat graphene nanoribbon due to the insufficient layer–layer interactions. Such interaction will lead to the AA stacking configuration, which provides larger interaction forces, and the average interlayer distance is about 3.35 Å. For a group of CNSs with a small deviation about the curvature radius, all the formations will result in the same internal length in order to maintain the AA stacking configuration. On the other hand, for the zigzag systems, the critical internal length is $N_{in} = 4$, and the optimal configuration is AB stacking with an average interlayer distance of ~3.2 Å. Such stacking configuration is the same as that in the bilayer graphene and nanoribbon. As compared with armchair CNSs, zigzag CNSs possess larger curvature radii, and layer–layer interactions. Therefore, zigzag CNSs can form with a smaller width ribbon.

The formation energy ($\Delta E$) is mainly dominated by the bending energy and the interlayer interaction. The competition between the bending energy and the interlayer interaction would determine the two critical structure parameters: the internal length and the ribbon width. Note that the $E_{for}$'s are greater than zero because the bending energy is always larger in magnitude than the interlayer interaction. For the armchair systems, three kinds of $N_y$-dependences of $\Delta E$ are found, and the $\Delta E$'s increase with the ribbon width. The relationship is attributed to the increased layer–layer interactions with the larger overlapping areas, yet the decreased bending energies with the larger curvature radius. Hence, the critical ribbon widths of $N_{in} = 7, 9$, and 11 are, respectively, 34, 43, and 47. As for the zigzag system, it presents similarities to the armchair type regarding the energy dependence. However, there are some differences since they are not rolled in the same manner. The zigzag systems possess larger curvatures and stronger layer–layer interactions owing to the AB stacking configuration. As the width increases, the energy decays faster in the AB stacked zigzag system than in the AA stacked armchair system so that the formation energy can easily and quickly reach the deeper negative levels. In conclusion, the lower total energy makes CNS more stable than the flat nanoribbon, and the critical ribbon widths of $N_{in} = 4$ and 9 are, respectively, 18 and 20.

**Author Contributions:** Conceptualization, S.-Y.L. and S.-L.C.; methodology, S.-L.C. and C.-R.C.; software, S.-Y.L., C.-R.C. and S.-L.C.; validation, S.-Y.L. and S.-L.C.; formal analysis, S.-Y.L. and S.-L.C.; writing—original draft preparation, S.-Y.L., C.-R.C., S.-L.C. and M.-F.L.; writing—review and editing, W.-B.L., H.-Y.L. and M.-F.L.; supervision, M.-F.L.; funding acquisition, S.-Y.L. All authors have read and agreed to the published version of the manuscript.

**Funding:** This work was supported by the National Science Council of Taiwan (Grant No. 107-2112-M-194-010-MY3).

**Institutional Review Board Statement:** Not applicable.

**Informed Consent Statement:** Not applicable.

**Data Availability Statement:** The data presented in this study are available on request from the corresponding author.

**Conflicts of Interest:** The authors declare no conflict of interest.

## References


1. Robertson, J. Diamond-like amorphous carbon. *Mater. Sci. Eng. R Rep.* **2002**, *37*, 129–281. [CrossRef]
2. Grill, A. Diamond-like carbon: State of the art. *Diam. Relat. Mater.* **1999**, *8*, 428–434. [CrossRef]
3. Graf, D.; Molitor, F.; Ensslin, K.; Stampfer, C.; Jungen, A.; Hierold, C.; Wirtz, L. Spatially resolved Raman spectroscopy of single-and few-layer graphene. *Nano Lett.* **2007**, *7*, 238–242. [CrossRef] [PubMed]
4. Meyer, J.C.; Geim, A.K.; Katsnelson, M.I.; Novoselov, K.S.; Booth, T.J.; Roth, S. The structure of suspended graphene sheets. *Nature* **2007**, *446*, 60–63. [CrossRef] [PubMed]
5. Gopinadhan, K.; Shin, Y.J.; Jalil, R.; Venkatesan, T.; Geim, A.K.; Neto, A.H.C.; Yang, H. Extremely large magnetoresistance in few-layer graphene/boron-nitride heterostructures. *Nat. Commun.* **2015**, *6*, 1–7. [CrossRef] [PubMed]
6. Ren, Z.; Huang, Z.; Xu, J.; Wang, J.; Bush, P.; Siegal, M.; Provencio, P. Synthesis of large arrays of well-aligned carbon nanotubes on glass. *Science* **1998**, *282*, 1105–1107. [CrossRef] [PubMed]
7. Wilder, J.W.; Venema, L.C.; Rinzler, A.G.; Smalley, R.E.; Dekker, C. Electronic structure of atomically resolved carbon nanotubes. *Nature* **1998**, *391*, 59–62. [CrossRef]
8. Liao, M.; Jiang, S.; Hu, C.R.; Zhang, R.; Kuang, Y.M.; Zhu, J.Z.; Zhang, Y.; Dong, Z. Tip Enhanced Raman Spectroscopic Imaging of Individual Carbon Nanotubes with Sub-nanometer Resolution. *Nano Lett.* **2016**, *16*, 4040–4046. [CrossRef]
9. Kosynkin, D.V.; Higginbotham, A.L.; Sinitskii, A.; Lomeda, J.R.; Dimiev, A.; Price, B.K.; Tour, J.M. Longitudinal unzipping of carbon nanotubes to form graphene nanoribbons. *Nature* **2009**, *458*, 872–876. [CrossRef]
10. Kim, K.; Sussman, A.; Zettl, A. Graphene nanoribbons obtained by electrically unwrapping carbon nanotubes. *ACS Nano* **2010**, *4*, 1362–1366. [CrossRef] [PubMed]
11. Chen, Y.C.; Cao, T.; Chen, C.; Pedramrazi, Z.; Haberer, D.; de Oteyza, D.G.; Fischer, F.R.; Louie, S.G.; Crommie, M.F. Molecular bandgap engineering of bottom-up synthesized graphene nanoribbon heterojunctions. *Nat. Nanotechnol.* **2015**, *10*, 156–160. [CrossRef] [PubMed]
12. Chung, H.C.; Chang, C.P.; Lin, C.Y.; Lin, M.F. Electronic and optical properties of graphene nanoribbons in external fields. *Phys. Chem. Chem. Phys.* **2016**, *18*, 7573–7616. [CrossRef] [PubMed]
13. Viculis, L.M.; Mack, J.J.; Kaner, R.B. A chemical route to carbon nanoscrolls. *Science* **2003**, *299*, 1361. [CrossRef] [PubMed]
14. Savoskin, M.V.; Mochalin, V.N.; Yaroshenko, A.P.; Lazareva, N.I.; Konstantinova, T.E.; Barsukov, I.V.; Prokofiev, I.G. Carbon nanoscrolls produced from acceptor-type graphite intercalation compounds. *Carbon* **2007**, *45*, 2797–2800. [CrossRef]
15. Shi, X.; Pugno, N.M.; Gao, H. Tunable core size of carbon nanoscrolls. *J. Comput. Theor. Nanosci.* **2010**, *7*, 517–521. [CrossRef]
16. Daff, T.D.; Collins, S.P.; Dureckova, H.; Perim, E.; Skaf, M.S.; Galvão, D.S.; Woo, T.K. Evaluation of carbon nanoscroll materials for post-combustion CO2 capture. *Carbon* **2016**, *101*, 218–225. [CrossRef]
17. Bacon, R. Growth, structure, and properties of graphite whiskers. *J. Appl. Phys.* **1960**, *31*, 283–290. [CrossRef]
18. Xie, X.; Ju, L.; Feng, X.; Sun, Y.; Zhou, R.; Liu, K.; Fan, S.; Li, Q.; Jiang, K. Controlled fabrication of high-quality carbon nanoscrolls from monolayer graphene. *Nano Lett.* **2009**, *9*, 2565–2570. [CrossRef]
19. Panchuk, R.; Prylutska, S.; Chumak, V.; Skorokhyd, N.; Lehka, L.; Evstigneev, M.; Prylutskyy, Y.I.; Berger, W.; Heffeter, P.; Scharff, P.; et al. Application of C60 fullerene-doxorubicin complex for tumor cell treatment in vitro and in vivo. *J. Biomed. Nanotechnol.* **2015**, *11*, 1139–1152. [CrossRef] [PubMed]
20. Chang, C.H.; Ortix, C. Theoretical prediction of a giant anisotropic magnetoresistance in carbon nanoscrolls. *Nano Lett.* **2017**, *17*, 3076–3080. [CrossRef]
21. Li, J.; Wang, L.; Jiang, W. Carbon microspheres produced by high energy ball milling of graphite powder. *Appl. Phys. A* **2006**, *83*, 385–388. [CrossRef]
22. Haddon, R.; Hebard, A.; Rosseinsky, M.; Murphy, D.; Duclos, S.; Lyons, K.; Miller, B.; Rosamilia, J.; Fleming, R.; Kortan, A.; et al. Conducting films of C60 and C70 by alkali-metal doping. *Nature* **1991**, *350*, 320–322. [CrossRef]
23. Kroto, H.W.; Allaf, A.; Balm, S. C60: Buckminsterfullerene. *Chem. Rev.* **1991**, *91*, 1213–1235. [CrossRef]
24. Mpourmpakis, G.; Tylianakis, E.; Froudakis, G.E. Carbon nanoscrolls: A promising material for hydrogen storage. *Nano Lett.* **2007**, *7*, 1893–1897. [CrossRef] [PubMed]
25. Coluci, V.; Braga, S.; Baughman, R.; Galvao, D. Prediction of the hydrogen storage capacity of carbon nanoscrolls. *Phys. Rev. B* **2007**, *75*, 125404. [CrossRef]
26. Zhao, J.; Yang, B.; Zheng, Z.; Yang, J.; Yang, Z.; Zhang, P.; Ren, W.; Yan, X. Facile preparation of one-dimensional wrapping structure: Graphene nanoscroll-wrapped of Fe3O4 nanoparticles and its application for lithium-ion battery. *ACS Appl. Mater. Interfaces* **2014**, *6*, 9890–9896. [CrossRef] [PubMed]
27. Shi, X.; Pugno, N.M.; Cheng, Y.; Gao, H. Gigahertz breathing oscillators based on carbon nanoscrolls. *Appl. Phys. Lett.* **2009**, *95*, 163113. [CrossRef]
28. Wang, J.; Hao, J.; Liu, D.; Qin, S.; Chen, C.; Yang, C.; Liu, Y.; Yang, T.; Fan, Y.; Chen, Y.; et al. Flower stamen-like porous boron carbon nitride nanoscrolls for water cleaning. *Nanoscale* **2017**, *9*, 9787–9791. [CrossRef]
29. Liu, Z.; Wang, J.; Ding, H.; Chen, S.; Yu, X.; Lu, B. Carbon nanoscrolls for aluminum battery. *ACS Nano* **2018**, *12*, 8456–8466. [CrossRef]
30. Uhm, T.; Na, J.; Lee, J.U.; Cheong, H.; Lee, S.W.; Campbell, E.; Jhang, S.H. Structural configurations and Raman spectra of carbon nanoscrolls. *Nanotechnology* **2020**, *31*, 315707. [CrossRef]

31. Braga, S.; Coluci, V.; Baughman, R.; Galvao, D. Hydrogen storage in carbon nanoscrolls: An atomistic molecular dynamics study. *Chem. Phys. Lett.* **2007**, *441*, 78–82. [CrossRef]
32. Huang, Y.; Li, T. Molecular mass transportation via carbon nanoscrolls. *J. Appl. Mech.* **2013**, *80*, 040903. [CrossRef]
33. Yan, M.; Wang, F.; Han, C.; Ma, X.; Xu, X.; An, Q.; Xu, L.; Niu, C.; Zhao, Y.; Tian, X.; et al. Nanowire templated semihollow bicontinuous graphene scrolls: Designed construction, mechanism, and enhanced energy storage performance. *J. Am. Chem. Soc.* **2013**, *135*, 18176–18182. [CrossRef] [PubMed]
34. Tojo, T.; Fujisawa, K.; Muramatsu, H.; Hayashi, T.; Kim, Y.A.; Endo, M.; Terrones, M.; Dresselhaus, M.S. Controlled interlayer spacing of scrolled reduced graphene nanotubes by thermal annealing. *RSC Adv.* **2013**, *3*, 4161–4166. [CrossRef]
35. Kresin, V.; Aharony, A. Fully collapsed carbon nanotubes. *Nature* **1995**, *377*, 1673–1686.
36. Qian, D.; Wagner, G.J.; Liu, W.K.; Yu, M.F.; Ruoff, R.S. Mechanics of carbon nanotubes. *Appl. Mech. Rev.* **2002**, *55*, 495–533. [CrossRef]
37. Son, Y.W.; Cohen, M.L.; Louie, S.G. Energy gaps in graphene nanoribbons. *Phys. Rev. Lett.* **2006**, *97*, 216803. [CrossRef]
38. Kane, C.L.; Mele, E. Size, shape, and low energy electronic structure of carbon nanotubes. *Phys. Rev. Lett.* **1997**, *78*, 1932. [CrossRef]
39. Shyu, F.L.; Lin, M.F. Electronic and optical properties of narrow-gap carbon nanotubes. *J. Phys. Soc. Jpn.* **2002**, *71*, 1820–1823. [CrossRef]
40. Chang, S.L.; Wu, B.R.; Yang, P.H.; Lin, M.F. Curvature effects on electronic properties of armchair graphene nanoribbons without passivation. *Phys. Chem. Chem. Phys.* **2012**, *14*, 16409–16414. [CrossRef] [PubMed]
41. Chang, S.L.; Lin, S.Y.; Lin, S.K.; Lee, C.H.; Lin, M.F. Geometric and electronic properties of edge-decorated graphene nanoribbons. *Sci. Rep.* **2014**, *4*, 1–8. [CrossRef] [PubMed]
42. Hwang, E.; Adam, S.; Sarma, S.D. Carrier transport in two-dimensional graphene layers. *Phys. Rev. Lett.* **2007**, *98*, 186806. [CrossRef] [PubMed]
43. Lai, Y.; Ho, J.; Chang, C.; Lin, M.F. Magnetoelectronic properties of bilayer Bernal graphene. *Phys. Rev. B* **2008**, *77*, 085426. [CrossRef]
44. Chang, S.L.; Wu, B.R.; Wong, J.H.; Lin, M.F. Configuration-dependent geometric and electronic properties of bilayer graphene nanoribbons. *Carbon* **2014**, *77*, 1031–1039. [CrossRef]
45. Lin, C.Y.; Wu, J.Y.; Ou, Y.J.; Chiu, Y.H.; Lin, M.F. Magneto-electronic properties of multilayer graphenes. *Phys. Chem. Chem. Phys.* **2015**, *17*, 26008–26035. [CrossRef] [PubMed]
46. Feng, J.; Qi, L.; Huang, J.Y.; Li, J. Geometric and electronic structure of graphene bilayer edges. *Phys. Rev. B* **2009**, *80*, 165407. [CrossRef]
47. Lin, S.Y.; Chang, S.L.; Shyu, F.L.; Lu, J.M.; Lin, M.F. Feature-rich electronic properties in graphene ripples. *Carbon* **2015**, *86*, 207–216. [CrossRef]
48. Son, Y.W.; Choi, S.M.; Hong, Y.P.; Woo, S.; Jhi, S.H. Electronic topological transition in sliding bilayer graphene. *Phys. Rev. B* **2011**, *84*, 155410. [CrossRef]
49. Tran, N.T.T.; Lin, S.Y.; Glukhova, O.E.; Lin, M.F. Configuration-induced rich electronic properties of bilayer graphene. *J. Phys. Chem. C* **2015**, *119*, 10623–10630. [CrossRef]
50. Kresse, G.; Joubert, D. From ultrasoft pseudopotentials to the projector augmented-wave method. *Phys. Rev. B* **1999**, *59*, 1758. [CrossRef]
51. Kresse, G.; Furthmüller, J. Efficient iterative schemes for ab initio total-energy calculations using a plane-wave basis set. *Phys. Rev. B* **1996**, *54*, 11169. [CrossRef]
52. Grimme, S. Semiempirical GGA-type density functional constructed with a long-range dispersion correction. *J. Comput. Chem.* **2006**, *27*, 1787–1799. [CrossRef] [PubMed]
53. Perdew, J.P.; Burke, K.; Ernzerhof, M. Generalized gradient approximation made simple. *Phys. Rev. Lett.* **1996**, *77*, 3865. [CrossRef] [PubMed]
54. Lin, S.Y.; Chang, S.L.; Tran, N.T.T.; Yang, P.H.; Lin, M.F. H–Si bonding-induced unusual electronic properties of silicene: A method to identify hydrogen concentration. *Phys. Chem. Chem. Phys.* **2015**, *17*, 26443–26450. [CrossRef] [PubMed]
55. Chang, S.L.; Wu, B.R.; Yang, P.H.; Lin, M.F. Geometric, magnetic and electronic properties of folded graphene nanoribbons. *RSC Adv.* **2016**, *6*, 64852–64860. [CrossRef]

*nanomaterials*

*Article*

# Ab Initio Theory of Photoemission from Graphene

Eugene Krasovskii [†,‡]

Departamento de Polímeros y Materiales Avanzados: Física, Química y Tecnología, Universidad del Pais Vasco/Euskal Herriko Unibertsitatea, 20080 Donostia/San Sebastián, Basque Country, Spain; eugene.krasovskii@ehu.eus
† Donostia International Physics Center (DIPC), 20018 Donostia/San Sebastián, Basque Country, Spain.
‡ IKERBASQUE, Basque Foundation for Science, 48013 Bilbao, Basque Country, Spain.

**Abstract:** Angle-resolved photoemission from monolayer and bilayer graphene is studied based on an ab initio one-step theory. The outgoing photoelectron is represented by the time-reversed low energy electron diffraction (LEED) state $\Phi^*_{\text{LEED}}$, which is calculated using a scattering theory formulated in terms of augmented plane waves. A strong enhancement of the emission intensity is found to occur around the scattering resonances. The effect of the photoelectron scattering by the underlying substrate on the polarization dependence of the photocurrent is discussed. The constant initial state spectra $I(\mathbf{k}_{||}, \hbar\omega)$ are compared to electron transmission spectra $T(E)$ of graphene, and the spatial structure of the outgoing waves is analyzed. It turns out that the emission intensity variations do not correlate with the structure of the $T(E)$ spectra and are caused by rather subtle interference effects. Earlier experimental observations of the photon energy and polarization dependence of the emission intensity $I(\mathbf{k}_{||}, \hbar\omega)$ are well reproduced within the dipole approximation, and the Kohn–Sham eigenstates are found to provide a quite reasonable description of the photoemission final states.

**Keywords:** graphene; angle-resolved photoemission; electron scattering; augmented plane waves

**Citation:** Krasovskii, E. Ab Initio Theory of Photoemission from Graphene. *Nanomaterials* **2021**, *11*, 1212. https://doi.org/10.3390/nano11051212

Academic Editor: Eugene Kogan

Received: 30 March 2021
Accepted: 30 April 2021
Published: 3 May 2021

**Publisher's Note:** MDPI stays neutral with regard to jurisdictional claims in published maps and institutional affiliations.

## 1. Introduction

Owing to the combination of high structural stability and unique electronic properties [1], graphene has become a paradigm two-dimensional material and a subject of numerous experimental and theoretical studies. The majority of research has addressed the vicinity of the Dirac point (DP), however, also the unbound states were discovered to exhibit fascinating phenomena, such as the electron-transmission slits at low kinetic energies [2–4] caused by the interlayer scattering and the scattering resonances due to the coupling of the in-plane and perpendicular motions at higher energies [5–7]. A detailed knowledge of the properties of unbound states is important for the interpretation of angle-resolved photoemission (ARPES), which is the most direct source of information about the occupied states. Graphene has been extensively studied with ARPES [8–19], and apparent final state effects were reported [9,13,14,18,19]. In particular, the circular dichroism [12,14,17] is of special interest owing to its close relation to the topological character of 2D states [20].

A characteristic feature of photoemission from graphite [21,22] and graphene [10,13,18] is the so-called "dark corridor", i.e., the suppression of emission with the $p$-polarized light from the occupied $\pi$ states along the $\bar{\Gamma}\bar{K}$ line in the second Brillouin zone (BZ) as a result of a destructive interference of the contributions to the photoemission matrix element from the two equivalent sublattices. In the monolayer graphene, the suppressed initial states are odd on reflection in the $\bar{\Gamma}\bar{K}$ line [18,23], so the dark corridor can be illuminated by the $s$-polarized light incident along $\bar{\Gamma}\bar{K}$ [13]. In Ref. [13] this was demonstrated experimentally, and, in addition, a strong photon-energy dependence of the emission intensity was revealed. These observations were analyzed using a multiple-scattering implementation [24] of the one-step theory of photoemission [25–29] based on a density-functional-derived one-particle potential. However, for a direct comparison between experiment and theory, the authors shifted the theoretical photon energies by 8.6 eV toward higher photon energies. The origin of such a large shift is unclear, especially in view of the fact that the LEED theory based on Kohn–Sham states describes the unoccupied continuum of both bulk graphite [30] and graphene [5,6,31–33] rather accurately. However, in Ref. [13] only two photon energies were studied, which may be insufficient for a conclusive comparison between experiment and theory. A more detailed measurement in the range $\hbar\omega = 42$–$55$ eV was reported in Ref. [18], and a rapid variation of the relative intensity for $s$ and $p$ light polarizations was observed.

A consistent and rigorous approach to photoemission is offered by the one-step theory [25–29], in which the photoexcitation and photoelectron transport to the detector (including elastic and inelastic scattering) are described by the time reversed LEED state $\Phi^*_{\mathrm{LEED}}$. Here, the one-step theory is applied to the monolayer and bilayer graphene with the aim to explain the experimentally observed features and analyze their relation to the properties of the relevant scattered waves. The final-state wave function $\Phi^*_{\mathrm{LEED}}$ is calculated using the augmented plane waves scattering formalism [34]. The present theory reproduces well both experiments [13,18] and reveals rapid variations of the character of the outgoing photoelectron wave with energy. These variations manifest themselves also in the electron transmission through the films and in the variations of the dwell time, i.e., the probability to find the scattered electron inside the film. However, the variations of these quantities do not correlate with each other, so the full knowledge of the wave function is necessary to describe the experiment, in particular, the lateral umklapp scattering proves to be essential. For a monolayer graphene, the question arises of how strongly the interaction with the substrate modifies the symmetry properties of the initial and the final states. Here, this question is addressed by comparing the symmetry of the emission from the monolayer and bilayer graphene. This estimate suggests that the reflection of the outgoing photoelectron from the underlying substrate may explain the experimentally observed symmetry breaking.

## 2. Computational Methodology and Approximations

According to the one-step theory of photoemission [25–29] the photocurrent $I((\mathbf{k}_{||}\omega)$ is proportional to the probability of the transition from the initial state $|\,\mathrm{i}\,\mathbf{k}_{||}\,\rangle$ to the time reversed LEED state $|\,\mathrm{f}\,\mathbf{k}_{||}\,\rangle$:

$$I(\omega, \mathbf{k}_{||}) \sim \sqrt{E_{\mathrm{f}} - E_{\mathrm{vac}}}\left|\langle\,\mathrm{f}\,\mathbf{k}_{||}\,|\hat{o}|\,\mathrm{i}\,\mathbf{k}_{||}\,\rangle\right|^2, \tag{1}$$

where $\langle\,\mathbf{r}\,|\,\mathrm{f}\,\mathbf{k}_{||}\,\rangle = \Phi^*_{\mathrm{LEED}}(\mathbf{r})$, and $\mathbf{k}_{||}$ is the crystal momentum parallel to the surface. In the dipole approximation the perturbation operator is $\hat{o} = -i\nabla \cdot \mathbf{e}$, where $\mathbf{e}$ is the light polarization vector. Thereby, the dielectric response of the electronic system is neglected. In principle, the related spatially inhomogeneous exciting field may lead to sharp structures in the photon energy dependence of the photoemission intensity. Such local field effects are known to be important below the plasma frequency, where the conditions for the excitation of the multipole plasmon may be met [35]. Here, the energies well above the plasmon are considered and, although the dielectric response may be tangible also at the higher energies, the experience with other materials [36] suggests that there one can hardly expect the local fields to give rise to sharp spectral features.

The LEED wave function $\Phi_{\mathrm{LEED}}(\mathbf{r})$ is a scattering solution for a plane wave incident from vacuum with the final state energy $E$. Inside the graphene layer it satisfies the Schrödinger equation with the Hamiltonian $\hat{H} = -\Delta + V(\mathbf{r}) - iV_{\mathrm{i}}$. Here an imaginary potential $-iV_{\mathrm{i}}$ is added to the crystal potential $V(\mathbf{r})$ to allow for the inelastic scattering of the outgoing electron. In photoemission from semi-infinite crystals, the absorbing potential simulates the surface sensitivity of photoemission and leads to a momentum broadening perpendicular to the surface. For finite-thickness films the interaction with the electronic system is limited to a thin layer. In the present calculation it is chosen to vanish outside a thin slab between $z = -1$ and $1$ a.u., see Figure 1. The results were found to be rather insensitive to $V_{\mathrm{i}}$ in the range from 1 to 4 eV: the absorbing potential smoothes the electron transmission curves $T(E)$ and constant initial state (CIS) spectra $I(\omega)$ and reduces the peak intensities of $I(\omega)$ (by around 20–30% per 1 eV increase in $V_i$). Otherwise $V_i$ does not affect the shape of the curves. Increasing $V_i$ from 1 to 4 eV leads to an increase by 20% of the peak value of the intensity ratio in Figure 4a. The present calculations are for $V_i = 1$ eV.

In the electron diffraction calculation the wave is incident from the right, and the space is divided into three regions: (i) left vacuum half-space $z < z_L$, which contains the transmitted plane waves, (ii) scattering region $z_L \leq z \leq z_R$, and (iii) right vacuum half-space $z > z_R$, which contains the incident plane wave and reflected waves. In the scattering region the wave function is a linear combination of the eigenfunctions $\psi$ of an auxiliary three-dimensional $z$-periodic crystal, which contains the scattering region as a part of the unit cell, Figure 1. The solution of the scattering problem consists in constructing a linear combinations of the basis functions $\psi$ that satisfies the Schrödinger equation in region (ii) and at the planes $z_L$ and $z_R$ matches the function and derivative of the plane-wave representations in regions (i) and (iii), respectively. This is achieved by the variational embedding method introduced in Ref. [34]. Thus, a Laue representation of the LEED state is constructed:

$$\Phi_{\text{LEED}}(\mathbf{r}_{||}, z) = \sum_{\mathbf{G}_{||}} \phi_{\mathbf{G}_{||}}(z) \exp[i(\mathbf{k}_{||} + \mathbf{G}_{||})\,\mathbf{r}_{||}], \tag{2}$$

which in the present calculation comprises 19 surface reciprocal vectors $\mathbf{G}_{||}$. The lattice constant of the auxiliary crystal along $z$ was $c = 15$ Å, and the basis set in region (ii) comprised the $\psi$ functions with energies up to about 40 eV above the highest energy of interest, which amounts to around 200 $\psi$ functions for the monolayer graphene. The Laue representation (2) is obtained by a straightforward expansion of the all-electron wave function in terms of 11,997 plane waves ($G \leq 11$ a.u.$^{-1}$). The potential $V(\mathbf{r})$ of the auxiliary crystal is determined self-consistently within the local density approximation by the full-potential augmented Fourier components method [37].

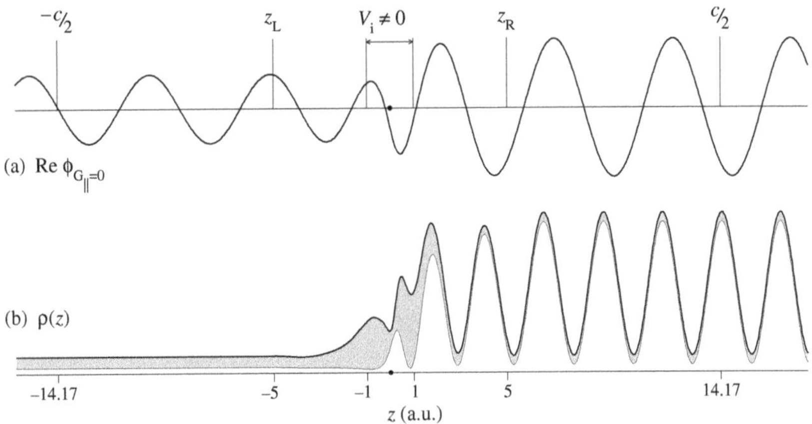


**Figure 1.** Wave function of the LEED state at $\mathbf{k}_{||} = 1.633$ Å$^{-1}$ along $\bar{\Gamma}\bar{K}$ and $E - E_F = 35$ eV. (a) Central beam $\phi_{\mathbf{G}_{||}=0}(z)$ of the Laue representation (2). (b) Density profile $\rho(z)$, see Equation (3). The graphene monolayer is at $z = 0$. The shaded area in graph (b) shows the contribution from the $\mathbf{G}_{||} \neq 0$ surface Fourier harmonics.

An example of the scattering solution for $\mathbf{k}_{||} = 1.633$ Å$^{-1}$ along $\bar{\Gamma}\bar{K}$ and $E = 35$ eV is presented in Figure 1. Figure 1b shows the density profile of this LEED state

$$\rho(z) = \int |\Phi_{\text{LEED}}(\mathbf{r}_{||}, z)|^2 \, d\mathbf{r}_{||} \tag{3}$$

and demonstrates that in the interior of the graphene layer the contribution from the $\mathbf{G}_{||} \neq 0$ harmonics strongly exceeds the $\mathbf{G}_{||} = 0$ contribution. It is the $\mathbf{G}_{||} \neq 0$ contribution that makes a single-plane-wave approximation for the final state $\Phi^*_{\mathrm{LEED}}$ unrealistic and misleading, see a detailed analysis in Ref. [38].

## 3. Results and Discussion

In this section, the calculation of photoemission form graphene is presented in the range $\hbar\omega = 20$ to 60 eV with the emphasis on the comparison with the experiments of Refs. [13,18]. The spectra are analyzed in terms of dipole transitions to the $\Phi^*_{\mathrm{LEED}}$ states for an all-electron Kohn–Sham potential. Detailed analysis of the monolayer and bilayer graphene is given in Sections 3.1 and 3.2, respectively.

### 3.1. Monolayer Graphene

Calculated polarization dependence of the photoemission from the monolayer and bilayer graphene is shown in Figure 2 for $\mathbf{k}_{||}$ along $\bar{\Gamma}\bar{K}$ around the DP for two photon energies $\hbar\omega = 35$ and 52 eV. The light incidence plane intersects the surface in the $\bar{\Gamma}\bar{K}$ line, and the angle of incidence is 18°, as in the experiment of Ref. [18]. For the monolayer graphene, the two branches have different parities under the reflection in $\bar{\Gamma}\bar{K}$ so the ascending branch ($B_2$ symmetry) is visible only in $p$ polarization and the descending branch ($A_2$) only in the $s$ polarization. A similar trend is observed in the graphene bilayer, only here the $\pi$ states are not parity eigenfunctions, so every state is visible in both polarizations, albeit with a striking intensity asymmetry. The absolute intensities and the asymmetry, however, depend on the photon energy, as seen from the comparison of Figure 2a–c,d–f.

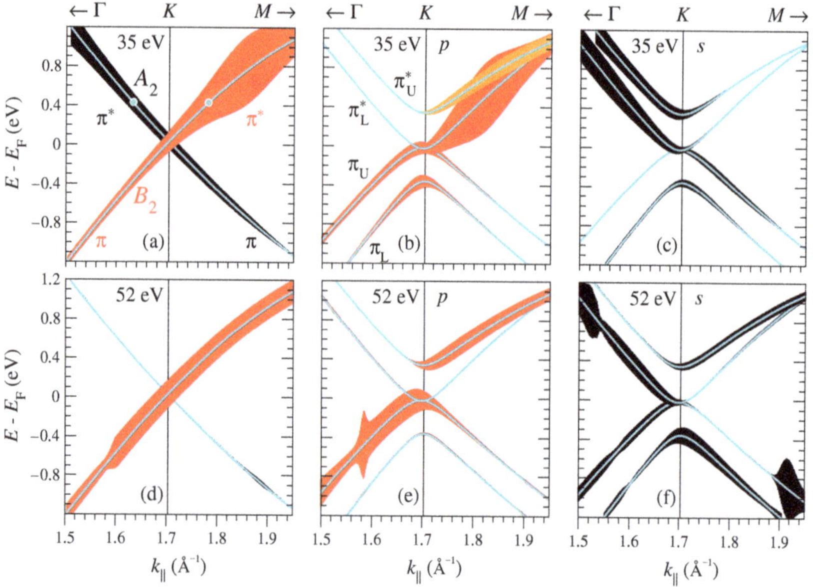

**Figure 2.** Crystal-momentum dependence of the photocurrent from graphene along $\bar{\Gamma}\bar{K}$: (**a,d**) monolayer; (**b,c,e,f**) bilayer. (**a–c**) $\hbar\omega = 35$ eV; (**d–f**) $\hbar\omega = 52$ eV. Light is incident along $\bar{\Gamma}\bar{K}$ at an angle of 18°. Intensity at $p$ polarization is shown by red and at $s$ polarization by black shading. The vertical extent of the shaded area is proportional to the relative intensity in the same graph (intensity is normalized differently in each of the graphs). The photon energy dependence of the intensity can be inferred from Figure 3. Because of the strict parity selection rules, for the monolayer graphene both polarizations are shown in the same graph (**a,d**). In graph (**a**) the two circles mark the initial states considered in Figure 4.

In agreement with the experimental observation of Ref. [13], in the monolayer graphene over a wide $\mathbf{k}_{\parallel}$ interval around $\bar{K}$ the intensity of the $s$ branch at 52 eV is an order of magnitude lower than at 35 eV. For each of the two photon energies 35 and 52 eV the intensity changes slowly and steadily with $\mathbf{k}_{\parallel}$, however, this is not the case for the $\hbar\omega$ interval between 35 and 52 eV, as illustrated by the intensity distribution $I(k_{\parallel}, \omega)$ in Figure 3. The $p$ branch manifests a sharp peak, which over the interval from 1.7 to 1.9 Å$^{-1}$ disperses from 40 to 33 eV and is followed by a minimum and a set of weaker structures at higher energies. The $s$ branch has two sharp maxima dispersing upwards: the one due to $\pi$ states (below the DP) in the second BZ around $\hbar\omega = 42$ eV and the one due to $\pi^*$ states (above the DP) in the first BZ around 38 eV. Apart from that, the $s$ branch of both $\pi$ and $\pi^*$ states manifests a sharp nondispersive dip at around 45.5 eV: the intensity drops by a factor of 5 over an interval of about 3 eV and then rapidly rises again.

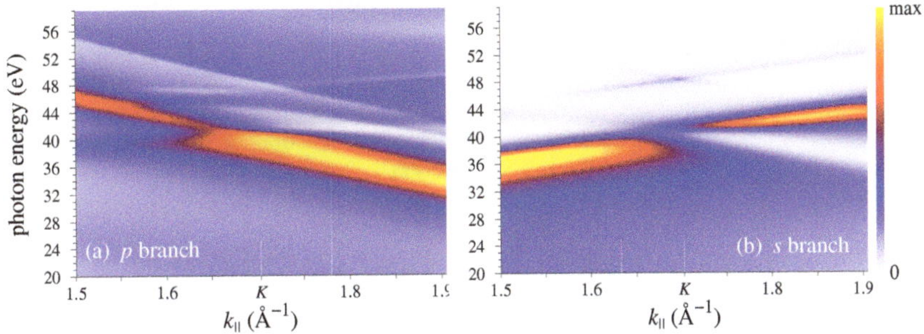

**Figure 3.** Photocurrent distribution in photon energy and crystal momentum from the $\pi$ and $\pi^*$ states for the same setup as in Figure 2. (**a**) $B_2$ states ($p$ branch). (**b**) $A_2$ states ($s$ branch). In each graph the two white vertical ticks show the $\bar{K}$ point and the $k_{\parallel}$ point presented in detail in Figure 4.

The constant initial state spectra for the $\pi^*$ states at $k_{\parallel} = 1.633$ and $1.780$ Å$^{-1}$ are shown in Figure 4b, and their ratio

$$R(\omega) = \frac{I_p(\omega)\cos^2\phi}{I_s(\omega)\sin^2\phi},\qquad(4)$$

where $\phi = 78°$ is the experimental polarization angle, is compared to the experiment [18] in Figure 4a. The minimum of the $s$ branch at $\hbar\omega = 45.5$ eV gives rise to a maximum in $R(\omega)$ very close in energy to the measured maximum at 46 eV. The calculated magnitude of $R(\omega)$ is two times lower than in the experiment, which can be considered a satisfactory agreement in view of the fact that it is related to a deep minimum in the denominator, i.e., to the cancellation effects in the momentum matrix element (1) for $s$ polarization. Naturally, in this situation the observables are especially sensitive to the accuracy of the wave functions, and an exact knowledge of all details is needed to achieve a perfect agreement. On the theoretical side, the discrepancy may arise from the neglect of the dielectric response (dipole approximation for the perturbation operator) and possibly from using the Kohn–Sham eigenfunctions for quasiparticles. Computational uncertainty related to the accuracy of the wave functions can hardly tangibly contribute to the discrepancy (judging by the convergence of the observables).

These results thereby establish the $R(\omega)$ peak to originate from the rapidly changing character of the final state wave function, and it is tempting to relate it to gross features of the scattered wave. In particular, because the initial states are confined to the graphene layer it is instructive to consider the spatial character of the LEED states as a function of energy, see Figure 5. The electron scattering by the graphene monolayer was first studied theoretically in Ref. [5], where the existence of scattering resonances was predicted that manifested themselves as rapid variations of the transmission probability $T(E)$ accompanied by a sharp enhancement of the density $\rho(z)$ at the graphene layer. Figure 5 shows that around the $\bar{K}$ point the resonances have rather complicated spatial structure, which strongly changes with $k_{||}$. Consequently, the probability to find the scattered electron at the graphene layer—the so-called dwell time $\tau(E)$—varies with energy. The $\tau(E)$ curves for $k_{||} = 1.633$ and $1.780$ Å$^{-1}$ are shown in Figure 4d (the probability density $\rho(z)$ was integrated from $z = -2$ to 2 a.u.). Both curves show rich structure, but the $\tau(E)$ variations do not correlate with those of the photocurrent, and although the dip in the $I_s(E)$ curve coincides with a minimum in the $\tau(E)$ curve, the former drops much deeper than the latter. Generally, the $\tau(E)$ variations are much weaker than the variations of the photocurrent, which points to the importance of the interference between different $\mathbf{G}_{||}$ contributions to the photoemission matrix element also for $p$ polarization.

Figure 4e shows the electron transmission spectra $T(E)$, i.e., the ratio of the transmitted current at $-\infty$ to the incident current from $+\infty$. The $T(E)$ curves show a minimum (at $E = 37$ eV for $k_{||} = 1.633$ Å$^{-1}$ and 34 eV for 1.780 Å$^{-1}$) followed by a maximum (at 42.5 and 39 eV, respectively), which is a signature of the scattering resonance [5]. The photoemission intensity peaks are located at $\hbar\omega = 37.3$ eV for $I_p$ and 38.1 eV for $I_s$, close to the inflection points of the respective $T(E)$ curves, $E = 37$ and 39 eV. Although it is not surprizing that the sharp enhancement of the intensity occurs in the resonance region, it cannot be directly related to the gross features of the final state, such as the transmission probability or density distribution.

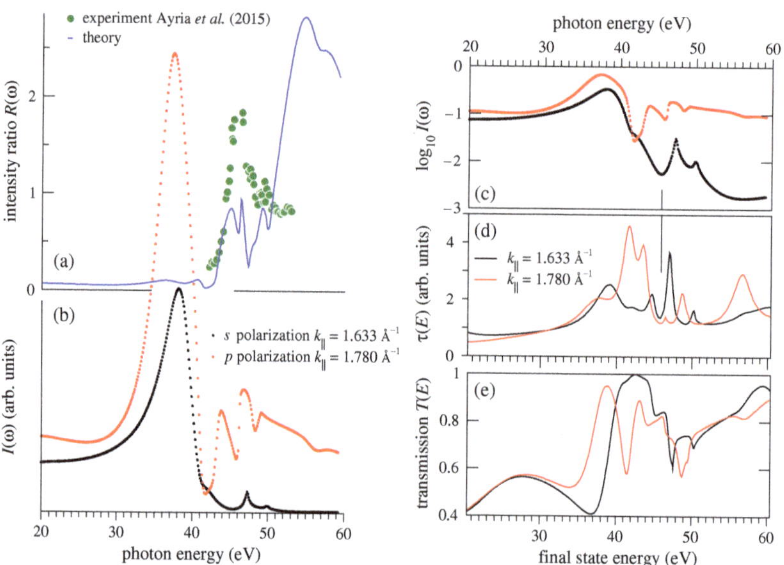

**Figure 4.** (**a**) Photon energy dependence of the relative intensity $R(\omega)$, see Equation (4), of the emission from the $B_2$ state ($I_p$) at $k_{||} = 1.780$ Å$^{-1}$ and $A_2$ state ($I_s$) at $k_{||} = 1.633$ Å$^{-1}$ for the polarization angle $\phi = 78°$ (4.3% of $p$ and 95.7% of $s$ polarization). Both initial states are located at about 0.4 eV above the DP. Full circles show the measurements of Ref. [18] (digitized from Figure 5 in that work). (**b**) Calculated constant initial state spectra $I_s(\omega)$ (black) and $I_p(\omega)$ (red). (**c**) Logarithmized intensities $\log_{10} I_s(\omega)$ (black) and $\log_{10} I_p(\omega)$ (red). (**d**) Dwell time $\tau(E)$: the probability to find the electron in the LEED state within a layer between $z = -2$ and 2 a.u., see Figure 1. The final state energy $E$ is relative to the DP. (**e**) Electron transmission $T(E)$ for $k_{||} = 1.633$ Å$^{-1}$ (black) and $k_{||} = 1.780$ Å$^{-1}$ (red). To facilitate the comparison, the $E$ range in graphs (**d**) and (**e**) is shifted by 0.4 eV (initial state energy) relative to the $\hbar\omega$ range in graph (**c**).

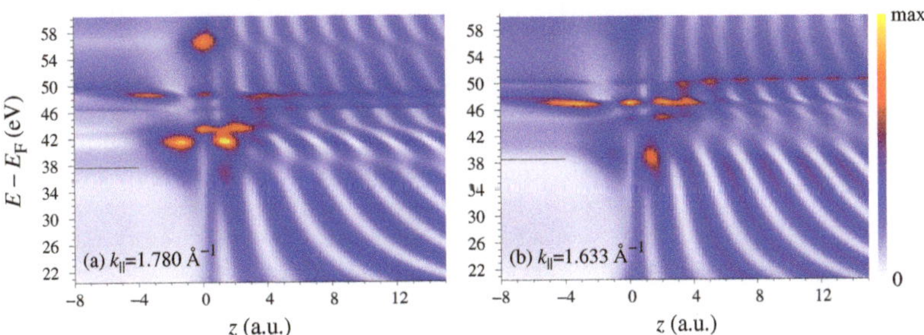

**Figure 5.** Energy dependence of the density distribution $\rho(z)$ in the LEED states: (**a**) $k_{||} = 1.780$ Å$^{-1}$ effects the $p$ branch emission, and (**b**) $k_{||} = 1.633$ Å$^{-1}$ the $s$ branch, see Figure 4b. The horizontal bars at 37.7 eV (**a**) and 38.5 eV (**b**) indicate the final states at the intensity peaks in Figure 4b.

*3.2. Bilayer Graphene: Relaxation of Parity Selection Rules*

Crystal momentum-photon energy distribution of the photocurrent for both light polarizations is presented in Figure 6 for the four bands around the $\bar{K}$ point: concave down bands $\pi_{\mathrm{L}}$ and $\pi_{\mathrm{U}}$ and concave up bands $\pi_{\mathrm{L}}^*$ and $\pi_{\mathrm{U}}^*$, see Figure 2b for notation. Similar to the monolayer graphene, the ascending branches are highlighted by the $p$-polarized light, while the descending ones by the $s$-polarized light. For both polarizations the CIS of each of the bands manifests a strong peak, which disperses downwards in $\hbar\omega$ with $k_{||}$ for $p$ polarization and upwards for $s$ polarization, compare Figures 3 and 6.

However, because the bilayer is not invariant under the reflection in the $\bar{\Gamma}\bar{K}$ line the parity selection rules are relaxed, and at certain $\hbar\omega$ the ascending and descending branches for a given light polarization may have comparable intensities. As seen in Figure 6, this happens when the intensity of the $p$- or $s$-highlighted branch drops off for reasons not related to the $\bar{\Gamma}\bar{K}$ reflection properties. For $p$ polarization this occurs, for example, around $\hbar\omega = 31$ eV, where the descending $\pi_{\mathrm{U}}^*$ branch turns out to have higher intensity than the ascending branch (Figure 6d). For $s$ polarization one can observe such asymmetry inversion for the $\pi_{\mathrm{L}}$ branch around 34 eV (Figure 6e).

In spite of the rather strong effect of the interlayer interaction on the $\pi$ states, the overall shape of the CIS curves is rather close for the monolayer and bilayer graphene, see Figure 7a,b. Let us now draw on these results to comment on the observation in Ref. [13] that in the monolayer graphene the emission from the $B_2$ band is visible also in the $s$-polarized light: this may be due to the scattering of the outgoing electron by the underlying substrate. It is reasonable to assume that the scattering by the substrate surface is comparable to the interlayer scattering in the bilayer graphene. To estimate its implications for the selection rules, we construct the matrix elements in Equation (1) between the initial states of the monolayer graphene (which are parity eigenfunctions) and the $\Phi_{\mathrm{LEED}}^*$ states of the bilayer graphene. This hybrid model yields the intensity distributions $I(k_{||}, \omega)$ very similar to those in Figure 3. As an example, the hybrid-model CIS curves for the $\pi^*$ states at $k_{||} = 1.780$ and $1.633$ Å$^{-1}$ are compared to the monolayer spectra in Figure 7c,d.

The extent to which the scattering by the second graphene layer relaxes the selection rules is revealed by Figure 7e–h, which compare the $k_{||}$ dependence of the emission from the $B_2$ and $A_2$ branch for both light polarizations by the hybrid model. For $p$ polarization (Figure 7e–g) the $B_2$ branch is about two orders of magnitude stronger than the $A_2$ branch. This is not surprising, as the dark corridor was also observed in photoemission from the bulk graphite [21,22]. The situation is somewhat different for $s$ polarization: again, for $\hbar\omega = 34$ and 35 eV the $A_2$ branch is two orders of magnitude more intense than the $B_2$ branch around the $\bar{K}$ point, but below the DP the intensities of the $A_2$ and $B_2$ branches become closer to each other in moving to lower energies, i.e., away from the $\bar{K}$ point, see Figure 7f. As we have seen for the two selected $k_{||}$, the intensity drop-off above the resonance is stronger for the $A_2$ branch than for the $B_2$ branch. Figure 7h demonstrates that this is the case over a wide $k_{||}$ interval around the $\bar{K}$ point and that at $\hbar\omega = 52$ eV the two branches are much closer in intensity than at 35 eV. This qualitatively agrees with the measurements of Ref. [13], where the overall contrast between the two branches was considerably stronger for 35 eV than for 52 eV. Furthermore, Figure 7f–h show that the contrast may be very sensitive to the photon energy: a variation of $\hbar\omega$ by 1 eV may change the intensity by a factor of 2. The hybrid model thus shows that the scattering of the photoelectron emitted from the graphene monolayer by the substrate may be sufficiently strong to break the symmetry of the photoexcitation. Another reason for the symmetry breaking is the spin–orbit interaction, as discussed in Ref. [13]. This effect is neglected in the present calculation.

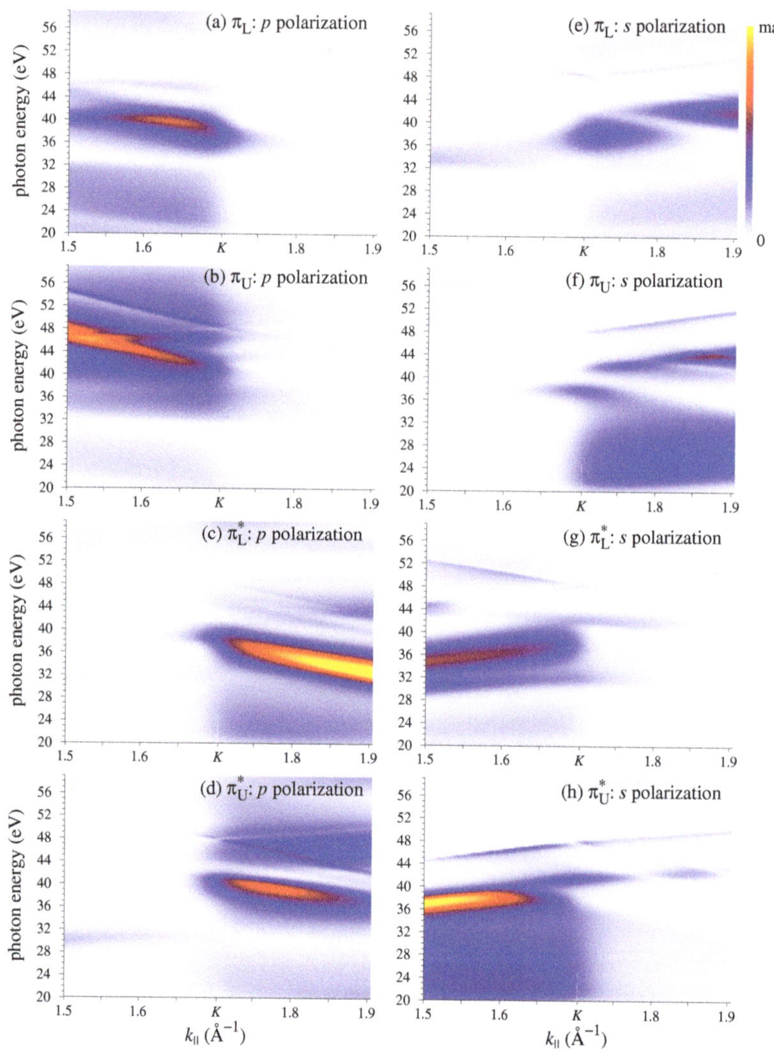

**Figure 6.** Photocurrent distribution in photon energy and crystal momentum from the $\pi$ and $\pi^*$ states of bilayer graphene, see Figure 2b for notation. (**a–d**) $p$ and (**e–h**) $s$ polarization. (Horizontal cross-sections of the maps at $\hbar\omega = 35$ and 52 eV are presented in Figure 2b,c,e,f).

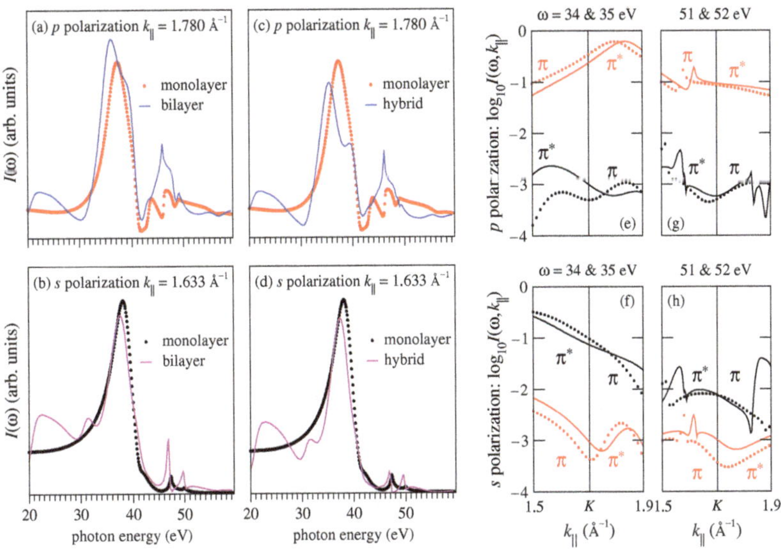

**Figure 7.** (**a**,**b**) Comparison of the CIS spectra for the monolayer (dots) and bilayer graphene (lines): (**a**) $p$ polarization, (**b**) $s$ polarization. The bilayer curves are a sum of the $\pi_L^*$ and $\pi_U^*$ spectra, see Figure 2b for notation. (**c**,**d**) Comparison of the monolayer (dots) and hybrid-model (lines) CIS spectra: (**c**) $p$ polarization, (**d**) $s$ polarization. The monolayer curves in graphs (**a**,**c**) and (**b**,**d**) are the same as the curves of the respective colors in Figure 4b. (**e**,**f**,**g**,**h**) Contrast between the $B_2$ and $A_2$ branches by the hybrid model. Crystal-momentum dependence of the intensity from the $\pi$ and $\pi^*$ states of the $B_2$ (red) and $A_2$ (black) branch for four photon energies: (**e**,**f**) $\hbar\omega = 34$ eV (lines) and 35 eV (dots); (**g**,**h**) $\hbar\omega = 51$ eV (lines) and 52 eV (dots) for $p$ polarization (**e**,**g**) and $s$ polarization (**f**,**h**).

## 4. Summary and Conclusions

The present application of the one-step theory of photoemission to the monolayer and bilayer graphene demonstrates a strong effect of the in-plane scattering of the outgoing photoelectron on the photoemission intensity. The continuum spectrum of graphene contains scattering resonances first discovered in Ref. [5] and interpreted as due to the coupling of the in-plane and perpendicular motions. At the $\bar{K}$ point the resonance is located around 38 eV above the DP, and the present theory predicts the photoemission from the Dirac cone to be strongly enhanced in the resonance region both for $p$ and for $s$ light polarization. Above the resonance the intensity drops more strongly for $s$ than for $p$ polarization, in agreement with the experiment [13]. In the interval up to about 15 eV above the resonance the scattering states have very complicated and rapidly changing structure, which is reflected both in the electron transmission and in the photoemission spectra, although no obvious correlation between the $T(E)$ and $I(\omega)$ curves is observed. (This means in particular that a single-plane-wave approximation for the final state would be completely inappropriate for graphene). The presence of this fine structure offers the possibility to relate the theoretically predicted spectral features to the measured ones and to verify the validity of the approximations involved, in particular, how accurately the density-functional derived potential simulates the excited states (it is known to underestimate the quasiparticle energies). The good agreement of the calculated energy dependence of the $I_p/I_s$ ratio (4) with the experiment [18] suggests that the self-energy shift is quite moderate (around 0.5 to 1 eV), as expected from previous experience [6,30–33], and that the Kohn–Sham quasiparticles are a good approximation for graphene.

The comparison of the monolayer and bilayer spectra is instructive in order to estimate the effect of the scattering by the substrate on the symmetry breaking in photoemission from the monolayer graphene. The true structure of the interface between the graphene monolayer and the substrate is very difficult to include in an ab initio calculation because the mismatch between the lattices of the substrate and graphene as well as the presence of the reconstructed buffer layer would require a huge supercell. Instead, we resorted to a hybrid model that combines the initial states of the monolayer graphene (which have $B_2$ or $A_2$ symmetry) with the final states of the bilayer (which are not symmetry eigenfunctions). Such a heuristic model is justified in view of the close similarity of the gross features of the monolayer and bilayer spectra. It shows that the relaxation of the selection rules is most important in the region of low intensity (above $\hbar\omega = 50$ eV for $s$ polarization) and that the symmetry breaking observed in Ref. [13] can be explained by the scattering from the substrate. Generally, at low intensities, the emission is very sensitive to this effect, which should be kept in mind in theoretically modeling this energy range with ideal free-standing graphene.

**Funding:** This research was funded by the Spanish Ministry of Science, Innovation and Universities (Project No. PID2019-105488GB-I00).

**Institutional Review Board Statement:** Not applicable.

**Informed Consent Statement:** Not applicable.

**Data Availability Statement:** Data is contained within the article.

**Conflicts of Interest:** The author declares no conflict of interest. The funders had no role in the design of the study; in the collection, analyses, or interpretation of data; in the writing of the manuscript, or in the decision to publish the results.

# References


1. Castro Neto, A.H.; Guinea, F.; Peres, N.M.R.; Novoselov, K.S.; Geim, A.K. The electronic properties of graphene. *Rev. Mod. Phys.* **2009**, *81*, 109–162. [CrossRef]
2. Hibino, H.; Kageshima, H.; Maeda, F.; Nagase, M.; Kobayashi, Y.; Yamaguchi, H. Microscopic thickness determination of thin graphite films formed on SiC from quantized oscillation in reflectivity of low-energy electrons. *Phys. Rev. B* **2008**, *77*, 075413. [CrossRef]
3. Srivastava, N.; Gao, Q.; Widom, M.; Feenstra, R.M.; Nie, S.; McCarty, K.F.; Vlassiouk, I.V. Low-energy electron reflectivity of graphene on copper and other substrates. *Phys. Rev. B* **2013**, *87*, 245414. [CrossRef]
4. Jobst, J.; Kautz, J.; Geelen, D.; Tromp, R.M.; van der Molen, S.J. Nanoscale measurements of unoccupied band dispersion in few-layer graphene. *Nat. Commun.* **2015**, *6*, 8926. [CrossRef] [PubMed]
5. Nazarov, V.U.; Krasovskii, E.E.; Silkin, V.M. Scattering resonances in two-dimensional crystals with application to graphene. *Phys. Rev. B* **2013**, *87*, 041405. [CrossRef]
6. Wicki, F.; Longchamp, J.N.; Latychevskaia, T.; Escher, C.; Fink, H.W. Mapping unoccupied electronic states of freestanding graphene by angle-resolved low-energy electron transmission. *Phys. Rev. B* **2016**, *94*, 075424. [CrossRef]
7. Krivenkov, M.; Marchenko, D.; Sánchez-Barriga, J.; Rader, O.; Varykhalov, A. Suppression of electron scattering resonances in graphene by quantum dots. *Appl. Phys. Lett.* **2017**, *111*, 161605. [CrossRef]
8. Ohta, T.; Bostwick, A.; Seyller, T.; Horn, K.; Rotenberg, E. Controlling the Electronic Structure of Bilayer Graphene. *Science* **2006**, *313*, 951–954.
9. Ohta, T.; Bostwick, A.; McChesney, J.L.; Seyller, T.; Horn, K.; Rotenberg, E. Interlayer Interaction and Electronic Screening in Multilayer Graphene Investigated with Angle-Resolved Photoemission Spectroscopy. *Phys. Rev. Lett.* **2007**, *98*, 206802. [CrossRef] [PubMed]
10. Bostwick, A.; Ohta, T.; Seyller, T.; Horn, K.; Rotenberg, E. Quasiparticle dynamics in graphene. *Nat. Phys.* **2007**, *3*, 36–40. [CrossRef]
11. Hwang, C.; Park, C.H.; Siegel, D.A.; Fedorov, A.V.; Louie, S.G.; Lanzara, A. Direct measurement of quantum phases in graphene via photoemission spectroscopy. *Phys. Rev. B* **2011**, *84*, 125422. [CrossRef]
12. Liu, Y.; Bian, G.; Miller, T.; Chiang, T.C. Visualizing Electronic Chirality and Berry Phases in Graphene Systems Using Photoemission with Circularly Polarized Light. *Phys. Rev. Lett.* **2011**, *107*, 166803. [CrossRef] [PubMed]
13. Gierz, I.; Henk, J.; Höchst, H.; Ast, C.R.; Kern, K. Illuminating the dark corridor in graphene: Polarization dependence of angle-resolved photoemission spectroscopy on graphene. *Phys. Rev. B* **2011**, *83*, 121408. [CrossRef]
14. Gierz, I.; Lindroos, M.; Höchst, H.; Ast, C.R.; Kern, K. Graphene Sublattice Symmetry and Isospin Determined by Circular Dichroism in Angle-Resolved Photoemission Spectroscopy. *Nano Lett.* **2012**, *12*, 3900–3904. [CrossRef] [PubMed]

15. Varykhalov, A.; Marchenko, D.; Sánchez-Barriga, J.; Scholz, M.R.; Verberck, B.; Trauzettel, B.; Wehling, T.O.; Carbone, C.; Rader, O. Intact Dirac Cones at Broken Sublattice Symmetry: Photoemission Study of Graphene on Ni and Co. *Phys. Rev. X* **2012**, *2*, 041017. [CrossRef]

16. Moreau, E.; Godey, S.; Wallart, X.; Razado-Colambo, I.; Avila, J.; Asensio, M.C.; Vignaud, D. High-resolution angle-resolved photoemission spectroscopy study of monolayer and bilayer graphene on the C-face of SiC. *Phys. Rev. B* **2013**, *88*, 075406. [CrossRef]

17. Hwang, C. Angle-resolved photoemission spectroscopy study on graphene using circularly polarized light. *J. Phys. Condens. Matter* **2014**, *26*, 335501. [CrossRef]

18. Ayria, P.; Nugraha, A.R.T.; Hasdeo, E.H.; Czank, T.R.; Tanaka, S.i.; Saito, R. Photon energy dependence of angle-resolved photoemission spectroscopy in graphene. *Phys. Rev. B* **2015**, *92*, 195148. [CrossRef]

19. Polley, C.M.; Johansson, L.I.; Fedderwitz, H.; Balasubramanian, T.; Leandersson, M.; Adell, J.; Yakimova, R.; Jacobi, C. Origin of the $\pi$-band replicas in the electronic structure of graphene grown on $4H$-SiC(0001). *Phys. Rev. B* **2019**, *99*, 115404. [CrossRef]

20. Schüler, M.; De Giovannini, U.; Hübener, H.; Rubio, A.; Sentef, M.A.; Werner, P. Local Berry curvature signatures in dichroic angle-resolved photoelectron spectroscopy from two-dimensional materials. *Sci. Adv.* **2020**, *6*.

21. Daimon, H.; Imada, S.; Nishimoto, H.; Suga, S. Structure factor in photoemission from valence band. *J. Electron Spectrosc. Relat. Phenom.* **1995**, *76*, 487–492.

22. Shirley, E.L.; Terminello, L.J.; Santoni, A.; Himpsel, F.J. Brillouin-zone-selection effects in graphite photoelectron angular distributions. *Phys. Rev. B* **1995**, *51*, 13614–13622. [CrossRef] [PubMed]

23. Kogan, E.; Nazarov, V.U.; Silkin, V.M.; Kaveh, M. Energy bands in graphene: Comparison between the tight-binding model and Ab Initio Calc. *Phys. Rev. B* **2014**, *89*, 165430. [CrossRef]

24. Braun, J. The theory of angle-resolved ultraviolet photoemission and its applications to ordered materials. *Rep. Prog. Phys.* **1996**, *59*, 1267. [CrossRef]

25. Adawi, I. Theory of the Surface Photoelectric Effect for One and Two Photons. *Phys. Rev.* **1964**, *134*, A788–A798. [CrossRef]

26. Mahan, G.D. Theory of Photoemission in Simple Metals. *Phys. Rev. B* **1970**, *2*, 4334–4350. [CrossRef]

27. Caroli, C.; Lederer-Rozenblatt, D.; Roulet, B.; Saint-James, D. Inelastic Effects in Photoemission: Microscopic Formulation and Qualitative Discussion. *Phys. Rev. B* **1973**, *8*, 4552–4569. [CrossRef]

28. Feibelman, P.J.; Eastman, D.E. Photoemission spectroscopy—Correspondence between quantum theory and experimental phenomenology. *Phys. Rev. B* **1974**, *10*, 4932–4947. [CrossRef]

29. Pendry, J.B. Theory of photoemission. *Surf. Sci.* **1976**, *57*, 679–705. [CrossRef]

30. Barrett, N.; Krasovskii, E.E.; Themlin, J.M.; Strocov, V.N. Elastic scattering effects in the electron mean free path in a graphite overlayer studied by photoelectron spectroscopy and LEED. *Phys. Rev. B* **2005**, *71*, 035427. [CrossRef]

31. de Jong, T.A.; Krasovskii, E.E.; Ott, C.; Tromp, R.M.; van der Molen, S.J.; Jobst, J. Intrinsic stacking domains in graphene on silicon carbide: A pathway for intercalation. *Phys. Rev. Mater.* **2018**, *2*, 104005. [CrossRef]

32. Geelen, D.; Jobst, J.; Krasovskii, E.E.; van der Molen, S.J.; Tromp, R.M. Nonuniversal Transverse Electron Mean Free Path through Few-layer Graphene. *Phys. Rev. Lett.* **2019**, *123*, 086802. [CrossRef]

33. Vilkov, O.Y.; Krasovskii, E.E.; Fedorov, A.V.; Rybkin, A.G.; Shikin, A.M.; Laubschat, C.; Budagosky, J.; Vyalikh, D.V.; Usachov, D.Y. Angle-resolved secondary photoelectron emission from graphene interfaces. *Phys. Rev. B* **2019**, *99*, 195421. [CrossRef]

34. Krasovskii, E.E. Augmented-plane-wave approach to scattering of Bloch electrons by an interface. *Phys. Rev. B* **2004**, *70*, 245322. [CrossRef]

35. Feibelman, P.J. Surface electromagnetic fields. *Prog. Surf. Sci.* **1982**, *12*, 287–407. [CrossRef]

36. Krasovskii, E.E.; Silkin, V.M.; Nazarov, V.U.; Echenique, P.M.; Chulkov, E.V. Dielectric screening and band-structure effects in low-energy photoemission. *Phys. Rev. B* **2010**, *82*, 125102. [CrossRef]

37. Krasovskii, E.E.; Starrost, F.; Schattke, W. Augmented Fourier components method for constructing the crystal potential in self-consistent band-structure calculations. *Phys. Rev. B* **1999**, *59*, 10504–10511. [CrossRef]

38. Krasovskii, E.E. Character of the outgoing wave in soft X-ray photoemission. *Phys. Rev. B* **2020**, *102*, 245139. [CrossRef]

 *nanomaterials*

*Article*

# Atomistic Band-Structure Computation for Investigating Coulomb Dephasing and Impurity Scattering Rates of Electrons in Graphene

Thi-Nga Do [1], Danhong Huang [2,*], Po-Hsin Shih [1], Hsin Lin [3] and Godfrey Gumbs [4]

[1] Department of Physics, National Cheng Kung University, Tainan 701, Taiwan; sofia90vn@gmail.com (T.-N.D.); PHShih@phys.ncku.edu.tw (P.-H.S.)

[2] US Air Force Research Laboratory, Space Vehicles Directorate (AFRL/RVSU), Kirtland Air Force Base, Albuquerque, NM 87117, USA

[3] Institute of Physics, Academia Sinica, Taipei 11529, Taiwan; nilnish@gmail.com

[4] Department of Physics and Astronomy, Hunter College of the City University of New York, 695 Park Avenue, New York, NY 10065, USA; ggumbs@hunter.cuny.edu

* Correspondence: danhong.huang@us.af.mil

**Abstract:** In this paper, by introducing a generalized quantum-kinetic model which is coupled self-consistently with Maxwell and Boltzmann transport equations, we elucidate the significance of using input from first-principles band-structure computations for an accurate description of ultra-fast dephasing and scattering dynamics of electrons in graphene. In particular, we start with the tight-binding model (TBM) for calculating band structures of solid covalent crystals based on localized Wannier orbital functions, where the employed hopping integrals in TBM have been parameterized for various covalent bonds. After that, the general TBM formalism has been applied to graphene to obtain both band structures and wave functions of electrons beyond the regime of effective low-energy theory. As a specific example, these calculated eigenvalues and eigen vectors have been further utilized to compute the Bloch-function form factors and intrinsic Coulomb diagonal-dephasing rates for induced optical coherence of electron-hole pairs in spectral and polarization functions, as well as the energy-relaxation time from extrinsic impurity scattering of electrons for non-equilibrium occupation in band transport.

**Keywords:** graphene; scattering; dephasing; relaxation time; band structure; tight-binding model

**Citation:** Do, T.-N.; Huang, D.; Shih, P.-H.; Lin, H.; Gumbs, G. Atomistic Band-Structure Computation for Investigating Coulomb Dephasing and Impurity Scattering Rates of Electrons in Graphene. *Nanomaterials* **2021**, *11*, 1194. https://doi.org/10.3390/nano11051194

Academic Editor: Francisco Javier García Ruiz

Received: 10 April 2021
Accepted: 28 April 2021
Published: 1 May 2021

**Publisher's Note:** MDPI stays neutral with regard to jurisdictional claims in published maps and institutional affiliations.



## 1. Introduction

Very recently, a generalized parameter-free quantum-kinetic model [1,2] based on many-body theory [3,4] has been developed, which is self-consistently coupled with Maxwell equations [5] for an interacting electromagnetic field and with Boltzmann transport equation [6] for a conduction current, as illustrated in Figure 1. Here, being an off-diagonal element in a density matrix, the induced quantum coherence for electron-hole pairs leads to a macroscopic optical polarization field [1] included in the Maxwell equations. Meanwhile, the modified electric field determined from the Maxwell equations can also change the microscopic quantum coherence [1] of electron-hole pairs. In this way, a self-consistent loop is constructed between electrons in the quantum-kinetic model and electric field in the Maxwell equations. This theory aims at enabling first-principles computations of ultra-fast dynamics for non-thermal photo-generated electron-hole pairs in undoped semiconductors [1,2]. At the same time, this a theory is also able to simultaneously describe electromagnetic, optical and electrical properties of crystal materials and their interplay all together. More importantly, the numerical output of this first-principles dynamics model can be utilized as an input for material optical and transport properties to be fed into a next-stage simulation software facilitated by finite-element methods, such as COMSOL

Multiphysics [7], for devices with various configurations. Consequently, device characteristics can be accurately predicted beyond the linear-response regime [3,4] for numerical bottom-up design and engineering. However, such a quantum-kinetic model itself requires an input from wave functions and band structures associated with different host materials in devices.

In Figure 1, we introduce the product of field frequency ($\omega$) with the carrier momentum-relaxation time ($\tau_p$). The situations with $\omega\tau_p \gg 1$ and $\omega\tau_p \ll 1$ correspond separately to optical and bias field regimes, while $\omega\tau_p \approx 1$ uniquely specifies the terahertz regime with dual optical and bias field characteristics. The bridging connection between the Maxwell [5] and semiconductor Bloch [8,9] equations is provided by the induced optical-polarization field $P(r,t)$ as a quantum-statistical average of the electric-dipole moment with the induced microscopic optical coherence $p_j(k,t)$ with $j$ the band index. The bridging connection between the Maxwell [5] and Boltzmann transport [6] equations, on the other hand, is fulfilled by the optically-induced magnetization field $M(r,t)$ as a quantum-statistical average of the induced microscopic magnetic-dipole moment $m_j(k,t)$ from spins or orbital angular momentum. Finally, the bridging connection between the semiconductor Bloch [8,9] and Boltzmann transport [6] equations is facilitated by the bias-induced macroscopic center-of-mass drift velocity $v_d(t)$ as a non-equilibrium quantum-statistical average of the microscopic electron group velocities $v_j(k)$ from multi-band dispersions for modifying optical-transition properties of driven carriers within the center-of-mass frame due to relative scattering motions of carriers.

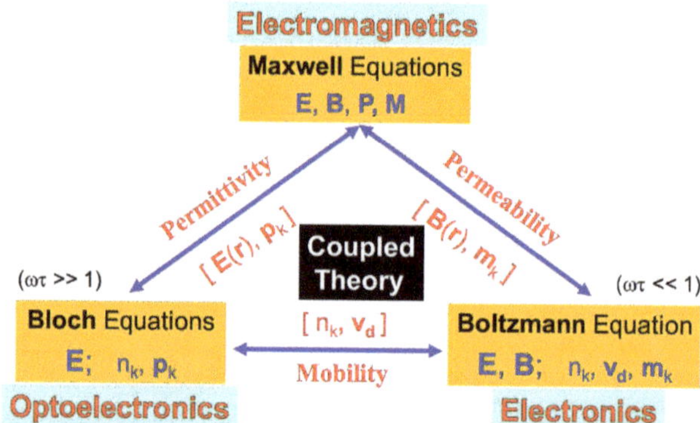


**Figure 1.** Illustration of a **Device Modeling & Simulation Triangle** for strong-coupling model applied to multi-functional electro-optical devices, where the device electromagnetic, opto-electronic and electronic characteristics are fully described by coupled Bloch, Maxwell, and Boltzmann equations all together.

The first-principles computation of electron Bloch wave function and band dispersion of a targeted material can be performed by employing the well-known Kohn-Sham density-functional theory [10]. Meanwhile, the tight-binding model [11–15] for solid crystals is usually considered as an alternative approach for computing electronic band structure using an approximate set of orbital wave functions based upon superposition of bond-orbital states for isolated atoms sitting at different lattice sites. In fact, this method is closely related to the linear combination of atomic orbitals method [16] adopted commonly in quantum chemistry. Such a real-space tight-binding model can be applied to a lot of solids, even including a magnetic field, Ref. [17] and it is proved giving rise to good qualitative results [18]. Moreover, this method can be combined with other models to produce better results whenever the tight-binding model fails. Here, we would like to emphasize that although the tight-binding model is only a one-electron model in nature, it indeed provides a basis for more advanced computations [11], such as the computation of surface states, application to various kinds of many-body problems, and quasi-particle calculation [19].

Historically, the family of carbon-based materials can be characterized into two distinct crystal forms, i.e., the isotropic diamond and anisotropic graphite. Recently, their allotropes, such as fullerenes and carbon nanotubes, entered into play and expanded to graphene, which is a unique material consisting of a two-dimensional lattice of carbon atoms with a honeycomb symmetry. Graphene stands for an physically interesting system [20,21], and becomes very promising for future device applications. On the atomic level, e.g., density-functional theory, electron certainly follows the Schrodinger equation. However, by using an approximate effective-mass Hamiltonian [22,23] for low-energy electrons near the $K$ or $K'$ valley, the quasi-particles are found to satisfy the relativistic Dirac equation for massless fermions. Today, the extensive investigations on various graphene systems have turned into a broad research field for qualitatively new two-dimensional systems [24]. Up to now, the basic properties of novel 2D allotropes of carbon, including graphene [22,23], graphene bilayer [25–27], multi-layer graphene [28,29], graphene on a silicon carbide substrate [30], are well known and the basis of graphene physics becomes well established.

In recent years, by using the low-energy Dirac Hamiltonian [4], we have extensively explored varieties of dynamical properties of electrons in graphene and other two-dimensional materials, including Landau quantization [18,31–35], many-body optical effects [36–41], band and tunneling transports [42–50], etc. In this paper, we particularly focus on the application of computed electronic states and band structures from a tight-binding model to the calculations of Coulomb and impurity scatterings of electrons in graphene on the basis of a many-body theory [3,4], where the former and latter determine the lineshape [1] of an absorption peak and the transport mobility [44], respectively.

The rest of paper is organized as follows. In Section 2, we present a general description of tight-binding model for novel two-dimensional materials. Section 3 is devoted to discuss the Slater-Koster approximation for bonding parameters and bonding integrals. We acquire the parameter values in Section 4 and obtain graphene wave functions and band structures. We study the Coulomb diagonal-dephasing rate of electron-hole pairs in undoped graphene in Section 5, as well as the impurity scattering rate of conduction electrons in Section 6, respectively. Finally, a brief summary is presented in Section 7 along with some remarks.

## 2. General Description of Tight-Binding Model

For completeness, we start with tight-binding model [14] for computing complete band structures of two-dimensional materials. The advantage of tight-binding model is easily incorporating a magnetic field through the so-called Peierls substitution in the phase of a hopping integral [51]. In quantum mechanics, the single-electron static Schrödinger equation is written as [52]

$$\hat{\mathcal{H}}\, \Phi_{\mathbf{k}}(\mathbf{r}) = E_{\mathbf{k}}\, \Phi_{\mathbf{k}}(\mathbf{r})\,, \tag{1}$$

where $\Phi_{\mathbf{k}}(\mathbf{r})$ is the Bloch wave function, $E_{\mathbf{k}}$ the eigen-energy, and $k$ is the wave vector of electrons within the first Brillouin zone of two-dimensional materials. The Hamiltonian operator $\hat{\mathcal{H}}$ in Equation (1) takes a general form

$$\hat{\mathcal{H}} = \hat{\mathcal{H}}_0 + V(\mathbf{r})\,, \tag{2}$$

in which the kinetic-energy operator $\hat{\mathcal{H}}_0$ is

$$\hat{\mathcal{H}}_0 = -\frac{\hbar^2}{2m_{\mathrm{e}}}\, \nabla_{\mathbf{r}}^2 \tag{3}$$

with free-electron mass $m_{\mathrm{e}}$, while the potential energy $V(\mathbf{r})$ for an electron within the lattice of two-dimensional materials is given by [11]

$$V(\mathbf{r}) = V_{\mathrm{ion}}(\mathbf{r}) + \Delta V_{\mathrm{L}}(\mathbf{r}) \tag{4}$$

with $V_{\mathrm{ion}}(\mathbf{r})$ and $\Delta V_{\mathrm{L}}(\mathbf{r})$ specifying the potentials of a single ion and that for the rest of ions, respectively. The Bloch wave function $\Phi_{\mathbf{k}}(\mathbf{r})$ of electrons in Equation (1) can be decomposed into a linear combination of a set of orbital wave functions $\{\phi_{\beta,\mathbf{k}}(r)\}$ within the first Brillouin zone, leading to [11]

$$\Phi_{\mathbf{k}}(r) = \sum_{\beta} C_{\beta}\, \phi_{\beta,\mathbf{k}}(r)\,, \tag{5}$$

where the index $\beta$ labels all the atomic orbitals of the lattice of two-dimensional materials. The expansion coefficient $C_{\beta}$ introduced in Equation (5) can be decided from

$$C_{\beta} = \int d^3r\, \phi^*_{\beta,\mathbf{k}}(r)\, \Phi_{\mathbf{k}}(r)\,, \tag{6}$$

where the orthonormal property for the set of orbital wave functions $\{\phi_{\beta,\mathbf{k}}(r)\}$ has been adopted.

Applying the method of linear combination of atomic orbitals (LCAO) of all ions on the lattice [16], we further express each orbital wave function $\phi_{\beta,\mathbf{k}}(r)$ in Equation (5) by a linear combination of bond-orbital states $\{\psi_{\beta}(r - R_j)\}$ within a unit cell in real space, namely

$$\phi_{\beta,\mathbf{k}}(r) = \frac{1}{\sqrt{N}}\sum_{j=1}^{N} \exp(ik\cdot R_j)\,\psi_{\beta}(r - R_j)\,, \tag{7}$$

where $j$ is the index for all bonded lattice ions, $R_j$ the lattice-ion position vector, and $N$ the total number of atoms within the unit cell. Here,

$$\psi_{\beta}(r - R_j) = \frac{1}{\sqrt{N}}\sum_{\mathbf{k}} \exp(-ik\cdot R_j)\,\phi_{\beta,\mathbf{k}}(r) \tag{8}$$

is termed as the localized Wannier function for the $\beta$ orbital of a bonded lattice ion at the site $R_j$, which satisfies the single-ion Schrödinger equation [16]

$$\left[\hat{\mathcal{H}}_0 + V_{\text{ion}}(r)\right]\psi_{\beta}(r - R_j) = \varepsilon_{j,\beta}\,\psi_{\beta}(r - R_j) \tag{9}$$

with $\varepsilon_{j,\beta}$ being the $\beta$th energy levels of electrons within an ion at the lattice site $R_j$.

Combining results in Equations (5) and (7), we acquire the following full LCAO expansion of a Bloch wave function [11]

$$\Phi_{\mathbf{k}}(r) = \frac{1}{\sqrt{N}}\sum_{j=1}^{N}\sum_{\beta} C_{\beta;j}(k)\,\psi_{\beta}(r - R_j) \tag{10}$$

with $C_{\beta;j}(k) = C_{\beta}\exp(ik\cdot R_j)$. At the same time, using Equation (5), we find from Equation (1) that

$$\sum_{\beta} C_{\beta} \int d^2r\, \phi^*_{\alpha,\mathbf{k}}(r)\,\hat{\mathcal{H}}\,\phi_{\beta,\mathbf{k}}(r) = E(k)\sum_{\beta} C_{\beta} \int d^2r\, \phi^*_{\alpha,\mathbf{k}}(r)\,\phi_{\beta,\mathbf{k}}(r)\,, \tag{11}$$

or equivalently, the following eigenvalue equation

$$\sum_{\beta} \mathcal{H}_{\alpha,\beta}(k)\,C_{\beta} = E(k)\sum_{\beta}\delta_{\alpha,\beta}\,C_{\beta} = E(k)\,C_{\alpha}\,. \tag{12}$$

As a result, the eigenvalue $E_n(k)$ can be determined from the secular determinant of Equation (12) for any given $k$, yielding

$$\mathcal{D}et\{\mathcal{H}_{\alpha,\beta}(k) - E_n(k)\,\delta_{\alpha,\beta}\} = 0\,, \tag{13}$$

and the orthonormal-eigenvectors $\{C_{n,\beta}\}$ are also obtained, corresponding to the eigenvalue $E_n(k)$ at given $k$, where the index $n$ labels different quantized energy bands of two-dimensional materials. Explicitly, using Equation (7), we obtain the Hamiltonian matrix elements in Equation (12) as [11]

$$\mathcal{H}_{\alpha\beta}(k) = \frac{1}{N} \sum_{i,j=1}^{N} \exp\left(ik \cdot R_{ij}\right) \mathcal{H}_{i\alpha,j\beta}, \tag{14}$$

in which $R_{ij} = R_j - R_i$, and

$$\mathcal{H}_{i\alpha,j\beta} = \int d^2r\, \psi_\alpha^*(r - R_i) \left[\hat{\mathcal{H}}_0 + V_{\text{ion}}(r)\right] \psi_\beta(r - R_j) + \int d^2r\, \psi_\alpha^*(r - R_i)\, \Delta V_{\text{L}}(r)\, \psi_\beta(r - R_j). \tag{15}$$

In fact, we know from Equation (9) that

$$\int d^2r\, \psi_\alpha^*(r - R_i) \left[\hat{\mathcal{H}}_0 + V_{\text{ion}}(r)\right] \psi_\beta(r - R_j) = \varepsilon_{j,\beta}\, \delta_{\alpha,\beta}\, \delta_{i,j}, \tag{16}$$

$$\int d^2r\, \psi_\alpha^*(r - R_i)\, \Delta V_{\text{L}}(r)\, \psi_\beta(r - R_j) \equiv \begin{cases} C_\Sigma\, \delta_{\alpha,\beta} & \text{if } i = j, \\ t_{\alpha\beta}(R_{ij}) & \text{if } i \neq j, \end{cases} \tag{17}$$

where $(\varepsilon_{j,\beta} + C_\Sigma)$ represents the site energy, and $t_{\alpha\beta}(R_{ij})$ is usually called the two-center (or hopping) integral [14].

As a final step, with the help from Equation (10), we arrive at the full expression for Hamiltonian matrix elements, given by

$$\int d^2r\, \Phi_{n',k'}^*(r)\, \hat{\mathcal{H}}\, \Phi_{n,k}(r) = \frac{1}{N} \sum_{j,j'=1}^{N} \sum_{\alpha,\beta} C_{n',\alpha}^* C_{n,\beta} \exp\left(ik \cdot R_j - ik' \cdot R_{j'}\right) \mathcal{H}_{j'\alpha,j\beta}$$

$$= \frac{(\varepsilon_{j,\beta} + C_\Sigma)}{N} \sum_{j=1}^{N} \sum_{\beta} C_{n',\beta}^* C_{n,\beta} \exp\left[i(k - k') \cdot R_j\right]$$

$$+ \frac{1}{N} \sum_{j,j'=1}^{N}{}' \sum_{\alpha,\beta} C_{n',\alpha}^* C_{n,\beta} \exp\left(ik \cdot R_j - ik' \cdot R_{j'}\right) t_{\alpha\beta}(R_{j'j}), \tag{18}$$

where the primed summation in the second term of the right-hand side of the last equation excludes the contribution from $j = j'$, and $C_{n,\beta}$ can be obtained from the calculated eigenvector from Equation (12). The matrix elements for other physical operators can be computed in a similar way.

## 3. Slater-Koster Approximation for Hopping Integrals

To seek for the feasibility of fast numerical computation, we introduce a parameterized process for the tight-binding model described in Section 2. For the Coulomb interaction between electron and ion within an atom, the potential field presents a spherical symmetry. Therefore, the energy levels labeled by the radial quantum number $n = 1, 2, \cdots$ will degenerate with the angular-momentum quantum number $\ell = 0, 1, \cdots, n - 1$, as well as the magnetic quantum number $m = -\ell, \cdots, 0, \cdots, \ell$ [52]. Consequently, there exists a total orbital degeneracy $n^2$ (excluding the spin-degeneracy). Customarily, we specify these orbitals by $\ell = 0, 1, 2, 3, \cdots$ for $\{s, p, d, f, \cdots\}$ orbitals.

In order to describe the chemical bonds between a pair of atoms inside a lattice, we often adopt the concept of overlapping electronic orbitals $\{s, p, d, f, \cdots\}$. To further specify the spatial direction of the chemical bonding between two atoms at the lattice sites $R_i$ and $R_j$, we have to rely on three directional cosines $\ell, m, n$, as defined in Figure 2.

Considering $s$ and $p$ orbitals as an example, we display their possible bonding potentials $V_{\ell,\ell';\sigma(\pi)}$ in Figure 3 for $s$, $p$ orbitals and four different configurations, including $\sigma$ and $\pi$ bonds. Meanwhile, we also list six different $\pi$, $\sigma$, $\delta$ bonding configurations in Figure 4 for $s$, $p$, $d$ orbitals.

To speed up numerical computations, the bonding potentials $V_{\ell,\ell';\sigma(\pi)}$ for $\ell, \ell' = s, p, d$ in Figures 3 and 4 are usually parameterized as: [53] $V_{\ell,\ell';\xi} = (\hbar^2/m_e d^2)\, \eta_{\ell\ell';\xi}$,

$V_{\ell,d;\xi} = (\hbar^2 r_d^{3/2} / m_e d^{7/2}) \, \eta_{\ell,d;\xi}$ and $V_{d,d;\xi} = (\hbar^2 r_d^3 / m_e d^5) \, \eta_{d,d;\xi}$, where $d$ and $r_d$ represent the bonding length and atomic radius, and $\xi = \sigma, \pi, \delta$ are for various bond configurations. Here, the dimensionless bonding parameters $\eta_{\ell,\ell';\xi}$ for different bonding types are listed in Table 1.

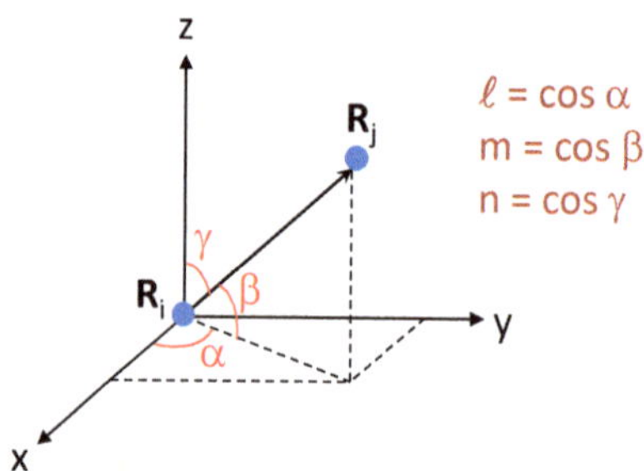


**Figure 2.** Illustration for three directional cosines $\ell$, $m$, $n$ in a three-dimensional position space for two atoms sitting at $r = R_i$ and $r = R_j$, respectively.

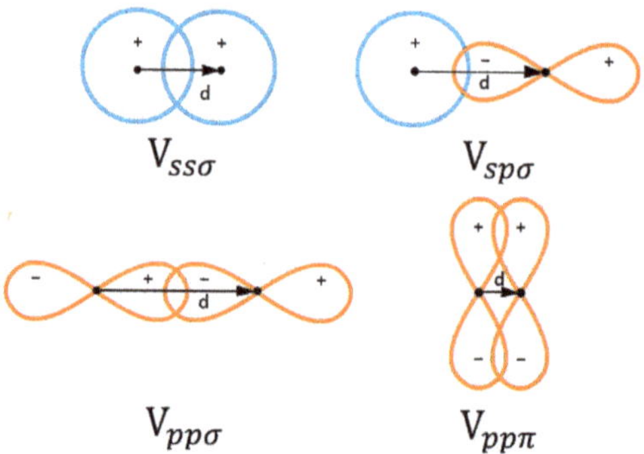


**Figure 3.** Illustrations for $\pi$ and $\sigma$ bonding of atomic $s$ and $p$ orbitals. Details on description of these bonding orbitals in this figure can be found in Ref. [11].

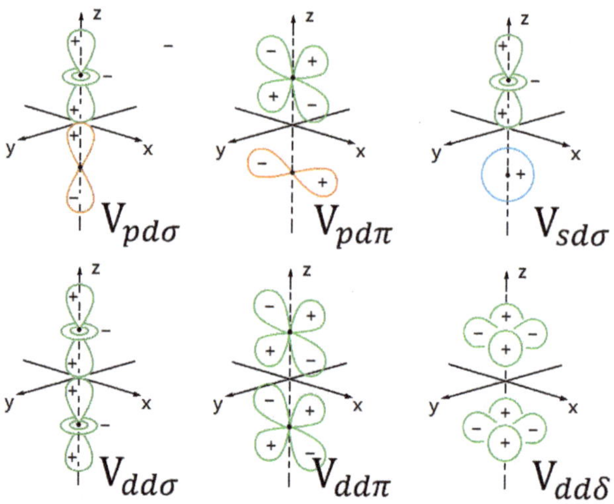


**Figure 4.** Illustrations for $\pi$, $\sigma$ and $\delta$ bonding of atomic $s$, $p$, and $d$ orbitals. Details on description of these bonding orbitals in this figure can be found in Ref. [11].

**Table 1.** Inter-atomic bonding parameters.

| Bonding Parameter | Value [11] |
|---|---|
| $\eta_{s,s;\sigma}$ | $-1.40$ |
| $\eta_{s,d;\sigma}$ | $-3.16$ |
| $\eta_{d,d;\sigma}$ | $-16.2$ |
| $\eta_{s,p;\sigma}$ | $1.84$ |
| $\eta_{p,d;\sigma}$ | $-2.95$ |
| $\eta_{d,d;\pi}$ | $8.75$ |
| $\eta_{p,p;\sigma}$ | $3.24$ |
| $\eta_{p,d;\pi}$ | $1.36$ |
| $\eta_{d,d;\delta}$ | $0$ |
| $\eta_{p,p;\pi}$ | $-0.81$ |

By using these parameterized bonding potentials $V_{\ell,\ell';\xi}$, $V_{\ell,d;\xi}$ and $V_{d,d;\xi}$, we are able to compute further the hopping integrals $t_{\alpha\beta}(R_{ij})$ based on the Slater-Koster approximation [14], and some commonly-used results are shown in Table 2.

**Table 2.** Expressions for Bonding Integrals.

| Bonding Integral | Expression [14] |
|---|---|
| $t_{s,s}$ | $V_{s,s;\sigma}$ |
| $t_{s,x}$ | $\ell\, V_{s,p;\sigma}$ |
| $t_{x,x}$ | $\ell^2\, V_{p,p;\sigma} + (1 - \ell^2)\, V_{p,p;\pi}$ |
| $t_{x,y}$ | $\ell m\, (V_{p,p;\sigma} - V_{p,p;\pi})$ |
| $t_{x,z}$ | $\ell n\, (V_{p,p;\sigma} - V_{p,p;\pi})$ |
| $t_{s,xy}$ | $\sqrt{3}\ell m\, V_{s,d;\sigma}$ |
| $t_{s,x^2-y^2}$ | $(\sqrt{3}/2)\, (\ell^2 - m^2)\, V_{s,d;\sigma}$ |
| $t_{3z^2-r^2}$ | $[(n^2 - (\ell^2 + m^2)/2]\, V_{s,d;\sigma}$ |
| $\vdots$ | $\vdots$ |

## 4. Tight-Binding Model for Graphene Band Structure

To seek for an application, we use the general theory, as developed in Sections 2 and 3, for novel two-dimensional graphene material in order to obtain its electronic wave functions and band structures for the full first Brillouin zone [54]. In this way, we are able to study scattering dynamics with respect to high-energy electrons in graphene resulted from Coulomb interactions between either pair of electrons or between electrons and ionized impurity atoms.

Monolayer graphene displays a hexagonal (or honeycomb) lattice structure of carbon atoms, as illustrated in Figure 5, where each carbon atom is connected by $\sigma$ covalent bonds with its three nearest neighbors. The electronic orbitals of a carbon atom are characterized as $1s^2\,2s^2\,2p^2$. However, the unique energy difference between the $2s$ and $2p$ orbitals favors the appearance of a mixed state of these two orbitals. The first-principles density-functional calculations reveal that it becomes energetically favorable to move an electron from the $2s$ orbital to the $2p$ orbital in this mixed state. Since the $2p$ orbitals include $2p_x$, $2p_y$, $2p_z$, as a result, each of these three $2p$ orbitals will accommodate one electron, leading to the $x$–$y$ orbitals within the plane of the lattice, as well as the $z$ orbital out of the lattice plane. Here, two electrons in the mixed $x$–$y$ orbitals form the higher-energy $\sigma$ bonds, while the remaining electron in the $z$ orbital leads to the lower-energy $\pi$ bonds, i.e., a side-on overlap of the $2p$-orbital wave functions. Consequently, these $\pi$-bond electrons give rise to the low-energy bands of graphene and will be studied exclusively based on a tight-binding model.

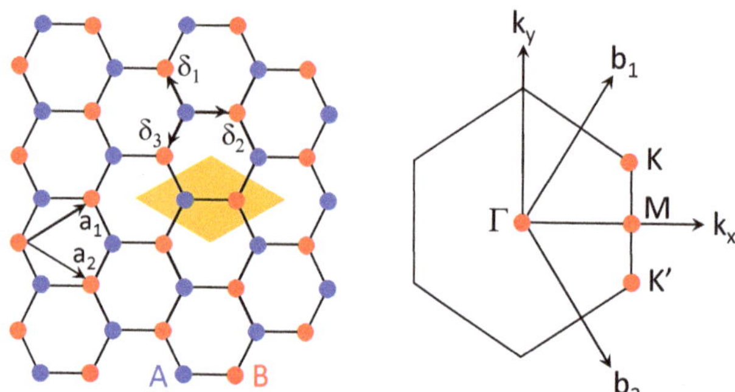

**Figure 5. (Left)** A diagram illustrating the hexagonal-lattice structure of a monolayer graphene with two sublattices $A$ (blue) and $B$ (red) with the Bravis lattice vectors $a_{1,2}$ and the nearest-neighbor lattice vectors $\delta_{1,2,3}$. **(Right)** the first Brillouin zone of graphene with labeled high-symmetry points $\Gamma$, $M$, $K$, $K'$ in the $k$–space with reciprocal-lattice vectors $b_{1,2}$. In the left panel, a unit cell is shown in the shaded region in yellow.

From Equation (7), we know the wave function for $\pi$-bond ($p_z$-orbital) electrons in graphene can be expressed as

$$\phi^{A,B}_{p_z,\mathbf{k}}(\mathbf{r}) = \frac{1}{\sqrt{N}} \sum_{j=1}^{N} \exp(i\mathbf{k} \cdot \mathbf{R}_j)\, \psi^{A,B}_{p_z}(\mathbf{r} - \mathbf{R}_j)\,, \tag{19}$$

where $\mathbf{k} \equiv (k_x, k_y)$, $\mathbf{R}_j \equiv (R_j^x, R_j^y) = m_j a_1 + n_j a_2$ represents the Bravis lattice-site vectors as indicated in Figure 5, and indexes $A$, $B$ refer to two sublattices of graphene. By including both sublattices $A$ and $B$, we have

$$\phi_{p_z,\mathbf{k}}(\mathbf{r}) = a_\mathbf{k}\, \phi^{A}_{p_z,\mathbf{k}}(\mathbf{r}) + b_\mathbf{k}\, \phi^{B}_{p_z,\mathbf{k}}(\mathbf{r})\,, \tag{20}$$

where $a_\mathbf{k}$ and $b_\mathbf{k}$ are two elements of the eigenvector corresponding to the eigenvalue equation with respect to two sublattices. Specifically, from Equations (11) and (20), we arrive at the matrix-form Schrödinger equation

$$\begin{bmatrix} \mathcal{H}_{AA}(\mathbf{k}) & \mathcal{H}_{AB}(\mathbf{k}) \\ \mathcal{H}_{BA}(\mathbf{k}) & \mathcal{H}_{BB}(\mathbf{k}) \end{bmatrix} \begin{bmatrix} a_\mathbf{k} \\ b_\mathbf{k} \end{bmatrix} = E_n(\mathbf{k}) \begin{bmatrix} \mathcal{S}_{AA}(\mathbf{k}) & \mathcal{S}_{AB}(\mathbf{k}) \\ \mathcal{S}_{BA}(\mathbf{k}) & \mathcal{S}_{BB}(\mathbf{k}) \end{bmatrix} \begin{bmatrix} a_\mathbf{k} \\ b_\mathbf{k} \end{bmatrix}\,, \tag{21}$$

where $\mathcal{S}_{\ell\ell'}(\mathbf{k}) = \langle \phi^{\ell}_{p_z,\mathbf{k}} | \phi^{\ell'}_{p_z,\mathbf{k}} \rangle$, $\mathcal{H}_{\ell\ell'}(\mathbf{k}) = \langle \phi^{\ell}_{p_z,\mathbf{k}} | \hat{\mathcal{H}} | \phi^{\ell'}_{p_z,\mathbf{k}} \rangle$, $\ell, \ell' = A$ or $B$, and $E_n(\mathbf{k})$ represents the eigen-energies of $\pi$-bond electrons with $n = 1$, 2 labeling two graphene low-energy bands determined by the secular determinant: $\mathcal{D}et\{ \mathcal{H}_{\ell\ell'}(\mathbf{k}) - E_n(\mathbf{k})\, \mathcal{S}_{\ell\ell'}(\mathbf{k}) \}_{2\times 2} = 0$.

As in Equation (7), we can rewrite the orbital wave function $\phi_{p_z,\mathbf{k}}(\mathbf{r})$ in Equation (20) approximately only by its near-neighbor decomposition, yielding

$$\phi_{p_z,\mathbf{k}}(\mathbf{r}) = \frac{1}{\sqrt{N_c}} \sum_{\ell \in A,B} \exp(i\mathbf{k} \cdot \mathbf{R}_\ell) \sum_{j=1}^{N_c} a_{j\mathbf{k}}\, \psi_{p_z}(\mathbf{r} - \mathbf{R}_\ell + \Delta_j)\,, \tag{22}$$

and then, the eigenvalue equation turns into $\mathcal{D}et\left\{ \mathcal{H}_{jj'}(\mathbf{k}) - E_s(\mathbf{k})\, \mathcal{S}_{jj'}(\mathbf{k}) \right\}_{N_c \times N_c} = 0$ with eigen-vectors $\{a_{j\mathbf{k}}\}_{N_c \times 1}$, where $N_c$ represents the number of near-neighbor atoms within a unit cell, $\Delta_j$ stands for the lattice vectors of the near-neighbor atoms relative to the sublattice site $R_\ell$, and $a_{j\mathbf{k}} = a_{\mathbf{k}} \exp(-i\mathbf{k} \cdot \Delta_j)$. Moreover, we find $\mathcal{H}_{jj'}(k) = \varepsilon_2 \, \mathcal{S}_{jj'}(k) + t_{jj'}(k)$, where $\varepsilon_2$ stands for the second energy level of electrons within a carbon atom,

$$\mathcal{S}_{jj'}(k) = \sum_{\ell \in A,B} \exp(ik \cdot R_\ell) \int d^2r\, \psi_{p_z}^*(r - R_\ell + \Delta_j)\, \psi_{p_z}(r - R_\ell + \Delta_{j'}) \tag{23}$$

is the overlap integral, while

$$t_{jj'}(k) = \sum_{\ell \in A,B} \exp(ik \cdot R_\ell) \int d^2r\, \psi_{p_z}^*(r - R_\ell + \Delta_j)\, \Delta V_L(r)\, \psi_{p_z}(r - R_\ell + \Delta_{j'}) \tag{24}$$

is the hopping integral.

For simplicity, we would omit the orbital index $p_z$ from now on. Without loss of generality, we can assume that the vectors that connect sublattice $A$ site to the equivalent site on the $B$ sublattice is $\delta_3$, as seen in Figure 5. As a result, the hopping and overlap amplitudes between the nearest neighbor (nn) and the next-nearest neighbor (nnn) can be computed explicitly from Equations (23) and (24), leading to

$$
\begin{aligned}
t_{AB}(k) &= t_{BA}^*(k) = \gamma^*(k)\, t_{nn}\,,\\[4pt]
t_{AA}(k) - C_p \mathcal{S}_{AA} &= t_{BB}(k) - C_p \mathcal{S}_{BB} = 2\, t_{nnn} \sum_{i=1}^{3} \cos(k \cdot a_i) = \left(|\gamma(k)|^2 - 3\right) t_{nnn}\,,\\[4pt]
\mathcal{S}_{AB}(k) &= \mathcal{S}_{BA}^*(k) = \gamma^*(k)\, s_{nn}\,,\\[4pt]
\mathcal{S}_{AA}(k) &= \mathcal{S}_{BB}(k) = 1 + \left(|\gamma(k)|^2 - 3\right) s_{nnn} \approx 1\,,
\end{aligned}
\tag{25}
$$

where $a_3 \equiv a_1 - a_2$, $\gamma(k) = 1 + \exp(ik \cdot a_1) + \exp(ik \cdot a_2)$, and the hopping and overlap integrals are calculated as

$$
\begin{aligned}
C_p &= \int d^2r\, \psi_A^*(r)\, \Delta V_L(r)\, \psi_A(r) = \int d^2r\, \psi_B^*(r)\, \Delta V_L(r)\, \psi_B(r)\,,\\[4pt]
t_{nn} &= \int d^2r\, \psi_A^*(r)\, \Delta V_L(r)\, \psi_B(r + \delta_3)\,,\\[4pt]
t_{nnn} &= \int d^2r\, \psi_A^*(r)\, \Delta V_L(r)\, \psi_A(r + a_1) = \int d^2r\, \psi_B^*(r)\, \Delta V_L(r)\, \psi_B(r + a_1)\,,\\[4pt]
s_{nn} &= \int d^2r\, \psi_A^*(r)\, \psi_B(r + \delta_3)\,,\\[4pt]
s_{nnn} &= \int d^2r\, \psi_A^*(r)\, \psi_A(r + a_1) = \int d^2r\, \psi_B^*(r)\, \psi_B(r + a_1)\,.
\end{aligned}
\tag{26}
$$

Particularly, the results for these tight-binding model parameters in Equation (26) for band structures are presented in Table 3, which have been computed from listed bonding parameters in Table 1 and bonding integrals in Table 2.

Finally, from the eigenvalue equation $\mathcal{D}et\{\, t_{\ell\ell'}(k) - E(k)\, \mathcal{S}_{\ell\ell'}(k)\,\}_{2\times 2} = 0$ in Equation (21) for $\ell, \ell' = A, B$, we obtain an explicit expression

$$E^2(k)\, \mathcal{D}et\{\overleftrightarrow{\mathcal{S}}\} - E(k)[\mathcal{S}_{AA} t_{BB} + \mathcal{S}_{BB} t_{AA} - \mathcal{S}_{AB} t_{BA} - \mathcal{S}_{BA} t_{AB}] + \mathcal{D}et\{t\} = 0\,, \tag{27}$$

where, by setting $s_{nnn} = 0$, we have three coefficients

$$
\begin{aligned}
\mathcal{D}et\{\overleftrightarrow{\mathcal{S}}\} &= 1 - s_{nn}^2\, |\gamma(k)|^2\,,\\[4pt]
\mathcal{D}et\{\overleftrightarrow{t}\} &= (|\gamma(k)|^2 - 3|)^2\, t_{nnn}^2 - t_{nn}^2\, |\gamma(k)|^2\,,\\[4pt]
\mathcal{S}_{AA} t_{BB} + \mathcal{S}_{BB} t_{AA} - \mathcal{S}_{AB} t_{BA} - \mathcal{S}_{BA} t_{AB} &= 2\Big[(|\gamma(k)|^2 - 3)\, t_{nnn} - t_{nn} s_{nn}\, |\gamma(k)|^2\Big]\,.
\end{aligned}
\tag{28}
$$

This leads to the explicit solution of Equation (27), namely [55]

$$E_\lambda(k) = (\varepsilon_2 + C_p) + \frac{(|\gamma(k)|^2 - 3|)\, t_{nnn} - t_{nn}s_{nn}\,|\gamma(k)|^2 + \lambda\sqrt{\mathcal{D}(k)}}{1 - s_{nn}^2|\gamma(k)|^2}\,, \tag{29}$$

where $\lambda = \pm 1$ correspond to valence $(-1)$ and conduction $(+1)$ bands, respectively, and

$$\begin{aligned}
\mathcal{D}(k) \;=\;& \left[(|\gamma(k)|^2 - 3)\, t_{nnn} - t_{nn}s_{nn}\,|\gamma(k)|^2\right]^2 - (1 - s_{nn}^2\,|\gamma(k)|^2) \\
\times\;& \left[(|\gamma(k)|^2 - 3|)^2\, t_{nnn}^2 - t_{nn}^2\,|\gamma(k)|^2\right] = |\gamma(k)|^2\left[(|\gamma(k)|^2 - 3)\, t_{nnn}\, s_{nn} + t_{nn}\right]^2 .
\end{aligned} \tag{30}$$

Table 3. Graphene structure and tight-binding model parameters.

| Parameter | Value [55] |
|:---:|:---:|
| $a_1$ | $(a/2)\,(3, \sqrt{3})$ |
| $a_2$ | $(a/2)\,(3, -\sqrt{3})$ |
| $a_3$ | $(a/2)\,(0, 2\sqrt{3})$ |
| $\delta_1$ | $(a/2)\,(1, \sqrt{3})$ |
| $\delta_2$ | $(a/2)\,(1, -\sqrt{3})$ |
| $\delta_3$ | $-a\,(1,0)$ |
| $K$ | $(2\pi/3\sqrt{3}a)\,(\sqrt{3}, 1)$ |
| $K'$ | $(2\pi/3\sqrt{3}a)\,(\sqrt{3}, -1)$ |
| $s_{nn}$ | $0.106$ |
| $s_{nnn}$ | $0.001$ |
| $t_{nn}$ | $-2.78\,\mathrm{eV}$ |
| $t_{nnn}$ | $-0.12\,\mathrm{eV}$ |
| $\varepsilon_2 + C_p$ | $-0.36\,\mathrm{eV}$ |

By using the result in Equations (29) and (30) can be rewritten as

$$\begin{aligned}
E_\lambda(k) \;=\;& (\varepsilon_2 + C_p) + \frac{(|\gamma(k)|^2 - 3)\, t_{nnn}[1 + \lambda|\gamma(k)|\,s_{nn}] - t_{nn}s_{nn}|\gamma(k)|^2 + \lambda\,|\gamma(k)|\,t_{nn}}{1 - s_{nn}^2|\gamma(k)|^2} \\
\approx\;& (\varepsilon_2 + C_p) + \frac{(|\gamma(k)|^2 - 3)\, t_{nnn} - t_{nn}s_{nn}|\gamma(k)|^2 + \lambda\, t_{nn}\,|\gamma(k)|}{1 - s_{nn}^2|\gamma(k)|^2} .
\end{aligned} \tag{31}$$

By setting $C_p + \varepsilon_2 = 0$ as the reference point for energy, the result in Equation (31) is plotted in Figure 6 by employing the graphene structural parameters listed in Table 3.

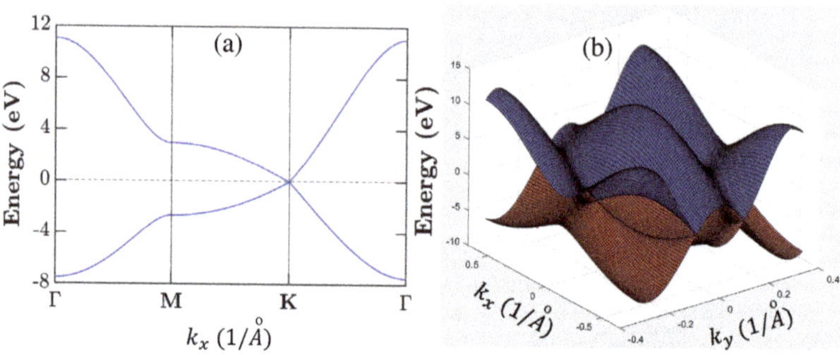


Figure 6. Calculated dispersion of energy bands for graphene. Panel (a) displays 2D plot for energy dispersion of graphene electrons. Panel (b) shows 3D plot for upper and lower bands touched at six Dirac points (three $K$ and three $K'$ valleys), at which the energy is set to be zero.

Furthermore, by using the result in Equation (31), two elements of the eigenvector, $a_{\mathbf{k}}^{\lambda}$ and $b_{\mathbf{k}}^{\lambda}$, are found to be

$$
\begin{aligned}
a_{\mathbf{k}}^{\lambda} &= \left\{ \frac{\gamma^*(k)\left[E_\lambda(k)\,s_{nn} - t_{nn}\right]}{\left(|\gamma(k)|^2 - 3\right)t_{nnn} - E_\lambda(k)} \right\} b_{\mathbf{k}}^{\lambda}, \\
b_{\mathbf{k}}^{\lambda} &= \frac{\left|\left(|\gamma(k)|^2 - 3\right)t_{nnn} - E_\lambda(k)\right|}{\sqrt{\left|\gamma^*(k)\left[t_{nn} - E_\lambda(k)\,s_{nn}\right]\right|^2 + \left|\left(|\gamma(k)|^2 - 3\right)t_{nnn} - E_\lambda(k)\right|^2}}.
\end{aligned}
\tag{32}
$$

As known experimentally, both the nearest-neighbor (nn) overlap and the next-nearest-neighbor (nnn) hopping integrals are much smaller than the nearest-neighbor (nn) hopping integral. By neglecting some constants, the dispersion in Equation (31) can be further simplified as

$$
E_\lambda(k) \approx 2\,t'_{nnn}\sum_{i=1}^{3}\cos(k\cdot a_i) + \lambda\,t_{nn}\left[3 + 2\sum_{i=1}^{3}\cos(k\cdot a_i)\right]^{1/2},
\tag{33}
$$

where $t'_{nnn} = t_{nnn} - s_{nn}t_{nn}$ is the corrected hopping amplitude.

## 5. Coulomb Diagonal-Dephasing Rate for Optical Coherence in Undoped Graphene

The quantum coherence of electrons is associated with the off-diagonal elements of their density matrix. The presence of an external field can induce coherence between two quantum states of electrons if the field frequency matches the energy separation between the two relevant electronic states. Dephasing refers to a physics mechanism which recovers classical behavior from a quantum system, and it quantifies the time required for electrons to lose their field-induced quantum coherence. Diagonal-dephasing rate connects to the ways in which coherence caused by perturbation decays over time, and then, the system goes back to the state before perturbation [1]. This is an important effect in molecular and atomic spectroscopy, and also in condense-matter physics of mesoscopic devices.

In order to demonstrate the significance of band-structure computation with a tight-binding model on dynamical properties of electrons in graphene, we first study Coulomb diagonal-dephasing (CDD) rate for induced optical polarization of thermally-excited electrons and holes around the Dirac point in an intrinsic (or undoped) graphene sample. For undoped graphene, conduction electrons can be introduced by a photo-excitation process [8], giving rise to equal number of electrons and holes $n_e = n_h \equiv n_0$, where $n_0$ represents the areal density of photo-excited carriers. For non-equilibrium photo-carriers under a transverse optical field, its induced optical coherence in steady states decays [1] with the sum of CDD rates $\Delta_e(k)$ and $\Delta_h(k)$ for electrons (e) and holes (h), respectively. These two rates determine the inhomogeneous line-shape of a resonant interband-absorption peak at $\hbar\omega = \varepsilon_{\mathbf{k}}^e + \varepsilon_{\mathbf{k}}^h$ for vertical transitions of electrons with their kinetic energies $\varepsilon_{\mathbf{k}}^{e,h}$ in valence and conduction bands.

As illustrated by Feynman diagrams [3] in Figure 7, the CDD rate $\Delta_e(k)$ of electrons is calculated as [1,44]

$$
\begin{aligned}
\Delta_e(k) &= \frac{8\pi}{\hbar \mathcal{A}^2}\sum_{\mathbf{k}_1,\mathbf{q}\neq 0}\left|V^{ee}_{\mathbf{k},\mathbf{k}_1;\mathbf{k}_1-\mathbf{q},\mathbf{k}+\mathbf{q}}\right|^2\left[\mathcal{L}_0(\varepsilon^e_{\mathbf{k}_1-\mathbf{q}} + \varepsilon^e_{\mathbf{k}+\mathbf{q}} - \varepsilon^e_{\mathbf{k}_1} - \varepsilon^e_{\mathbf{k}}, \Gamma_e)\right. \\
&\quad \times \left.\left\{ f^e_{\mathbf{k}_1-\mathbf{q}}\,f^e_{\mathbf{k}+\mathbf{q}}\left(1 - f^e_{\mathbf{k}_1}\right) + \left(1 - f^e_{\mathbf{k}_1-\mathbf{q}}\right)\left(1 - f^e_{\mathbf{k}+\mathbf{q}}\right)f^e_{\mathbf{k}_1}\right\}\right] \\
&\quad + \frac{8\pi}{\hbar \mathcal{A}^2}\sum_{\mathbf{k}_1,\mathbf{q}\neq 0}\left|V^{he}_{\mathbf{k},\mathbf{k}_1;\mathbf{k}_1-\mathbf{q},\mathbf{k}-\mathbf{q}}\right|^2\left[\mathcal{L}_0(\varepsilon^e_{\mathbf{k}_1-\mathbf{q}} + \varepsilon^h_{-(\mathbf{k}-\mathbf{q})} - \varepsilon^e_{\mathbf{k}_1} - \varepsilon^h_{-\mathbf{k}}, \Gamma_{eh})\right. \\
&\quad \times \left.\left\{ f^e_{\mathbf{k}_1-\mathbf{q}}\,f^h_{-(\mathbf{k}-\mathbf{q})}\left(1 - f^e_{\mathbf{k}_1}\right) + \left(1 - f^e_{\mathbf{k}_1-\mathbf{q}}\right)\left(1 - f^h_{-(\mathbf{k}-\mathbf{q})}\right)f^e_{\mathbf{k}_1}\right\}\right],
\end{aligned}
\tag{34}
$$

where both spin and valley degeneracies are included, $\mathcal{A}$ represents the surface area of graphene sample, the first and second terms correspond to the left and right panels of Figure 7, and both scattering-in and scattering-out contributions [44] are taken into consideration in these two terms. Moreover, $\varepsilon_{\mathbf{k}}^{\mathrm{e,h}}$ in Equation (34) stands for the kinetic energy of electrons (e) or holes (h), and $f_{\mathbf{k}}^{\mathrm{e,h}} = \{1 + \exp[(\varepsilon_{\mathbf{k}}^{\mathrm{e,h}} - \mu_{\mathrm{e,h}})/k_{\mathrm{B}}T]\}^{-1}$ is the Fermi function for thermal-equilibrium photo-carriers with their chemical potentials $\mu_{\mathrm{e,h}}$ at temperature $T$. Here, $\mu_{\mathrm{e,h}}$ are separately determined by following two equations for given $T$, i.e.,

$$n_{\mathrm{e,h}} = \frac{4}{\mathcal{A}} \sum_{\mathbf{k}} \frac{1}{1 + \exp[(\varepsilon_{\mathbf{k}}^{\mathrm{e,h}} - \mu_{\mathrm{e,h}})/k_{\mathrm{B}}T]} \, , \tag{35}$$

where both spin and valley degeneracies are included and $\mu_{\mathrm{e}} = \mu_{\mathrm{h}}$ in our case. Furthermore, in Equation (34), $\mathcal{L}_0(a,b) = (b/\pi)/(a^2 + b^2)$ is the Lorentzian line-shape function, $\Gamma_{\mathrm{e,h}}$ are inverse lifetime of unperturbed electrons or holes, and $\Gamma_{\mathrm{eh}} = (\Gamma_{\mathrm{e}} + \Gamma_{\mathrm{h}})/2$.

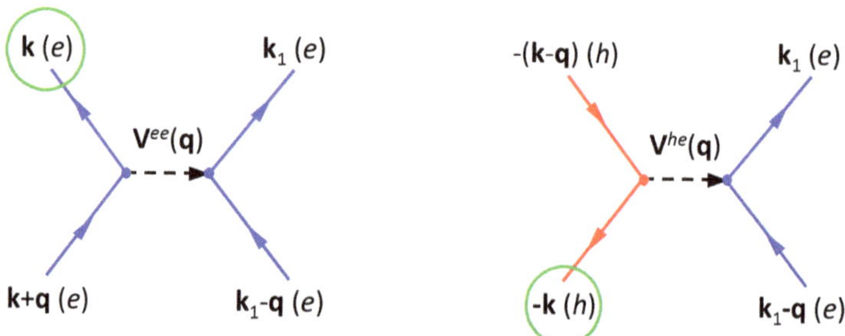


**Figure 7.** Feynman diagrams for CDD rate $\Delta_{\mathrm{e}}(k)$ of electrons in Equation (34). (**left**) Coulomb coupling between pair of electrons in one inelastic-scattering event; (**right**) Coulomb coupling between an electron and a hole in another inelastic-scattering event.

In addition, we have introduced in Equation (34), as well as in Equation (40) below, the Coulomb-interaction matrix elements, given by [56]

$$\begin{aligned}
V_{\mathbf{k},\mathbf{k}_1;\mathbf{k}_1-\mathbf{q},\mathbf{k}+\mathbf{q}}^{\mathrm{ee}} &= u_{\mathrm{c}}(q)\, \mathcal{F}_{\mathbf{k},\mathbf{k}+\mathbf{q}}^{(c)}(q)\, \mathcal{F}_{\mathbf{k}_1,\mathbf{k}_1-\mathbf{q}}^{(c)}(-q) \, , \\
V_{\mathbf{k},\mathbf{k}_1;\mathbf{k}_1-\mathbf{q},\mathbf{k}-\mathbf{q}}^{\mathrm{he}} &= u_{\mathrm{c}}(q)\, \mathcal{F}_{\mathbf{k},\mathbf{k}-\mathbf{q}}^{(v)}(q)\, \mathcal{F}_{\mathbf{k}_1,\mathbf{k}_1-\mathbf{q}}^{(c)}(-q) \, , \\
V_{\mathbf{k},\mathbf{k}_1;\mathbf{k}_1-\mathbf{q},\mathbf{k}+\mathbf{q}}^{\mathrm{hh}} &= u_{\mathrm{c}}(q)\, \mathcal{F}_{\mathbf{k},\mathbf{k}+\mathbf{q}}^{(v)}(q)\, \mathcal{F}_{\mathbf{k}_1,\mathbf{k}_1-\mathbf{q}}^{(v)}(-q) \, , \\
V_{\mathbf{k},\mathbf{k}_1;\mathbf{k}_1-\mathbf{q},\mathbf{k}-\mathbf{q}}^{\mathrm{eh}} &= u_{\mathrm{c}}(q)\, \mathcal{F}_{\mathbf{k},\mathbf{k}-\mathbf{q}}^{(c)}(q)\, \mathcal{F}_{\mathbf{k}_1,\mathbf{k}_1-\mathbf{q}}^{(v)}(-q) \, ,
\end{aligned} \tag{36}$$

where $u_{\mathrm{c}}(q) = e^2/[2\epsilon_0\epsilon_{\mathrm{r}}(q + q_0)]$ in Equation (36) is the two-dimensional Fourier transformed Coulomb potential $\sim 1/r$ including static screening, $\epsilon_0$ represents the vacuum permittivity, and $\epsilon_{\mathrm{r}} = 2.4$ is the average dielectric constant of the host material. Additionally, $q_0$ stands for the inverse Thomas-Fermi screening length, and can be given by a semi-classical model as [57]

$$q_0 = \left(\frac{e^2}{8\epsilon_0\epsilon_{\mathrm{r}}k_{\mathrm{B}}T}\right)\frac{4}{\mathcal{A}}\sum_{\mathbf{k}}\left[\cosh^{-2}\left(\frac{\varepsilon_{\mathbf{k}}^{\mathrm{e}} - \mu_{\mathrm{e}}}{2k_{\mathrm{B}}T}\right) + \cosh^{-2}\left(\frac{\varepsilon_{\mathbf{k}}^{\mathrm{h}} - \mu_{\mathrm{h}}}{2k_{\mathrm{B}}T}\right)\right] , \tag{37}$$

where both spin and valley degeneracies have been included.

Furthermore, the introduced $\mathcal{F}_{\mathbf{k},\mathbf{k}'}^{(s)}(q)$ in Equation (36) with $s = c, v$ represents the Bloch-function form factor, calculated as [57]

$$\mathcal{F}_{\mathbf{k},\mathbf{k}'}^{(s)}(\boldsymbol{q}) = \int d^2\boldsymbol{r}\,[\Phi_{\mathbf{k}}^{(s)}(\boldsymbol{r})]^*\,\exp(i\boldsymbol{q}\cdot\boldsymbol{r})\,\Phi_{\mathbf{k}'}^{(s)}(\boldsymbol{r}) = \frac{1}{N_c}\left\{[a_{\mathbf{k}}^{(s)}]^\dagger\otimes a_{\mathbf{k}'}^{(s)}\right\}$$

$$\times\ \sum_{j,j'=1}^{N_c}\exp[-i(\boldsymbol{q}-\boldsymbol{k})\cdot\boldsymbol{\Delta}_j - i\boldsymbol{k}'\cdot\boldsymbol{\Delta}_{j'}]\sum_{\ell,\ell'\in A,B}\exp[i(\boldsymbol{k}'-\boldsymbol{k}+\boldsymbol{q})\cdot\boldsymbol{R}_\ell]\,\mathcal{W}_s(\boldsymbol{q},\boldsymbol{R}_{\ell'\ell}+\boldsymbol{\Delta}_{jj'})\,, \tag{38}$$

where the Bloch functions $\Phi_{\mathbf{k}}^{c,v}(\boldsymbol{r})$ in Equations (5) and (22) have been employed. In Equation (38), $N_c$ represents the number of near-neighbor atoms within a unit cell, $\boldsymbol{\Delta}_j$ stands for the lattice vectors of the near-neighbor atoms relative to the sublattice site $\boldsymbol{R}_\ell$, and $a_{\mathbf{k}}^{(s)}$ are two column eigenvectors in Equation (32) for $s = c,\,v$. The Wannier-function structure factor $\mathcal{W}_s(\boldsymbol{q},\boldsymbol{R}_{\ell'\ell}+\boldsymbol{\Delta}_{jj'})$ in Equation (38) is defined as

$$\mathcal{W}_s(\boldsymbol{q},\boldsymbol{R}_{\ell'\ell}+\boldsymbol{\Delta}_{jj'}) = \int d^2\boldsymbol{r}\,[\psi_{p_z}^{(s)}(\boldsymbol{r})]^*\,\exp(i\boldsymbol{q}\cdot\boldsymbol{r})\,\psi_{p_z}^{(s)}(\boldsymbol{r}-\boldsymbol{R}_{\ell'\ell}-\boldsymbol{\Delta}_{jj'})\,, \tag{39}$$

where $\boldsymbol{R}_{\ell'\ell} = \boldsymbol{R}_{\ell'} - \boldsymbol{R}_\ell$ and $\boldsymbol{\Delta}_{jj'} = \boldsymbol{\Delta}_j - \boldsymbol{\Delta}_{j'}$. In fact, Equations (34) and (36)–(39) are the key results in this paper for connecting the calculated tight-binding wave functions and band structures to a quantum-statistical theory for graphene optical properties.

Similarly, as illustrated by Feynman diagrams [3] in Figure 8, the CDD rate $\Delta_h(\boldsymbol{k})$ of holes takes the form [1,44]

$$\begin{aligned}
\Delta_h(\boldsymbol{k}) =\ & \frac{8\pi}{\hbar\mathcal{A}^2}\sum_{\mathbf{k}_1,\mathbf{q}\neq 0}\left|V_{\mathbf{k},\mathbf{k}_1;\mathbf{k}_1-\mathbf{q},\mathbf{k}+\mathbf{q}}^{\mathrm{hh}}\right|^2\left[\mathcal{L}_0(\varepsilon_{-(\mathbf{k}_1-\mathbf{q})}^{\mathrm{h}}+\varepsilon_{-(\mathbf{k}+\mathbf{q})}^{\mathrm{h}}-\varepsilon_{-\mathbf{k}_1}^{\mathrm{h}}-\varepsilon_{-\mathbf{k}}^{\mathrm{h}},\Gamma_{\mathrm{h}})\right.\\
& \times\ \left.\left\{f_{-(\mathbf{k}_1-\mathbf{q})}^{\mathrm{h}}\,f_{-(\mathbf{k}+\mathbf{q})}^{\mathrm{h}}(1-f_{-\mathbf{k}_1}^{\mathrm{h}})+(1-f_{-(\mathbf{k}_1-\mathbf{q})}^{\mathrm{h}})(1-f_{-(\mathbf{k}+\mathbf{q})}^{\mathrm{h}})\,f_{-\mathbf{k}_1}^{\mathrm{h}}\right\}\right]\\
& + \frac{8\pi}{\hbar\mathcal{A}^2}\sum_{\mathbf{k}_1,\mathbf{q}\neq 0}\left|V_{\mathbf{k},\mathbf{k}_1;\mathbf{k}_1-\mathbf{q},\mathbf{k}-\mathbf{q}}^{\mathrm{eh}}\right|^2\left[\mathcal{L}_0(\varepsilon_{\mathbf{k}-\mathbf{q}}^{\mathrm{e}}+\varepsilon_{-(\mathbf{k}_1-\mathbf{q})}^{\mathrm{h}}-\varepsilon_{\mathbf{k}}^{\mathrm{e}}-\varepsilon_{-\mathbf{k}_1}^{\mathrm{h}},\Gamma_{\mathrm{eh}})\right.\\
& \times\ \left.\left\{f_{-(\mathbf{k}_1-\mathbf{q})}^{\mathrm{h}}\,f_{\mathbf{k}-\mathbf{q}}^{\mathrm{e}}(1-f_{-\mathbf{k}_1}^{\mathrm{h}})+(1-f_{-(\mathbf{k}_1-\mathbf{q})}^{\mathrm{h}})(1-f_{\mathbf{k}-\mathbf{q}}^{\mathrm{e}})\,f_{-\mathbf{k}_1}^{\mathrm{h}}\right\}\right].
\end{aligned} \tag{40}$$

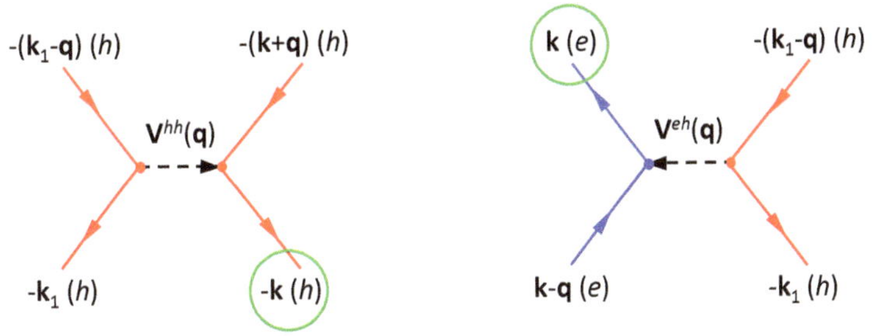

**Figure 8.** Feynman diagrams for CDD rate $\Delta_h(\boldsymbol{k})$ of holes in Equation (40). (**left**) Coulomb coupling between pair of holes in one inelastic-scattering event; (**right**) Coulomb coupling between a hole and an electron in another inelastic-scattering event.

Computationally, the $\pi$-electron band structure of graphite can be obtained by employing the nearest-neighbor tight-binding model [58,59]. For graphene, the reciprocal lattice in the wave-vector space also acquires the hexagonal symmetry, same as that in real lattice. Moreover, the low energy bands are found linear and isotropic near the corners of the first Brillouin zone or $K$ point. Such $K$-point linear bands become essential for the low-energy (or small wave-number) excitation of electrons. The calculated energy dispersions by diagonalizing the $2\times 2$ Hamiltonian matrix are given by [58,59]

$$\varepsilon_{\mathbf{k}}^{\mathrm{e,h}} = \pm\frac{3}{2}\gamma_0 bk \equiv \hbar v_{\mathrm{F}}k\,, \tag{41}$$

where $\gamma_0 = 2.4\,\text{eV}$ is the hopping integral between the nearest-neighbor atoms, $b = 1.42\,\text{Å}$ is the C–C bond length, and signs $\pm$ represents conduction $(+)$ and hole $(-)$ bands, respectively. Meanwhile, the corresponding spinor-type Bloch wave functions are found to be

$$\phi_{p_z,\mathbf{k}}^{(c,v)}(\mathbf{r}) = \frac{1}{\sqrt{2}}\left[U_{\mathbf{k}}^{(1)}(\mathbf{r}) \mp e^{i\theta_{\mathbf{k}}}\, U_{\mathbf{k}}^{(2)}(\mathbf{r})\right], \tag{42}$$

where as shown in Equation (20), $U_{\mathbf{k}}^{(1)}(\mathbf{r})$ and $U_{\mathbf{k}}^{(2)}(\mathbf{r})$ are two sublattice Bloch functions built from the superposition of the periodic $2p_z$ orbitals, Ref. [59] and $\theta_{\mathbf{k}} = \tan^{-1}(k_y/k_x)$ is the angle between the wave vector $\mathbf{k}$ and $x$-axis. As in Equation (22), we can further express the $2p_z$ atomic orbital by means of a generalized hydrogen-like wave function, given by [60]

$$\psi_{p_z}(\mathbf{r}) = C_0\, r\, \cos\theta\, e^{-Z^* r/2a_0}, \tag{43}$$

where $C_0$ is a normalization factor, $a_0$ the Bohr radius, and an effective nucleus charge number $Z^*$ is 3.18.

In particular, the structure factor introduced in Equation (38) can be calculated explicitly as

$$\begin{aligned}
\mathcal{F}_{\mathbf{k}',\mathbf{k}}^{(s)}(\mathbf{q}) &= \delta_{\mathbf{k}',\mathbf{k}+\mathbf{q}} \int d^2r\, [\Phi_{\mathbf{k}+\mathbf{q}}^{(s)}(\mathbf{r})]^* \exp(i\mathbf{q}\cdot\mathbf{r})\, \Phi_{\mathbf{k}}^{(s)}(\mathbf{r}) \\
&\approx \delta_{\mathbf{k}',\mathbf{k}+\mathbf{q}} \langle \phi_{p_z,\mathbf{k}+\mathbf{q}}^{(s)}(\mathbf{r})\,|\, \exp(i\mathbf{q}\cdot\mathbf{r})\,|\, \phi_{p_z,\mathbf{k}}^{(s)}(\mathbf{r})\rangle,
\end{aligned} \tag{44}$$

where $s = c, v$ for Bloch wave function. Moreover, the Bloch-function structure factor in Equation (44) takes the form [60]

$$\begin{aligned}
&\langle \phi_{p_z,\mathbf{k}+\mathbf{q}}^{(s)}(\mathbf{r})\,|\, \exp(i\mathbf{q}\cdot\mathbf{r})\,|\, \phi_{p_z,\mathbf{k}}^{(s)}(\mathbf{r})\rangle \\
&= \frac{1}{N_A + N_B} \sum_{\mathbf{R}=\mathbf{R}_A,\mathbf{R}_B} \langle \psi_{p_z}(\mathbf{r}-\mathbf{R})\,|\, \exp[i\mathbf{q}\cdot(\mathbf{r}-\mathbf{R})]\,|\, \psi_{p_z}(\mathbf{r}-\mathbf{R})\rangle \frac{1}{2}\left[1 \pm \frac{\gamma(\mathbf{k}+\mathbf{q})\,\gamma^*(\mathbf{k})}{|\gamma(\mathbf{k}+\mathbf{q})\,\gamma(\mathbf{k})|}\right],
\end{aligned} \tag{45}$$

where tight-binding function $\psi_{p_z}(\mathbf{r})$ is given by Equation (43), and the signs $(\pm)$ correspond to conduction $(+)$ and valence $(-)$ bands, respectively [59].

For intrinsic graphene, we have chemical potential $\mu_e = \mu_h = 0$ [61]. However, there is still a finite intrinsic areal density $n_i \approx (\pi/6)\,(k_B T/\hbar v_F)^2$ due to thermal excitation of electrons and holes at finite temperatures $T$. In fact, we find $f_{\mathbf{k}}^e = f_{\mathbf{k}}^h = 1/2$ at the $K$ valley or $k = 0$. Here, the calculated CDD rates from Equations (34) and (40), respectively, for electrons $\Delta_e(\mathbf{k})$ and holes $\Delta_h(\mathbf{k})$ are presented in Figure 9a at $T = 77\,\text{K}$ and in Figure 9b at $T = 300\,\text{K}$. Since $f_{\mathbf{k}}^{e,h} \sim \exp(-\varepsilon_{\mathbf{k}}^{e,h}/k_B T)$ as $\varepsilon_{\mathbf{k}}^{e,h} \gg k_B T$, the thermal occupations of electron and hole states will be limited mostly to wave numbers close to the $K$ valley due to their lower kinetic energies $\varepsilon_{\mathbf{k}}^{e,h}$ around $k = 0$, as seen in Figure 6.

The Coulomb diagonal-dephasing rates $\Delta_{e,h}(\mathbf{k})$ presented in Figure 9a,b quantifies an amplitude-decay process of induced electron-hole optical coherence with wave vector $\mathbf{k}$ by an optical field towards the state before external perturbation. Furthermore, the Coulomb off-diagonal-dephasing rates $\Lambda_{e,h}(\mathbf{k},\mathbf{q})$ reveals deformations of induced optical-polarization waves with different wave vectors $\mathbf{k}+\mathbf{q}$ [8].

Considering the fact that major occupations of electrons and holes are accumulated around $k = 0$, we have $f_{\mathbf{k}}^{e,h} \approx 0$ only if $k$ is large. As a result, we find from Equation (34) that $f_{\mathbf{k}_1-\mathbf{q}}^e f_{\mathbf{k}+\mathbf{q}}^e (1 - f_{\mathbf{k}_1}^e) \ll 1$ at $k = 0$ since we require $1 - f_{\mathbf{k}_1}^e \approx 1$ for large $k_1$, $f_{\mathbf{k}+\mathbf{q}}^e \approx 1$ for small $q$, and $f_{\mathbf{k}_1-\mathbf{q}}^e \approx 1$ for both large $q$ and $k_1$, which, however, cannot be satisfied simultaneously. Similar conclusion can also be drawn for the second term in Equation (34), where we find $f_{\mathbf{k}_1-\mathbf{q}}^e f_{-(\mathbf{k}-\mathbf{q})}^h (1 - f_{\mathbf{k}_1}^e) \ll 1$. Combining these two facts together, we expect that a dip will occur at $k = 0$ for the Coulomb diagonal-dephasing rate $\Delta_e(\mathbf{k})$, as seen in Figure 9a. Moreover, the observed anisotropic energy dispersion in Figure 6a along the $K$-$M$ and $K$-$\Gamma$ directions directly leads to a staircase-like feature in Figure 9a for both $\Delta_e(\mathbf{k})$

and $\Delta_h(k)$. As temperature $T$ is raised from 77 K in Figure 9a to 300 K in Figure 9b, the thermally-excited areal densities of electrons and holes are increased with $T^2$; therefore, the Coulomb interaction ($\propto T^4$) between electrons and holes, as well as the Coulomb interaction among electrons or holes, will be enhanced greatly. Consequently, we find that both $\Delta_e(k)$ and $\Delta_h(k)$ are enhanced by a factor of 2.4, in addition to amplified depth of the dip at $k = 0$. Furthermore, different structural factors in Equation (39), corresponding to + signs for conduction and valence bands, give rise to a slightly larger value of $\Delta_h(k)$ in comparison with that of $\Delta_e(k)$, as well as different dispersion features around the $K$ valley for $\Delta_e(k)$ and $\Delta_h(k)$. These two computed Coulomb diagonal-dephasing rates can be physically applied to the spectral [32] and polarization [1,36] functions in order to study transport and optical properties of graphene material.

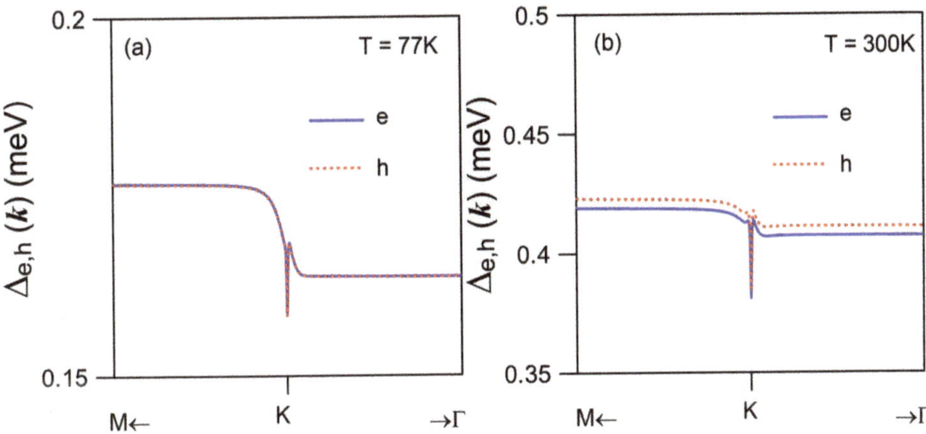

**Figure 9.** Calculated Coulomb diagonal-dephasing rates $\Delta_{e,h}(k)$ for electrons (e, blue solid curves) from Equation (34) and $\Delta_h(k)$ for holes (h, red dashed curves) from Equation (40) as functions of wave number $k$ (with respect to $k = 0$ at the $K$ valley) at temperatures $T = 77$ K in (**a**) and $T = 300$ K in (**b**), where $\epsilon_r = 2.4$ and $\Gamma_e = \Gamma_h = 0.01$ meV are assumed.

### 6. Carrier Energy-Relaxation Rate in Doped Graphene

In condensed-matter physics, the microscopic energy-relaxation time usually refers to a measure of the time it requires for one electron in the system to be significantly affected by the presence of other electrons, lattice vibrations, and randomly-distributed ionized impurity atoms in the system through an either scattering-in or scattering-out process mediated by electron-electron, electron-phonon and electron-impurity interactions, respectively. Since the microscopic energy-relaxation time is assigned to a specific electronic state, we are able to define a thermally-averaged energy-relaxation time through the diagonal density-matrix elements of electrons for all electronic states. In this way, one can reveal unique temperature dependence of this macroscopic energy-relaxation time and utilize it for simplifying the well-known Boltzmann transport equation within the relaxation-time approximation [44].

By going beyond the intrinsic graphene samples, we would like to investigate further the impurity scattering of electrons in extrinsic (or doped) graphene materials. In parallel with the discussion on scattering rates in Section 5, we present here the calculations for intraband-scattering of electrons by randomly-distributed impurities. Results for intraband-scattering of holes can be obtained in a similar way.

By using the detailed-balance condition, the microscopic energy-relaxation time $\tau_{\text{rel}}(k)$ of electrons in the presence of randomly-distributed ionized impurities can be calculated according to [44]

$$\frac{1}{\tau_{\text{rel}}(k)} = \mathcal{W}_{\text{in}}(k) + \mathcal{W}_{\text{out}}(k) , \tag{46}$$

where the scattering-in rate for electrons in the final $|k\rangle$ state is

$$\mathcal{W}_{\text{in}}(k) = \frac{4\pi n_{\text{im}}}{\hbar \mathcal{A}} \sum_{q \neq 0} \left\{ \left| U^{\text{im}}_{k, k-q}(q) \right|^2 f^{\text{e}}_{k-q} \mathcal{L}_0(\varepsilon^{\text{e}}_k - \varepsilon^{\text{e}}_{k-q}, \Gamma_{\text{e}}) \right.$$
$$\left. + \left| U^{\text{im}}_{k, k+q}(q) \right|^2 f^{\text{e}}_{k+q} \mathcal{L}_0(\varepsilon^{\text{e}}_k - \varepsilon^{\text{e}}_{k+q}, \Gamma_{\text{e}}) \right\}, \tag{47}$$

whereas the scattering-out rate for electrons in the initial $|k\rangle$ state takes the form

$$\mathcal{W}_{\text{out}}(k) = \frac{4\pi n_{\text{im}}}{\hbar \mathcal{A}} \sum_{q \neq 0} \left\{ \left| U^{\text{im}}_{k+q, k}(q) \right|^2 (1 - f^{\text{e}}_{k+q}) \mathcal{L}_0(\varepsilon^{\text{e}}_{k+q} - \varepsilon^{\text{e}}_k, \Gamma_{\text{e}}) \right.$$
$$\left. + \left| U^{\text{im}}_{k-q, k}(q) \right|^2 (1 - f^{\text{e}}_{k-q}) \mathcal{L}_0(\varepsilon^{\text{e}}_{k-q} - \varepsilon^{\text{e}}_k, \Gamma_{\text{e}}) \right\}. \tag{48}$$

Here, $n_{\text{im}}$ represents the areal density of ionized impurity atoms in the crystal, and $|U^{\text{im}}_{k, k'}(q)|^2$ comes from the randomly-impurity scattering of electron in the second-order Born approximation [48,62]. Explicitly, the random impurity-interaction matrix elements are calculated as

$$\left| U^{\text{im}}_{k, k'}(q) \right|^2 = Z^{*2} |u_{\text{c}}(q)|^2 \left| \int d^2 r \, [\Phi^{(c)}_k(r)]^* \exp(iq \cdot r) \, \Phi^{(c)}_{k'}(r) \right|^2 = Z^{*2} |u_{\text{c}}(q)|^2 \left| \mathcal{F}^{(c)}_{k, k'}(q) \right|^2, \tag{49}$$

where $Z^*$ is the charge number of ionized impurity atoms.

Substituting Equation (49) back into Equation (47), we obtain

$$\mathcal{W}^{\text{in}}_k = \frac{4\pi n_{\text{im}} Z^{*2}}{\hbar \mathcal{A}} \sum_{q \neq 0} \left\{ f^{\text{e}}_{k-q} \mathcal{L}_0(\varepsilon^{\text{e}}_k - \varepsilon^{\text{e}}_{k-q}, \Gamma_{\text{e}}) |u_{\text{c}}(q)|^2 \left| \mathcal{F}^{(c)}_{k, k-q}(q) \right|^2 \right.$$
$$\left. + f^{\text{e}}_{k+q} \mathcal{L}_0(\varepsilon^{\text{e}}_k - \varepsilon^{\text{e}}_{k+q}, \Gamma_{\text{e}}) |u_{\text{c}}(q)|^2 \left| \mathcal{F}^{(c)}_{k, k+q}(q) \right|^2 \right\}, \tag{50}$$

$$\mathcal{W}^{\text{out}}_k = \frac{4\pi n_{\text{im}} Z^{*2}}{\hbar \mathcal{A}} \sum_{q \neq 0} \left\{ (1 - f^{\text{e}}_{k+q}) \mathcal{L}_0(\varepsilon^{\text{e}}_{k+q} - \varepsilon^{\text{e}}_k, \Gamma_{\text{e}}) |u_{\text{c}}(q)|^2 \left| \mathcal{F}^{(c)}_{k+q, k}(q) \right|^2 \right.$$
$$\left. + (1 - f^{\text{e}}_{k-q}) \mathcal{L}_0(\varepsilon^{\text{e}}_{k-q} - \varepsilon^{\text{e}}_k, \Gamma_{\text{e}}) |u_{\text{c}}(q)|^2 \left| \mathcal{F}^{(c)}_{k-q, k}(q) \right|^2 \right\}. \tag{51}$$

Using the inverse microscopic energy-relaxation time in Equation (46), we can further calculate the macroscopic thermally-averaged energy-relaxation time $\tau_{\text{rel}}(T)$ as a function of temperature $T$, yielding [44]

$$\frac{1}{\tau_{\text{rel}}(T)} = \frac{4}{n_{\text{e}} \mathcal{A}} \sum_k \left[ \frac{1}{\tau_{\text{rel}}(k)} \right] f^{\text{e}}_k. \tag{52}$$

Actually, the results in Equation (46) and in Equations (50)–(52) demonstrate the approach for relating the computed tight-binding wave functions and band structures to graphene transport properties described by a many-body scattering theory. This calculated relaxation time in Equation (52) can be employed for building up different orders of moment equations [63] based on semi-classical Boltzmann transport equation [6] under the relaxation-time approximation [44]. Here, the zeroth-order moment equation [63] grantees the conservation of conduction electrons and allows us to find the chemical potential of electrons, as in Equation (35), for given areal doping density and temperature. Moreover, the first-order moment equation [63] makes it possible to find transport mobility and conductivity [64] for bias-field driven conduction electrons.

For doped graphene, we have Fermi energy $E_{\text{F}} = \hbar v_{\text{F}} \sqrt{\pi n_0}$ at low temperatures, Ref. [61] where $n_0$ represents the areal electron density from doping, i.e., $n_0 = n_{\text{im}}$ for completely ionized doping atoms. For low temperatures with $k_BT \ll E_F$, we have $f_k^e = \Theta(E_F - \varepsilon_k^e)$ or $f_k^e = \Theta(k_F - k)$, where $\Theta(x)$ is a unity step function and $k_F = \sqrt{\pi n_0}$ is the Fermi wave number.

Physically, the Coulomb diagonal-dephasing rates $\Gamma(k) = \Delta_e(k) + \Delta_h(k)$ in Figure 9 describes a decay process of induced electron-hole optical coherence, which is induced by an optical field over time, towards the state before perturbation. On the other hand, the electron energy-relaxation rate $1/\tau_{rel}(T)$, determined by Equations (46) and (52), reflects the time, which is a quantum-statistical average over all occupied states of electrons, needed for recovering from a non-equilibrium-state occupation after an external perturbation to an initial thermal-equilibrium-state occupation before external perturbation via an elastic electron-impurity scattering process. Therefore, these two rates, as shown by Figures 9 and 10, respectively, represent two fundamentally different microscopic physics mechanisms.

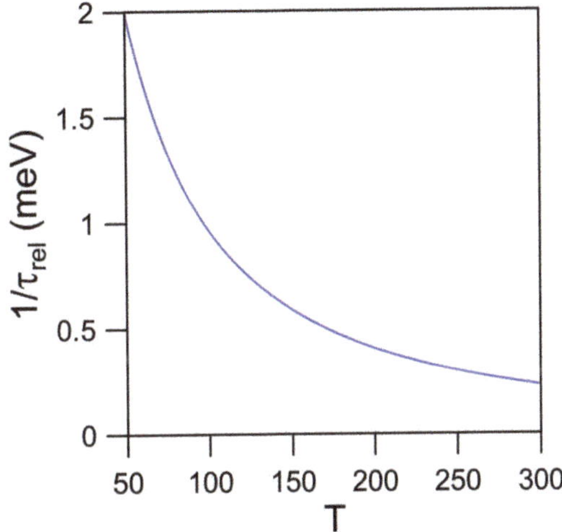

**Figure 10.** Calculated average energy-relaxation rate $1/\tau_{rel}(T)$ from Equation (52) as a function of temperature $T$ due to elastic scattering of doped electrons with impurities in graphene material, where $\epsilon_r = 2.4$, $\Gamma_e = \Gamma_h = 0.01$ meV, $Z^* = 1$, doped electron areal density $n_0 = 1 \times 10^{11}$ cm$^{-2}$, and impurity areal density $n_{im} = n_0$ are assumed.

As seen from Figure 10, we find the electron energy-relaxation rate $1/\tau_{rel}(T)$ reduces with increasing temperature $T$ due to enhanced screening effect on Coulomb interaction $u_c(q)$ between two electrons or the rising of $q_0$ in Equation (37) with $T$, which implies that we have to wait a longer time $\tau_{rel}(T)$ for our system returning to its initial thermal-equilibrium state at an elevated temperature. Furthermore, using the second-order Boltzmann moment equation [44], we would emphasize that this average energy-relaxation time $\tau_{rel}(T)$, as determined from Equations (46) and (52), is directly associated with the mobility of transport electrons limited by elastic scattering from existence of impurities in the system.

### 7. Conclusions and Remarks

In conclusion, by introducing a generalized first-principles quantum-kinetic model coupled self-consistently with Maxwell and Boltzmann transport equations, we demonstrate the importance to incorporate inputs from first-principles band-structure computations for accurately describing non-equilibrium optical and transport properties of electrons in graphene. Generally speaking, the physical properties of an active material in a device are determined by both underlined band structures of involved materials and non-equilibrium responses to various external impulses.

In this study, we initialize with the tight-binding model for investigating band structures of solid covalent crystals by means of localized Wannier orbital functions, and further parameterize the hopping integrals in the tight-binding model for different covalent bonds. After that, we apply the general tight-binding-model formalism to graphene in order to acquire both band structures and wave functions of electrons within the whole first Brillouin zone of two-dimensional materials. For illustrating their significance, we utilize them to explore the intrinsic electron-hole Coulomb diagonal-dephasing rates used for spectral and polarization functions of graphene materials, and meanwhile, the energy-relaxation rate from extrinsic elastic scattering by impurities for transport mobility of doped electrons in graphene.

Theoretically, our current theory is capable of first-principles calculations of ultra-fast dynamics for non-thermal photo-generated electron-hole pairs. Simultaneously, this a theory also enables to describe electromagnetic, optical and electrical properties of semiconductor materials all together, as well as their interplay. Technologically, in combination with first-principles band-structure computations, the numerical output of current first-principles dynamics model can be used as an input for material optical and transport properties and put into a next-step simulation software, such as COMSOL Multiphysics, for a target device. Consequently, device characteristics can be predicted accurately for numerical bottom-up design and engineering.

**Author Contributions:** Conceptualization, D.H. and G.G.; computations, P.-H.S. and T.-N.D.; writing—review and editing, D.H. and G.G., supervision, D.H., G.G. and H.L.; project administration, D.H. All authors have read and agreed to the published version of the manuscript.

**Funding:** This research was funded by Air Force Research Laboratory (AFRL), Air Force Office of Scientific Research (AFOSR), and Ministry of Science and Technology (Taiwan).

**Institutional Review Board Statement:** Not applicable.

**Informed Consent Statement:** Not applicable.

**Data Availability Statement:** The datasets generated during and/or analyzed during the current study are available from the corresponding author on reasonable request.

**Acknowledgments:** D.H. would like to acknowledge the financial support from the Air Force Office of Scientific Research (AFOSR). G.G. would like to acknowledge Grant No. FA9453-21-1-0046 from the Air Force Research Laboratory (AFRL). P.-H. Shih and T.-N. Do would like to thank the Ministry of Science and Technology of Taiwan and Tay-Rong Chang from the Department of Physics of National Cheng Kung University for the support through the Grant No. MOST 110-2636-M-006-002.

**Conflicts of Interest:** There are no conflict of interest to declare.

## References

1. Gulley, J.R.; Huang, D.H. Self-consistent quantum-kinetic theory for interplay between pulsed-laser excitation and nonlinear carrier transport in a quantum-wire array. *Opt. Expr.* **2019**, *27*, 17154–17185. [CrossRef]
2. Huang, D.H.; Easter, M.M.; Gumbs, G.; Maradudin, A.A.; Lin, S.-Y.; Cardimona, D.A.; Zhang, X. Controlling quantum-dot light absorption and emission by a surface-plasmon field. *Opt. Expr.* **2014**, *22*, 27576–27605. [CrossRef]
3. Gumbs, G.; Huang, D.H. *Properties of Interacting Low-Dimensional Systems*; John Wiley & Sons: Hoboken, NJ, USA, 2011.
4. Kotov, V.N.; Uchoa, B.; Pereira, V.M.; Guinea, F.; Neto, A.H.C. Electron-Electron Interactions in Graphene: Current Status and Perspectives. *Rev. Mod. Phys.* **2012**, *84*, 1067. [CrossRef]
5. Jackson, J.D. *Classical Electrodynamics*, 3rd ed.; John Wiley & Sons: Hoboken, NJ, USA, 1999.
6. Ziman, J.M. *Principles of the Theory of Solids*, 2nd ed.; Cambridge University Press: Cambridge, UK, 1972.
7. Pryor, R.W. *Multiphysics Modeling Using COMSOL*; Jones and Bartlell Publishers: Burlington, MA, USA, 2011.
8. Lindberg, M.; Koch, S.W. Effective Bloch equations for semiconductors. *Phys. Rev. B* **1988**, *38*, 3342. [CrossRef]
9. Haug, H.; Koch, S.W. *Quantum Theory of the Optical and Electronic Properties of Semiconductors*, 4th ed.; World Scientific Publishing Co.: Singapore, 2004.
10. Kohn, W.; Sham, L. Self-Consistent Equations Including Exchange and Correlation Effects. *Phys. Rev.* **1965**, *140*, A1133. [CrossRef]
11. Harrison, W.A. *Electronic Structure and the Properties of Solids*; Dover Publications: Mineola, NY, USA, 1989.
12. Ashcroft, N.W.; Mermin, N.D. *Solid State Physics*; Thomson Learning: Toronto, ON, Canada, 1976.
13. Davies, J.H. *The Physics of Low-Dimensional Semiconductors: An Introduction*; Cambridge University Press: Cambridge, UK, 1998.

14. Slater, J.C.; Koster, G.F. Simplified LCAO Method for the Periodic Potential Problem. *Phys. Rev.* **1954**, *94*, 1498. [CrossRef]
15. Goringe, C.M.; Bowler, D.R.; Hernández, E. Tight-binding modelling of materials. *Rep. Progr. Phys.* **1997**, *60*, 1447. [CrossRef]
16. Huheey, J.E.; Keiter, E.A.; Keiter, R.L. *Inorganic Chemistry: Principles of Structure and Reactivity*, 4th ed.; Pearson: London, UK, 1993.
17. Roeland, L.W.; Cock, G.J. High-field magnetization of Tb single crystals. *J. Phys. C Solid State Phys.* **1975**, *8*, 3427. [CrossRef]
18. Shih, P.-H.; Do, T.-N.; Gumbs, G.; Huang, D.H.; Pham, H.D.; Lin, M.-F. Rich Magnetic Quantization Phenomena in AA Bilayer Silicene. *Nat. Sci. Rep.* **2019**, *9*, 14799. [CrossRef] [PubMed]
19. Altland, A.; Simons, B. *Condensed Matter Field Theory*, 2nd ed.; Cambridge University Press: Cambridge, UK, 2010.
20. Geim, A.K.; Novoselov, K.S. The rise of graphene. *Nat. Mater.* **2007**, *6*, 183–191 [CrossRef]
21. Neto, A.H.C.; Guinea, F.; Peres, N.M.R.; Novoselov, K.S.; Geim, A.K. The electronic properties of graphene. *Rev. Mod. Phys.* **2009**, *81*, 109. [CrossRef]
22. Novoselov, K.S.; Geim, A.K.; Morozov, S.V.; Jiang, D.; Katsnelson, M.I.; Grigorieva, I.V.; Dubonos, S.V.; Firsov, A.A. Two-dimensional gas of massless Dirac fermions in graphene. *Nature* **2005**, *438*, 197–200. [CrossRef] [PubMed]
23. Zhang, Y.B.; Tan, Y.W.; Stormer, H.L.; Kim, P. Experimental observation of the quantum Hall effect and Berry's phase in graphene. *Nature* **2005**, *438*, 201–204. [CrossRef] [PubMed]
24. Geim, A.K. Graphene: Status and Prospects. *Science* **2009**, *324*, 1530–1534. [CrossRef] [PubMed]
25. Novoselov, K.S.; McCann, E.; Morozov, S.V.; Fal'ko, V.I.; Katsnelson, K.I.; Zeitler, U.; Jiang, D.; Schedin, F.; Geim, A.K. Unconventional quantum Hall effect and Berry's phase of $2\pi$ in bilayer graphene. *Nat. Phys.* **2006**, *2*, 177–180. [CrossRef]
26. McCann, E.; Fal'ko, V.I. Landau-Level Degeneracy and Quantum Hall Effect in a Graphite Bilayer. *Phys. Rev. Lett.* **2006**, *96*, 086805. [CrossRef]
27. Cao, Y.; Fatemi, V.; Fang, S.; Watanabe, K.; Taniguchi, T.; Kaxiras, E.; Jarillo-Herrero, P. Unconventional superconductivity in magic-angle graphene superlattices. *Nature* **2018**, *556*, 43–50. [CrossRef] [PubMed]
28. Novoselov, K.S.; Geim, A.K.; Morozov, S.V.; Jiang, D.; Zhang, Y.; Dubonos, S.V.; Grigorieva, I.V.; Firsov, A.A. Electric field effect in atomically thin carbon films. *Science* **2004**, *306*, 666–669. [CrossRef]
29. Craciun, M.F.; Russo, S.; Yamamoto, M.; Oostinga, J.B.; Morpurgo, A.F.; Tarucha, S. Trilayer graphene is a semimetal with a gate-tunable band overlap. *Nat. Nanotech.* **2009**, *4*, 383–388. [CrossRef] [PubMed]
30. Berger, C.; Song, Z.; Li, T.; Li, X.; Ogbazghi, A.Y.; Feng, R.; Dai, Z.; Marchenkov, A.N.; Conrad, E.H.; First, P.N.; et al. Ultrathin Epitaxial Graphite: 2D Electron Gas Properties and a Route toward Graphene-based Nanoelectronics. *J. Phys. Chem. B* **2004**, *108*, 19912–19916. [CrossRef]
31. Shih, P.-H.; Do, T.-N.; Gumbs, G.; Huang, D.; Pham, T.P.; Lin, M.-F. Magneto-transport properties of B-, Si- and N-doped graphene. *Carbon* **2020**, *160*, 211–218. [CrossRef]
32. Do, T.-N.; Gumbs, G.; Shih, P.-H.; Huang, D.; Lin, M.-F. Valley- and spin-dependent quantum Hall states in bilayer silicene. *Phys. Rev. B* **2019**, *100*, 155403. [CrossRef]
33. Huang, D.; Iurov, A.; Gumbs, G.; Zhemchuzhna, L. Effects of site asymmetry and valley mixing on Hofstadter-type spectra of bilayer graphene in a square-scatter array potential. *J. Phys. Condens. Matter* **2019**, *31*, 125503. [CrossRef]
34. Do, T.-N.; Gumbs, G.; Shih, P.-H.; Huang, D.H.; Chiu, C.-W.; Chen, C.-Y.; Lin, M.-F. Peculiar optical properties of bilayer silicene under the influence of external electric and magnetic fields. *Nat. Sci. Rep.* **2019**, *9*, 624. [CrossRef]
35. Shih, P.-H.; Do, T.-N.; Huang, B.-L.; Gumbs, G.; Huang, D.; Lin, M.-F. Magneto-electronic and optical properties of Si-doped graphene. *Carbon* **2019**, *144*, 608–614. [CrossRef]
36. Iurov, A.; Gumbs, G.; Huang, D.H. Many-body effects and optical properties of single and double layer $\alpha$-T$_3$ lattices. *J. Phys. Condens. Matter* **2020**, *32*, 415303. [CrossRef]
37. Balassis, A.; Dahal, D.; Gumbs, G.; Iurov, A.; Huang, D.H.; Roslyak, O. Magnetoplasmons for the $\alpha$-T$_3$ model with filled Landau levels. *J. Phys. Condens. Matter* **2020**, *32*, 485301. [CrossRef]
38. Iurov, A.; Huang, D.H.; Gumbs, G.; Pan, W.; Maradudin, A.A. Effects of optical polarization on hybridization of radiative and evanescent field modes. *Phys. Rev. B* **2017**, *96*, 081408. [CrossRef]
39. Iurov, A.; Gumbs, G.; Huang, D.H. Exchange and correlation energies in silicene illuminated by circularly polarized light. *J. Mod. Opt.* **2017**, *64*, 913–920. [CrossRef]
40. Iurov, A.; Gumbs, G.; Huang, D.H.; Zhemchuzhna, L. Controlling plasmon modes and damping in buckled two-dimensional material open systems. *J. Appl. Phys.* **2017**, *121*, 084306. [CrossRef]
41. Iurov, A.; Gumbs, G.; Huang, D.H. Temperature-dependent collective effects for silicene and germanene. *J. Phys. Condens. Matter* **2017**, *29*, 135602. [CrossRef]
42. Anwar, F.; Iurov, A.; Huang, D.H.; Gumbs, G.; Sharma, A.K. Interplay between effects of barrier tilting and scatterers within a barrier on tunneling transport of Dirac electrons in graphene. *Phys. Rev. B* **2020**, *101*, 115424. [CrossRef]
43. Iurov, A.; Zhemchuzhna, L.; Dahal, D.; Gumbs, G.; Huang, D.H. Quantum-statistical theory for laser-tuned transport and optical conductivities of dressed electrons in $\alpha$-T$_3$ materials. *Phys. Rev. B* **2020**, *101*, 035129. [CrossRef]
44. Huang, D.H.; Iurov, A.; Xu, H.-Y.; Lai, Y.-C.; Gumbs, G. Interplay of Lorentz-Berry forces in position-momentum spaces for valley-dependent impurity scattering in $\alpha$-T$_3$ lattices. *Phys. Rev. B* **2019**, *99*, 245412. [CrossRef]
45. Iurov, A.; Gumbs, G.; Huang, D.H. Peculiar electronic states, symmetries, and Berry phases in irradiated $\alpha$-T$_3$ materials. *Phys. Rev. B* **2019**, *99*, 205135. [CrossRef]

46. Iurov, A.; Gumbs, G.; Huang, D.H. Temperature- and frequency-dependent optical and transport conductivities in doped buckled honeycomb lattices. *Phys. Rev. B* **2018**, *98*, 075414. [CrossRef]
47. Iurov, A.; Zhemchuzhna, L.; Gumbs, G.; Huang, D.H. Exploring interacting Floquet states in black phosphorus: Anisotropy and bandgap laser tuning. *J. Appl. Phys.* **2017**, *122*, 124301. [CrossRef]
48. Xu, H.; Huang, L.; Huang, D.H.; Lai, Y.-C. Geometric valley Hall effect and valley filtering through a singular Berry flux. *Phys. Rev. B* **2017**, *96*, 045412. [CrossRef]
49. Roslyak, O.; Iurov, A.; Gumbs, G.; Huang, D.H. Unimpeded tunneling in graphene nanoribbons. *J. Phys. Condens. Matter* **2010**, *22*, 165301. [CrossRef]
50. Iurov, A.; Gumbs, G.; Roslyak, O.; Huang, D.H. Anomalous photon-assisted tunneling in graphene. *J. Phys. Condens. Matter* **2010**, *24*, 015303. [CrossRef] [PubMed]
51. Shyu, F.L.; Chang, C.P.; Chen, R.B.; Chiu, C.W.; Lin, M.F. Magnetoelectronic and optical properties of carbon nanotubes. *Phys. Rev. B* **2003**, *67*, 045405. [CrossRef]
52. Lifshiz, L.D.L.E.M. *Quantum Mechanics: Non-Relativistic Theory*, 3rd ed.; Pergamon Press: Oxford, UK, 1977.
53. Froyen, S.; Harrison, W.A. Elementary prediction of linear combination of atomic orbitals matrix elements. *Phys. Rev. B* **1979**, *20*, 2420. [CrossRef]
54. Saito, R.; Dresselhaus, M.S.; Dresselhaus, G. *Physical Properties of Carbon Nanotubes*; Imperial College Press: London, UK, 1998.
55. Kundu, R. Tight-Binding Parameters for Graphene. *Mod. Phys. Lett. B* **2011**, *25*, 163. [CrossRef]
56. Huang, D.H.; Iurov, A.; Gao, F.; Gumbs, G.; Cardimona, D.A. Many-Body Theory of Proton-Generated Point Defects for Losses of Electron Energy and Photons in Quantum Wells. *Phys. Rev. Appl.* **2018**, *9*, 024002. [CrossRef]
57. Huang, D.H.; Manasreh, M.O. Intersubband transitions in strained $In_{0.07}Ga_{0.93}As/Al_{0.40}Ga_{0.60}As$ multiple quantum wells and their application to a two-colors photodetector. *Phys. Rev. B* **1996**, *54*, 5620–5628. [CrossRef] [PubMed]
58. Wallace, P.R. The Band Theory of Graphite. *Phys. Rev.* **1947**, *71*, 622 [CrossRef]
59. Ho, J.H.; Chang, C.P.; Chen, R.B.; Lin, M.F. Electron decay rates in a zero-gap graphite layer. *Phys. Lett. A* **2006**, *357*, 401. [CrossRef]
60. Shung, K.W.-K. Dielectric function and plasmon structure of stage-1 intercalated graphite. *Phys. Rev. B* **1986**, *34*, 979. [CrossRef] [PubMed]
61. Zebrev, G.I. *Graphene Field Effect Transistors: Diffusion-Drift Theory*; Mikhailov, S., Ed.; IntechOpen: London, UK, 2011; Chapter 23; pp. 475–498.
62. Lyo, S.K.; Huang, D.H. Multisublevel magnetoquantum conductance in single and coupled double quantum wires. *Phys. Rev. B* **2001**, *64*, 115320. [CrossRef]
63. Huang, D.H.; Gumbs, G.; Roslyak, O. Optical modulation effects on nonlinear electron transport in graphene in terahertz frequency range. *J. Mod. Opt.* **2011**, *58*, 1898–1907. [CrossRef]
64. Backes, D.; Huang, D.; Mansell, R.; Lanius, M.; Kampmeier, J.; Ritchie, D.; Mussler, G.; Gumbs, G.; Grützmacher, J.; Narayan, V. Disentangling surface and bulk transport in topological-insulator $p - n$ junctions. *Phys. Rev. B* **2017**, *96*, 125125. [CrossRef]

MDPI

St. Alban-Anlage 66

4052 Basel

Switzerland

Tel. +41 61 683 77 34

Fax +41 61 302 89 18

www.mdpi.com

*Nanomaterials* Editorial Office

E-mail: nanomaterials@mdpi.com

www.mdpi.com/journal/nanomaterials